From Sky to Soil

Abhijit Sarkar · Pardeep Singh · Chandan Barman ·
Robert Popek
Editors

From Sky to Soil

The Ecological Journey of Particulate Pollution and Pathways to Sustainable Airscapes

Springer

Editors
Abhijit Sarkar
Department of Botany
University of Gour Banga
Malda, West Bengal, India

Pardeep Singh
Department of Environmental Studies
PGDAV College, University of Delhi
New Delhi, Delhi, India

Chandan Barman
Department of Botany
University of Gour Banga
Malda, West Bengal, India

Robert Popek
Centre for Climate Research SGGW
Warsaw University of Life Sciences
Warsaw, Poland

ISBN 978-3-032-31990-6 ISBN 978-3-032-31991-3 (eBook)
https://doi.org/10.1007/978-3-032-31991-3

This Springer imprint is published by the registered company Springer Nature Switzerland AG
The registered company address is: Gewerbestrasse 11, 6330 Cham, Switzerland

Contents

Chapter 1
Physico-Chemical Nature of Particulate Pollution: A Real-Time Global Threat

Zuzanna Zając, Katarzyna Gładysz, Mariia Pismanik, and Robert Popek

Abstract Particulate matter (PM) represents a heterogeneous mixture of microscopic solid and liquid particles suspended in the atmosphere, posing an acute global threat to health and ecosystems. Classified by aerodynamic diameter, PM includes coarse PM_{10} (< 10 μm, penetrating upper airways), fine $PM_{2.5}$ (< 2.5 μm, reaching deep lung alveoli), and submicron PM_1 (< 1 μm, entering bloodstream), with smaller fractions dominating particle numbers despite low mass. Sources span natural (sea spray, dust storms, volcanoes) and anthropogenic (transport, industry, biomass burning) origins, distinguishing primary direct emissions from secondary formations via gas-to-particle conversion of precursors like SO_2, NO_x, and VOCs. Physically, PM exhibits diverse morphology (spherical to irregular agglomerates), variable density, high specific surface area enabling toxin adsorption, and optical properties influencing climate through scattering (cooling) or absorption (warming by black carbon). Chemically, PM comprises inorganics (sulfates, nitrates, ammonium), organics (carbonaceous compounds, PAHs), and toxic trace metals (Pb, Cd, As), undergoing atmospheric aging via oxidation, increasing oxidation states and toxicity. Beyond chronic diseases, PM demands real-time response due to immediate cardiovascular and respiratory risks, burdening global health per WHO guidelines. This chapter elucidates PM's physico-chemical dynamics, urging urgent mitigation.

Keywords Particulate matter (PM) · Aerodynamic diameter · Primary emissions · Secondary aerosols · Physico-chemical properties

Z. Zając · K. Gładysz · M. Pismanik · R. Popek (✉)
Centre for Climate Research SGGW, Warsaw University of Life Sciences-SGGW (WULS-SGGW), Warsaw, Poland
e-mail: robert_popek@sggw.edu.pl

Z. Zając
e-mail: zuzanna_zajac1@sggw.edu.pl

K. Gładysz
e-mail: katarzyna_gladysz@sggw.edu.pl

M. Pismanik
e-mail: mariia_pismanik@sggw.edu.pl

A. Sarkar et al. (eds.), *From Sky to Soil*, https://doi.org/10.1007/978-3-032-31991-3_1

1.1 Introduction to Particulate Matter Pollution

Particulate matter (PM) is defined as a heterogeneous mixture of microscopic solid particles that remain suspended in the air. This mixture contains both organic and inorganic particles, such as dust, pollen, soot, and smoke (Popek et al., 2025). It is worth nothing that this definition also includes liquid droplets (aerosols). Due to their negligible mass, PM particles can remain suspended in the air for a very long time (Sicard et al., 2023).

1.1.1 Classification of Particulate Matter

The basic criterion for the classification of suspended dust is the aerodynamic diameter of the particles (Fig. 1.1). This parameter determines both the time a particle remains in the atmosphere and the depth of its penetration into the human respiratory system (Lee et al., 2017).

The following fractions are distinguished:

A. Coarse **PM_{10}**—particles with an aerodynamic diameter of less than **10 μm**. This fraction can freely penetrate the upper respiratory system. It includes most naturally occurring dust, such as soil particles, crushed rocks, and plant pollen (Lin et al., 2022).
B. Fine **$PM_{2.5}$**—this is a respirable fraction containing particles with a diameter of less than **2.5 μm**. Due to its significantly smaller size, this dust is able to overcome the barriers of the upper respiratory system and reach deeper into the bronchi, bronchioles, and alveoli (Feng et al. 2016). Combustion products and secondary atmospheric aerosols dominate here (Popek et al., 2022).
C. Submicron **PM_1**—this fraction includes particles with a diameter not exceeding **1 μm**. Although they constitute a negligible part of the total mass of suspended

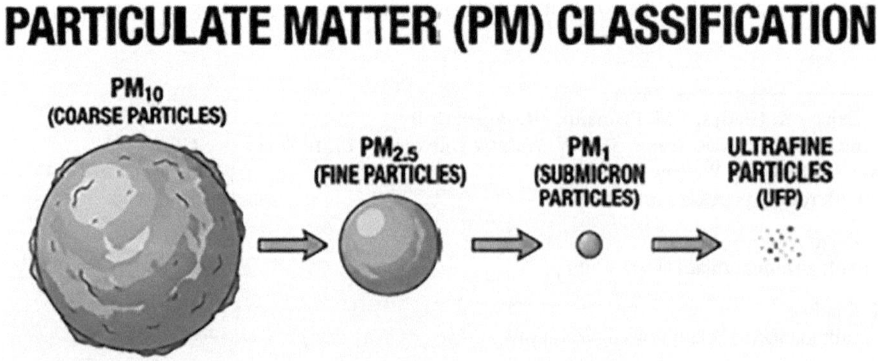

Fig. 1.1 Classification of particulate matter according to particle size

dust, they dominate in terms of the number of particles in the air (Das et al., 2023; Dzierżanowski et al., 2011). Due to their size, which is similar to viruses or large protein molecules, they behave almost like gases in the body. They have the ability to pass from alveoli directly into the bloodstream and can be transported with the blood to all internal organs (Kim et al., 2015).

1.1.2 The Scale of the Problem Globally

Air pollution caused by particulate matter is currently classified as one of the most significant environmental threats to public health. According to the latest reports from the World Health Organization, it is estimated that the vast majority of countries do not meet air quality standards (WHO, 2021). Exposure to elevated concentrations of particulate matter results in millions of premature deaths each year, mainly attributed to ischemic heart disease, stroke, chronic obstructive pulmonary disease, lung cancer, and acute respiratory infections (Lippmann & Chen, 2009). In addition to its devastating health effects, this phenomenon places a huge burden on the global economy (IPCC, 2023). It includes direct healthcare costs, losses resulting from reducing labor productivity, and the negative impact of pollution on ecosystems and agricultural yields (Harrison & Yin., 2000).

1.1.3 Particulate Matter Pollution as a Real-Time Hazard

Public and scientific discussions often focus on the long-term effects of inhaling pollutants, such as reduced life expectancy and the development of chronic diseases after years of exposure. However, **it should be clearly stated that particulate matter also poses a real-time threat** (Choi et al., 2022). Treating air pollution as a real-time threat changes the health protection paradigm from long-term prevention to the need for immediate response, warning humanity, and emergency risk management (Kim et al., 2021).

1.2 Sources of Particulate Matter Emissions

The origin of particulate matter suspended in the atmosphere is extremely diverse. The key distinction here is between natural and anthropogenic (human-induced) sources, as well as between primary and secondary emissions (Fig. 1.2).

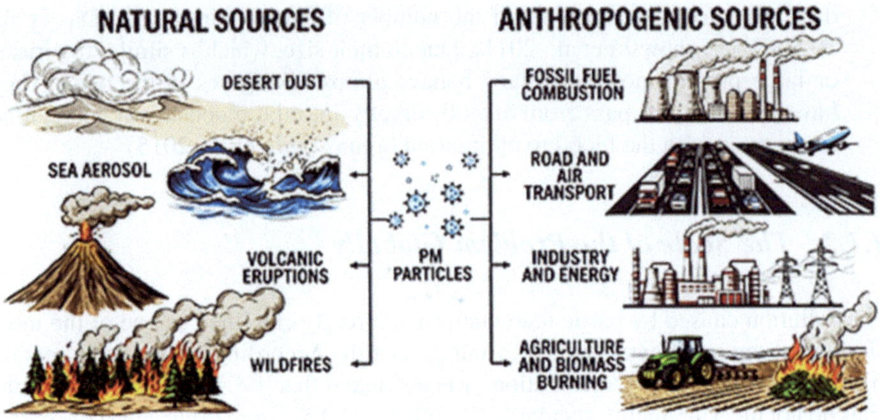

Fig. 1.2 Natural and anthropogenic sources of particulate matter

1.2.1 Natural Sources

Natural processes emit huge amounts of particles into the atmosphere, which form the background for pollution caused by civilization. These include

1. Sea aerosol—one of the dominant natural components on a global scale. It is formed as a result of mechanical disruption of the ocean surface (waves, bursting air bubbles). In addition to salt (NaCl), the aerosol contains sulfates and organic substances (May et al., 2018).
2. Desert and soil dust—mineral particles are carried by the wind from areas with sparse plant cover and can then be transported thousands of kilometers, e.g., dust from the Sahara (Shahsavani et al., 2020).
3. Volcanic eruptions—these are violent, localized sources of emissions of huge amounts of ash and gases, affecting the atmosphere even on a planetary scale (Mandal et al., 2023).
4. Fires—natural fires are a source of carbon and ash particles (Butwin et al., 2019).

1.2.2 Anthropogenic Sources

Human activity is responsible for the emission of particulate matter with the most harmful chemical composition, usually concentrated in densely populated areas. The main sectors include

1. Road and air transport—particles are generated directly in the engine and form just behind the exhaust pipe through the nucleation and condensation of hot exhaust gases and as a result of mechanical abrasion of tires, brake linings, and road surfaces, as well as the stirring up of dust already lying on the road (Popek et al., 2017; Roy et al., 2024a).

2. Industry and energy—include the combustion of fossil fuels in power plants and combined heat and power plants, as well as specific technological processes (metallurgical, cement, chemical) that emit dust containing heavy metals (Cheriyan & Choi, 2020).
3. Municipal sector and agriculture—the combustion of soil fuels and biomass (straw, agricultural waste) in households is the main source of so-called low emissions. In agriculture, ammonia from fertilizers and livestock farming, which is a precursor to secondary particles, is also a significant source (Klimont et al., 2017).

1.2.3 Primary Versus Secondary Emissions

When analyzing air pollution, it is crucial to distinguish the moment when pollution becomes a solid particle. This determines not only the chemical composition of the dust, but also the strategy for its reduction.

Primary emissions refer to particles that are emitted directly into the atmosphere in solid or liquid form. Their physical form is already established when they leave the emission source (chimney, exhaust pipe, ground surface) (Huang et al., 2019). They are emitted directly from anthropogenic and natural sources. **Secondary emissions** are formed in the atmosphere from gaseous substrates (precursors). The abovementioned process, known as gas-to-particle conversion, involves the transformation of gaseous pollutants (SO_2, NO_x, NH_3, VOCs) into solid or liquid matter (Huang et al., 2019) (Fig. 1.3). The mechanism has two modes of action:

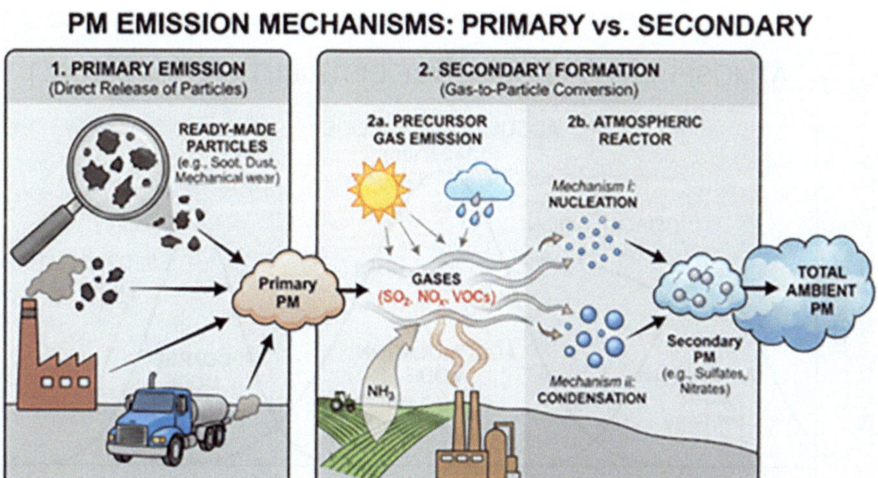

Fig. 1.3 Mechanism of primary and secondary PM emissions

A. Nucleation—the formation of new, microscopic particles from the gas phase (Kim et al., 2002).
B. Condensation—the deposition of chemical reaction products on the surface of existing primary particles (Parvez et al., 2017).

1.3 Physical Characteristics of Particles

It is essential to understand that the physical properties of particulate matter are fundamental to assessing its impact on human health and the climate. The key parameters determining the behavior of particles in the environment include size (aerodynamic diameter), morphology, density, specific surface area, and optical properties (Perrone et al., 2018).

1.3.1 Particle Size and Dimensions

Size is the main physical parameter, determining how long a particle remains in the atmosphere and how deeply it penetrates the respiratory system (Reijnders et al., 2018). The distribution of particles sizes in given air volume is heterogenous (Fig. 1.4). There are three dominant ranges:

A. Nucleation mode—particles < 0.1 μm, formed from gases.
B. Accumulation mode—particles 0.1–2.5 μm, formed as a result of the coagulation of smaller particles.

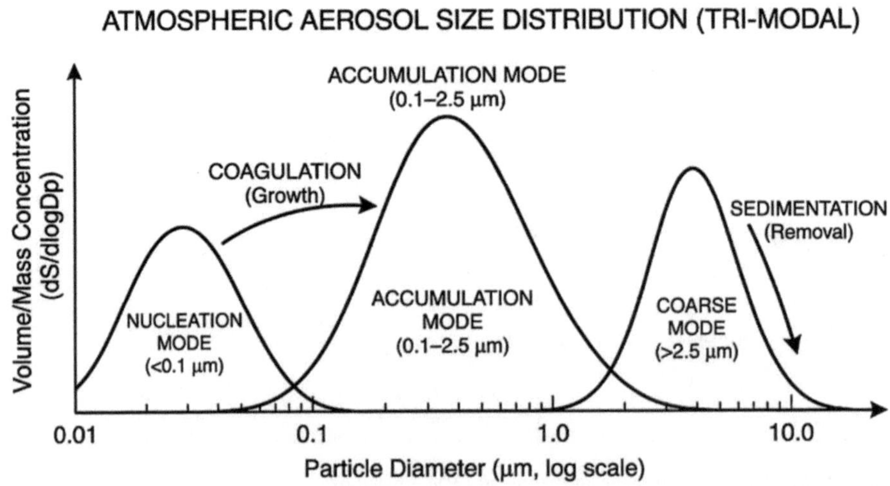

Fig. 1.4 Particulate matter size distribution diagram (trimodal model)

C. Coarse mode—particles > 2.5 μm, originating mainly from mechanical processes (Reijnders et al., 2018).

The distribution of suspended particle sizes is also crucial when selecting air purification technology. In biomass combustion installations, the effectiveness of dust removal devices is highly correlated with dust granulation: cyclones effectively remove fractions above 2.5 μm, while electrostatic precipitators can capture submicron particles (< 1 μm) with an efficiency of up to 99% (Zhang et al., 2010).

1.3.2 Shape, Density, and Specific Surface Area

The structure of solid particles is highly diverse and depends directly on the mechanism of their formation. Solid particles can be spherical, fibrous, or irregular in shape. Studies using scanning electron microscopy (SEM) have shown that fine dust fractions (e.g., from biomass combustion) tend to form agglomerates, clusters of smaller grains bound together by physical forces (Zeb et al., 2018). Coarse particles are more often characterized by separate, irregular edges.

Specific surface area is a parameter inversely proportional to particle size. Fine and ultrafine dusts, despite their low mass, have a huge specific surface area (Okuda, 2013). This promotes adsorption, these particles act like a "sponge," binding toxic chemicals (e.g., PAHs, heavy metals) to their surface and transporting them into the body (Okuda, 2013).

As opposed to homogeneous liquids, the particle density is strongly variable. Black carbon particles often have a porous, "fluffy" structure and very low effective density (< 1 g/cm^3), while mineral dusts or metal oxides are much denser (> 2–5 g/cm^3) (Dong et al., 2020). This difference is crucial because density together with size determines how quickly a particle falls through the air and where it settles in the lung. For this reason, the concept of aerodynamic diameter is used, which normalizes the density of different particles to the density of water (1 g/cm^3), allowing their behavior to be compared (Dong et al., 2020).

1.3.3 Optical Characteristics

The interaction of dust with solar radiation is one of the main factors influencing climate change and atmospheric visibility. Particles (e.g., sulfates, nitrates, sea salt) have **the ability to scatter solar radiation in all directions** (Chen et al., 2018). This leads to the so-called atmospheric cooling effect, reflecting energy back into space and reducing visibility (the phenomenon of fogging). Elementa: carbon plays a special role in the phenomenon of **light absorption**. Due to their dark color, these particles strongly absorb solar radiation, converting it into heat (Song et al., 2015).

1.3.4 *The Dynamics of Particles in the Atmosphere*

Suspended particulate matter is not a static component of air; it undergoes constant dynamic changes that determine how long it remains suspended. **Sedimentation** is the dominant process for removing large particles. Gravity dominates air resistance, causing them to fall rapidly to the ground near the emission source (Kravchishina et al., 2018). **Diffusion** affects the smallest particles. They move chaotically, colliding with gas molecules (Zhu et al., 2024). On the other hand, **coagulation** is a process in which smaller particles collide with each other, permanently combining to form larger agglomerates (Fig. 1.5). This process leads to a gradual shift from nucleation to accumulation, which changes the size distribution of dust transport from the source (Avino & Manigrasso., 2017).

1.4 Chemical Characterization of Particulate Matter

The chemical characterization of particulate matter (PM) provides critical insight into its emission sources, environmental behavior, and potential impacts on air quality, climate, and human health. The chemical composition of PM is heterogeneous and varies depending on emission source, atmospheric processing, and particle size. In urban and industrialized regions, fine particulate matter ($PM_{2.5}$) is generally dominated by inorganic ions, organic compounds, and trace elements.

Inorganic components are an important part of $PM_{2.5}$ mass. The dominant components include sulfate (SO_4^{2-}), nitrate (NO_3^-), and ammonium (NH_4^+), which

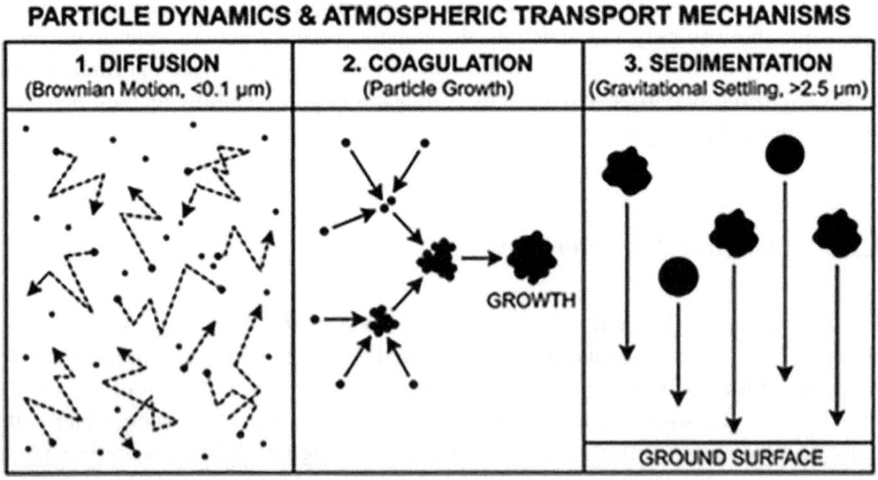

Fig. 1.5 The visualization of the main mechanisms of suspended dust particle dynamics: diffusion, coagulation, and sedimentation

together often constitute a significant portion of fine particulate matter in polluted environments (Lee et al., 2025). These ions play a key role in determining the hygroscopicity, acidity, and mass concentration, thereby affecting both atmospheric visibility and human exposure.

The organic fraction of particulate matter includes organic carbon (OC) and elemental carbon (EC). Organic carbon originates from both primary emissions, such as biomass burning and vehicular exhaust, and secondary formation from gaseous precursors (Alves et al., 2025). Elemental carbon, commonly associated with black carbon, is a product of incomplete combustion and has strong light-absorbing properties, contributing to positive radiative forcing (Duc et al., 2023). In addition, particulate matter contains hazardous organic species, including polycyclic aromatic hydrocarbons (PAHs) and semi-volatile organic compounds, many of which exhibit toxic and carcinogenic properties (Abdel-Shafy & Mansour, 2016; Yang et al., 2021).

Trace elements and heavy are another important component of PM, especially in urban environments. These elements originate mainly from anthropogenic activities such as industrial processes, fossil fuel combustion, vehicular exhaust, and non-exhaust sources including brake and tire wear, as well as resuspension of contaminated road dust (Querol et al., 2007; Wiseman et al., 2021). Metals such as lead (Pb), cadmium (Cd), arsenic (As), nickel (Ni), chromium (Cr), and mercury (Hg) are of particular concern due to their persistence, toxicity, and ability to induce oxidative stress and inflammatory responses in biological systems (Abd Elnabi et al., 2023).

1.5 Physico-Chemical Transformations in the Atmosphere

After being emitted into the atmosphere, particulate matter does not remain chemically or physically static, but undergoes continuous physico-chemical changes, collectively referred to as aerosol aging (Attah et al., 2025). These processes modify particle size, chemical composition, morphology, mixing state, and optical properties, thereby affecting atmospheric lifetime, transport behavior, toxicity, and climate impact (Leskinen et al., 2023).

Secondary aerosol formation is the main process regulating PM diversity. Secondary inorganic aerosols, including sulfate, nitrate, and ammonium, are formed by the atmospheric oxidation of gaseous precursors: sulfur dioxide (SO_2), nitrogen oxides (NO_x), and ammonia (NH_3) (Perraud et al., 2012). In parallel, secondary organic aerosols (SOA) are produced through the oxidation of volatile organic compounds (VOCs) emitted from anthropogenic and biogenic sources. These oxidation processes are initiated by highly reactive atmospheric oxidants, including hydroxyl radicals (OH), ozone (O_3), and nitrate radicals (NO_3) (Palm et al., 2017). The resulting reaction products typically exhibit lower volatility and partition into the particle phase, increasing particulate matter mass and altering chemical composition.

Photochemical reactions driven by solar radiation dominate the formation and transformation of aerosols during the day, thereby controlling the production of secondary pollutants and shaping daily changes in the composition of particulate matter. Under conditions of high relative humidity, low temperature, or limited sunlight (e.g., night or winter), heterogeneous and multiphase reactions occurring on particle surfaces or in the aqueous phases of aerosols become increasingly important (Kim et al., 2018). These processes can significantly increase sulfate production, promote efficient nitrate formation, and accelerate the oxidative aging of organic aerosols.

As aerosols age, organic components undergo functionalization and fragmentation reactions that increase their oxygen content and modify physico-chemical properties: hygroscopicity, water uptake, and optical behavior (Cappa et al., 2011). These transformations affect particle toxicity, cloud condensation nucleus activity, and radiative forcing. Aging processes can also alter the chemical form, solubility, and bioavailability of trace elements associated with particles, affecting their reactivity in the environment and relevance to health (Binda et al., 2024).

Meteorological conditions have a strong influence on transformation pathways and PM concentrations. Temperature affects reaction kinetics and gas–particle partitioning, particularly for semi-volatile compounds such as ammonium nitrate and organic compounds (Chen et al., 2020). Relative humidity regulates the water content of aerosols and facilitates aqueous-phase chemistry, while atmospheric stability and wind speed influence the dispersion and accumulation of pollutants (Gopinath et al., 2022).

Lastly, physico-chemical aging allows fine particles to remain suspended in the atmosphere for longer periods of time, enabling transport over long distances ranging from hundreds to thousands of kilometers (Al-Abadleh, 2021). During transport, aerosols become increasingly oxidized, internally mixed, and hygroscopic, which increases their interactions with radiation, clouds, and contributes to transboundary air pollution and regional climate impacts.

1.6 Real-Time Measurement and Monitoring Methods

1.6.1 Traditional Reference Methods

The manual gravimetric method is the basis for air quality monitoring and is considered the "gold standard" in European Union (according to PN-EN 12341) (European Committee for Standardization, 2014) and US (US EPA) (U.S. Environmental Agency, 2024) legislation. It serves as a reference point for all other measurement techniques. The procedure is based on direct measurement of the mass of suspended particulate matter. Atmospheric air is sucked in by a selection head, which separates PM fractions at a constant flow rate for 24 h. The dust settles on a pre-weighed filter. After exposure, the filter is sent to the laboratory, where it is conditioned under strictly defined humidity and temperature conditions (to eliminate the influence of water) and then reweighed on an analytical microbalance (Patel and Agarwal, 2022). The main advantage of this method is its precision and the possibility of subjecting the

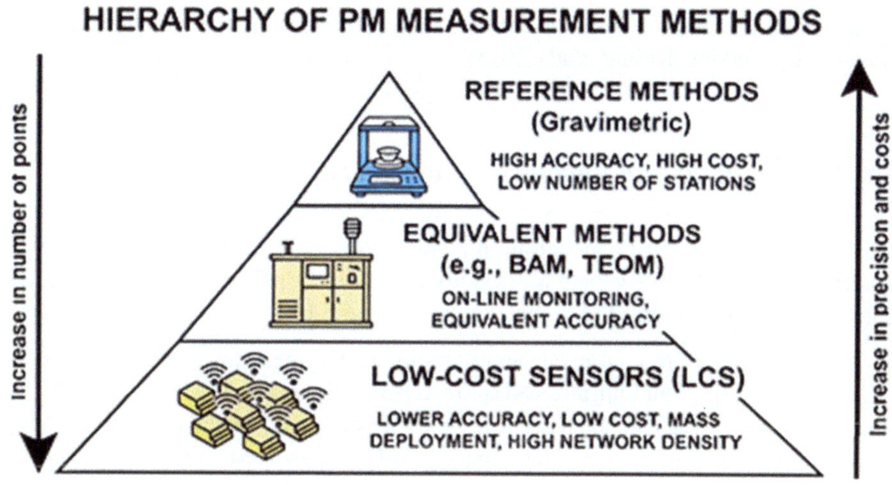

Fig. 1.6 Hierarchy of measurement methods

collected dust to further chemical analysis. However, it has significant operational limitations, as the results are available with a delay of several days, and the time resolution is limited to a daily average (Fig. 1.6). This makes it impossible to detect sudden peaks in concentration and to quickly inform the public about smog hazards (Noble et al., 2001).

1.6.2 Low-Cost Sensors and Online Monitoring

The need to obtain information in real time has led to the development of two groups of devices—professional automatic monitors and low-cost networks. In state monitoring networks (e.g., GIOŚ in Poland), devices operating in continuous mode, which have proven to be equivalent to the gravimetric method, are standard. The most common technology here is beta radiation absorption (BAM—Beta Attenuation Monitor). The principle of operation of BAM is based on Beer–Lambert's law. Air is sucked through a filter tape, through which a beam of beta radiation (emitted by an isotope source) is passed through it (Raja et al., 2017). Dust particles deposited on the tape absorb electrons—the greater the mass of particulate matter, the weaker the signal reaching the detector. An alternative is oscillating microbalances (TEOM), which measure the change in the frequency of vibration of an element under the influence of mass (Amaral et al., 2015).

The last decade has brought a revolution in the form of miniature sensors based on the nephelometric (light scattering) method. A laser diode illuminates the measuring chamber, and dust particles crossing the beam scatter the light, which is recorded by a photodetector. Despite its lower accuracy, this technology has made it possible to densify measurement networks and create high-resolution spatial smog maps in the Smart City concept (Kumar et al., 2015).

1.6.3 Chemical and Spectroscopic Analysis

Quantitative data on dust concentration alone are insufficient for comprehensive air quality assessment. To evaluate air toxicity and identify pollution sources, detailed chemical analysis, referred to as speciation, is required.

1. Inorganic ions (IC)—ion chromatography technique allows the detection of sulfates (SO_4^{2-}) and nitrates (NO_3^-) (Karthikeyan et al., 2007).
2. Heavy and trace metals (ICP-MS)—to detect toxic metals (such as lead, cadmium, arsenic, nickel), the sample must be dissolved in strong acids. The resulting solution is analyzed using inductively coupled plasma mass spectrometry. This is an extremely sensitive technique, capable of detecting even trace amounts of metals (Bacon et al., 2022).
3. Elemental composition—X-ray fluorescence (XRF) is a destruction-free method that allows for rapid determination of heavy metals and other elements (Chiari et al., 2018; Roy et al., 2024b).

1.6.4 Challenges and Uncertainties Associated with Measurements

Each sampling technique is subject to certain errors that must be taken into account when interpreting the results. The most important of these is the **hygroscopicity of PM** (Won et al., 2021). Particulate matter suspended in humid air absorbs water, increasing its diameter and mass. Optical methods do not measure mass directly, but estimate it based on the amount of scattered light. To convert the signal into a result, the device's algorithm must make certain assumptions about the density and color of PM. The problem is that sensors are usually factory calibrated on standard dust, which has different properties than local suspended dust.

1.7 Particulate Matter Pollution as a Global Threat

1.7.1 Worldwide and Regional Trends

High-income countries have successfully decoupled economic growth from PM emissions through technological regulation. In contrast, low- and middle-income countries face a "double burden" of pollution from traditional biomass burning and rapid industrialization. Analysis of satellite data confirms that while $PM_{2.5}$ levels are stabilizing in Europe and North America, they are rising in South Asia and parts of Africa, creating dangerous conditions for millions of people (Southerland et al., 2022).

1.7.2 WHO and EU Standards Exceeded

The revision of the World Health Organization's (2021) Air Quality Guidelines is a significant shift in thinking, indicating that there is no safe level of exposure to $PM_{2.5}$. The reduction of annual $PM_{2.5}$ standards to 5 $\mu g/m^3$ has resulted in most cities around the world failing to meet these requirements. In the European context, this highlights the differences between the current EU limit values (which are lenient) and the WHO health recommendations, which call for a more rigorous policy approach (WHO 2021).

1.7.3 The Impact of Urbanization and Climate Change

Urbanization exacerbates pollution through the "canyon effect," in which tall buildings trap pollutants at street level. This phenomenon exacerbates the urban heat island effect. Ernst et al. (2022) analyze the interaction between temperature and pollution, where high temperatures cause plants to close their stomata to conserve water, which paradoxically reduces their ability to absorb pollutants and trap PM during periods of intense pollution. Climate change further threatens global food security, with increased levels of ozone and particulate matter estimated to significantly reduce global crop yields (e.g., wheat, rice) by interfering with photosynthesis (Manono et al., 2025).

1.8 Challenges and Future Directions for Research

1.8.1 Gaps in Knowledge About Ultrafine Particles (UFPs)

Ultrafine particles (UFPs, $< 0.1\ \mu m$) represent a critical focus in current aerobiology and toxicology research. Conventional monitoring systems primarily assess particle mass (PM_{10}, $PM_{2.5}$), thereby overlooking the risks associated with the high number concentration and active surface area of UFPs, despite their minimal mass. Schraufnagel (2020) notes that the elevated surface-to-volume ratio of UFPs renders them highly biologically reactive. The absence of standardized assessment methods—specifically, the measurement of particle number versus mass—complicates reliable risk analysis. Notably, UFPs are capable of traversing biological barriers, including the placental barrier, and can accumulate on the fetal side (Bové et al., 2019).

1.8.2 *Real-Time Data Integration and New Analytical Techniques*

The future of air quality monitoring lies in integrated measurement systems, rather than individual solutions. This requires combining

A. **Low-cost sensors (LCS)**, which require advanced calibration methods.
B. **Biomonitoring**: The use of living organisms to assess environmental quality. Recent studies suggest that certain insect species (e.g., *Bombus terrestris*) or plant leaves can serve as passive biomonitoring samples for particulate matter (Moniuszko et al., 2025).

1.8.3 *The Importance of an Interdisciplinary Approach*

Solving the air quality crisis requires a combination of different disciplines.

A. **Plant Physiology and Urban Planning**: The selection of plant species for cities cannot be random. We need species that are not only effective filters (high accumulation) but also physiologically resistant (high tolerance) to harsh urban conditions (Gładysz et al., 2025; Popek et al., 2024; Przybysz et al., 2014).
B. **Medicine and Engineering**: Cooperation in the design of filtration systems inspired by biological barriers.
C. **Data Science**: Model complex interactions between climate change, plant phenology, and pollutant dispersion.

Authors' Declaration on Language Tools We declare that English language and grammar correction tools like Gemini, developed by Google LLC, USA, were used solely for proofreading and polishing the manuscript. These tools did not contribute to developing or generating any scientific content.

References

Abd Elnabi, M. K., Elkaliny, N. E., Elyazied, M. M., Azab, S. H., Elkhalifa, S. A., Elmasry, S., Mouhamed, M. S., Shalamesh, E. M., Alhorieny, N. A., Abd Elaty, A. E., Elgendy, I. M., Etman, A. E., Saad, K. E., Tsigkou, K., Ali, S. S., Kornaros, M., & Mahmoud, Y.A.-G. (2023). Toxicity of heavy metals and recent advances in their removal: A review. *Toxics, 11*(7), 580. https://doi.org/10.3390/toxics11070580

Abdel-Shafy, H. I., & Mansour, M. S. M. (2016). A review on polycyclic aromatic hydrocarbons: Source, environmental impact, effect on human health and remediation. *Egyptian Journal of Petroleum, 25*(1), 107–123. https://doi.org/10.1016/j.ejpe.2015.03.011

Al-Abadleh, H. A. (2021). Aging of atmospheric aerosols and the role of iron in catalyzing brown carbon formation. *Environmental Science: Atmospheres, 1*, 297–345. https://doi.org/10.1039/D1EA00038A

Alves, C. A., Vicente, E. D., Nunes, T., Cipoli, Y., Charres, I., Yubero, E., Galindo, N., Ryšavý, J., & Leitão, A. A. (2025). Elemental and carbonaceous composition of PM_{10} and its oxidative potential in schools in Luanda. *Atmospheric Environment, 364*, 121640. https://doi.org/10.1016/j.atmosenv.2025.121640

Amaral, S. S., De Carvalho, J. A., Jr., Costa, M. A. M., & Pinheiro, C. (2015). An overview of particulate matter measurement instruments. *Atmosphere, 6*(9), 1327–1345. https://doi.org/10.3390/atmos6091327

Attah, R., Kaur, K., Reilly, C. A., Deering-Rice, C. E., & Kelly, K. E. (2025). The effects of photochemical aging and interactions with secondary organic aerosols on cellular toxicity of combustion particles. *Journal of Aerosol Science, 183*, 106473. https://doi.org/10.1016/j.jaerosci.2024.106473

Avino, P., & Manigrasso, M. (2017). Dynamic of submicrometer particles in urban environment. *Environmental Science and Pollution Research, 24*, 13908–13920. https://doi.org/10.1007/s11356-016-6752-8

Bacon, J. R., Butler, O. T., Clairns, W. R., Cavoura, O., Cook, J. M., Davidson, C. M., & Metz-Kraus, R. (2022). Atomic spectrometry update—A review of advances in environmental analysis. *Journal of Analytical Atomic Spectrometry, 33*(1), 8–56. https://doi.org/10.1039/D4JA90056A

Binda, G., Kalčíková, G., Allan, I. J., Hurley, R., Rødland, E., Spanu, D., & Nizzetto, L. (2024). Microplastic aging processes: Environmental relevance and analytical implications. *TrAC—Trends in Analytical Chemistry, 172*, 117566. https://doi.org/10.1016/j.trac.2024.117566

Bové, H., Bongaerts, E., Slenders, E., Bijnens, E. M., Saenen, N. D., Gyselaers, W., Van Eyken, P., Plusquin, M., Roeffaers, M. B. J., Ameloot, M., & Nawrot, T. S. (2019). Ambient black carbon particles reach the fetal side of human placenta. *Nature Communications, 10*, 3866. https://doi.org/10.1038/s41467-019-11654-3

Butwin, M. K., Lowis, S., Pfeffer, M. A., & Thorsteinsson, T. (2019). The effects of volcanic eruptions on the frequency of particulate matter suspension event in Iceland. *Journal of Aerosol Science, 128*, 99–113. https://doi.org/10.1016/j.jaerosci.2018.12.004

Cappa, C. D., Che, D. L., Kessler, S. H., Kroll, J. H., & Wilson, K. R. (2011). Variations in organic aerosol optical and hygroscopic properties upon heterogeneous OH oxidation. *Journal of Geophysical Research: Atmospheres, 116*(D15), D15204. https://doi.org/10.1029/2011JD015918

Chen, D., Liu, X., Han, J., Jiang, M., Xu, Y., & Xu, M. (2018). Measurements of particulate matter concentration by the light scattering method: Optimization of the detection angle. *Fuel Processing Technology, 179*, 124–134. https://doi.org/10.1016/j.fuproc.2018.06.016

Chen, Z., Chen, D., Zhao, C., Kwan, M. P., Cai, J., Zhuang, Y., Zhao, B., Wang, X., Chen, B., Yang, J., Li, R., He, B., Gao, B., Wang, K., & Xu, B. (2020). Influence of meteorological conditions on $PM_{2.5}$ concentrations across China: A review of methodology and mechanism. *Environment International, 139*, 105558. https://doi.org/10.1016/j.envint.2020.105558

Cheriyan, D., & Choi, J. (2020). A review of research on particulate matter pollution in the construction industry. *Journal of Cleaner Production*, 254. https://doi.org/10.1016/j.jclepro.2020.120077

Chiari, M., Yubero, E., Calzolai, G., Lucarelli, F., Crespo, J., Galindo, N., Nicolas, J. F., Giannoni, M., & Nava, S. (2018). Comparison of PIXE and XRF analysis of airborne particulate matter samples collected on Teflon and quartz fibre filters. *Nuclear Instruments and Methods in Physics Research Section B: Beam Interactions with Materials and Atoms, 417*, 128–132. https://doi.org/10.1016/j.nimb.2017.07.031

Choi, J. H., Khamraev, K., & Cheriyan, D. (2022). Hybrid health risk assessment model using real-time particulate matter, biometrics, and benchmark device. *Journal of Cleaner Production, 350*. https://doi.org/10.1016/j.jclepro.2022.131443

Das, S., Roy, A., Masiwal, R., Mandal, M., Popek, R., Chakraborty, M., Sarkar, A., et al. (2023). Comprehensive analysis of PM_1 composition in the Eastern Indo-Gangetic Basin: A three-year urban study. *Sustainability, 15*(20), 14894. https://doi.org/10.3390/su152014894

Dong, J., Chi, Y., Ephraim, A., Nzihou, A., & Millán, L. M. R. (2020). Particulate matter. In *Handbook on characterization of biomass, biowaste and related by-products*. Springer. ISBN 9783030350208.

Duc, H. N., Azzi, M., Zhang, Y., Kirkwood, J., White, S., Trieu, T., Riley, M., Salter, D., Chang, L.T.-C., Capnerhurst, J., Ho, J., Gunashanhar, G., & Monk, K. (2023). Black carbon emissions, transport and effect on radiation forcing modelling during the summer 2019–2020 wildfires in Southeast Australia. *Atmosphere, 14*(4), 699. https://doi.org/10.3390/atmos14040699

Dzierżanowski, K., Popek, R., Gawrońska, H., Sæbo, A., & Gawroński, S. W. (2011). Deposition of particulate matter of different size fractions on leaf surfaces and in waxes of urban forest species. *International Journal of Phytoremediation, 13*(10), 1037–1046. https://doi.org/10.1080/15226514.2011.552929

Ernst, M., Le Mentec, S., Louvrier, M., Loubet, B., Personne, E., & Stella, P. (2022). Impact of urban greening on microclimate and air quality in the urban canopy layer: Identification of knowledge gaps and challenges. *Frontiers in Environmental Science, 10*, 924742. https://doi.org/10.3389/fenvs.2022.92474

European Committee for Standardization. (2014). *Ambient air—Standard gravimetric measurement method for the determination of the PM_{10} or $PM_{2.5}$ mass concentration of suspended particulate matter (EN 12341:2014).*

Feng, S., Gao, D., Liao, F., Zhou, F., & Wang, X. (2016). The health effects of ambient $PM_{2.5}$ and potential mechanisms. *Ecotoxicology and Environmental Safety, 128*, 67–74. https://doi.org/10.1016/j.ecoenv.2016.01.030

Gładysz, K., Wrochna, M., & Popek, R. (2025). Tracking particulate matter accumulation on green roofs: A study at Warsaw University Library. *Air, 3*(1), 4. https://doi.org/10.3390/air3010004

Gopinath, A. K., Raj, S. S., Kommula, S. M., Jose, C., Panda, U., Bishambu, Y., Ojha, N., Ravikrishna, R., Liu, P., & Gunthe, S. S. (2022). Complex interplay between organic and secondary inorganic aerosols with ambient relative humidity implicates the aerosol liquid water content over India during wintertime. *Journal of Geophysical Research: Atmospheres, 127*(13). https://doi.org/10.1029/2021JD036430

Harrison, R. M., & Yin, J. (2000). Particulate matter in the atmosphere: Which particle properties are important for its effects on health? *Science of the Total Environment, 249*(1–3), 85–101. https://doi.org/10.1016/S0048-9697(99)00513-6

Huang, R. J., Wang, Y., Cao, J., Lin, C., Duan, J., Chen, Q., Li, Y., Gu, Y., Yan, J., Xu, W., Frohlich, R., Canonaco, F., Bozzetti, C., Ovadnevaite, J., Ceburnis, D., Canagaratna, M. R., Jayne, J., Worsnop, D. R., El-Haddad, I., Prevot, A. S. H., et al. (2019). Primary emissions versus secondary formation of fine particulate matter in the most polluted city (Shijiazhuang) in North China. *Atmospheric Chemistry and Physics, 19*, 2283–2298. https://doi.org/10.5194/acp-19-2283-2019

Intergovernmental Panel on Climate Change (IPCC). (2023). Short-lived climate forcers. In *Climate change 2021—The physical science basis: Working group I contribution to the sixth assessment report of the intergovernmental panel on climate change* (pp. 817–922). Cambridge University Press (Chapter). https://doi.org/10.1017/9781009157896.008

Karthikeyan, S., See, S. W., & Balasubramanian, R. (2007). Simultaneous determination of inorganic anions and selected organic acids in airborne particulate matter by ion chromatography. *Analytical Letters, 40*(4), 793–804. https://doi.org/10.1080/00032710601017920

Kim, D., Gautam, M., & Gera, D. (2002). Parametric studies on the formation of diesel particulate matter via nucleation and coagulation modes. *Journal of Aerosol Science, 13*, 1609–1621. https://doi.org/10.1016/S0021-8502(02)00119-2

Kim, H., Zhang, Q., & Heo, J. (2018). Influence of intense secondary aerosol formation and long-range transport on aerosol chemistry and properties in the Seoul Metropolitan Area during springtime: Results from KORUS-AQ. *Atmospheric Chemistry and Physics, 18*, 7149–7168. https://doi.org/10.5194/acp-18-7149-2018

Kim, J., Go, T., & Lee, S. J. (2021). Accurate real-time monitoring of high particulate matter concentration based on holographic speckles and deep learning. *Journal of Hazardous Materials*, 409. https://doi.org/10.1016/j.jhazmat.2020.124637

Kim, K. H., Kabir, E., & Kabir, S. (2015). A review on the human health impact of airborne particulate matter. *Environment International, 71*, 136–143. https://doi.org/10.1016/j.envint.2014.10.005

Klimont, Z., Kupiainen, K., Heyes, C., Purohit, P., Cofala, J., Rafaj, P., Borhen-Kleefeld, J., & Schopp, W. (2017). Global anthropogenic emissions of particulate matter including black carbon. *Atmospheric Chemistry and Physics, 17*, 8681–8723. https://doi.org/10.5194/acp-17-8681-2017

Kravchishina, M. D., Lisitsyn, A. P., Klyuvithin, A. A., Novigatsky, A. N., Politova, N. V., & Shevchenko, V. P. (2018). Suspended particulate matter as a main source and proxy of the sedimentation processes. *The Handbook of Environmental Chemistry, 82*, 13–48. https://doi.org/10.1007/698_2018_353

Kumar, P., Morawska, L., Martani, C., Biskos, G., Neophytou, M., Di Sabatino, S., Bell, M., Norford, L., & Britter, R. (2015). The rise of low-cost sensing for managing air pollution in cities. *Environment International, 75*, 199–205. https://doi.org/10.1016/j.envint.2014.11.019

Lee, E. S., Lee, T. J., Park, M. B., Park, D. S., & Kim, D. S. (2017). Characteristics of particulate matter concentration and classification of contamination patterns in the Seoul metropolitan subway tunnels. *Journal of Korean Society for Atmospheric Environment*. https://doi.org/10.5572/KOSAE.2017.33.6.593

Lee, H. M., Bae, M. S., Kim, J. H., et al. (2025). Characteristics and chemical composition of $PM_{2.5}$ in agricultural regions: A case study in Jeonbuk-do, South Korea. *Air Quality, Atmosphere & Health, 18*, 3567–3580. https://doi.org/10.1007/s11869-025-01849-7

Leskinen, J., Hartikainen, A., Väätäinen, S., Ihalainen, M., Virkkula, A., Mesceriakovas, A., Tiitta, P., Miettinen, M., Lamberg, H., Czech, H., Yli-Pirilä, P., Tissari, J., Jakobi, G., Zimmermann, R., & Sippula, O. (2023). Photochemical aging induces changes in the effective densities, morphologies, and optical properties of combustion aerosol particles. *Environmental Science & Technology, 57*(13), 5137–5148. https://doi.org/10.1021/acs.est.2c04151

Lin, Y.-C., Shih, H.-S., & Lai, C.-Y. (2022). Classification of air quality zones and fine particulate matter sensitives areas by risk assessment approach. *Environmental Research, 215*, 1. https://doi.org/10.1016/j.envres.2022.114208

Lippmann, M., & Chen, L. C. (2009). Health effects of concentrated ambient air particulate matter (CAPs) and its components. *Critical Reviews in Toxicology, 39*(10), 865–913. https://doi.org/10.3109/10408440903300080

Mandal, M., Popek, R., Przybysz, A., Roy, A., Das, S., & Sarkar, A. (2023). Breathing fresh air in the city: Implementing avenue trees as a sustainable solution to reduce particulate pollution in urban agglomerations. *Plants, 12*(7), 1545. https://doi.org/10.3390/plants12071545

Manono, B. O., Sadiq, F. K., Sadiq, A. A., Matsika, T. A., & Tanko, F. (2025). Impacts of air quality on global crop yields and food security: An integrative review and future outlook. *Air, 3*(3), 24. https://doi.org/10.3390/air3030024

May, N. W., Gunsch, M. J., Olson, N. E., Bondy, A. L., Kirpes, R. M., Bertman, S. B., China, S., Laskin, A., Hopke, P. K., Ault, P. K., & Pratt, K. A. (2018). Unexpected contributions of sea spray and lake spray aerosol to inland particulate matter. *Environmental Science & Technology Letters I, 5*(7), 405–412. https://doi.org/10.1021/acs.estlett.8b00254

Moniuszko, H., Przybysz, A., Borański, M., Splitt, A., Jachuła, J., & Popek, R. (2025). Buff-tailed bumblebee, an underrated indicator of air pollution: A comparison of particulate matter accumulation by *Bombus terrestris* L. and *Apis mellifera* L. *Environmental Toxicology and Chemistry, 44*(1), 282–293. https://doi.org/10.1093/etojnl/vgae020

Noble, C. A., Vanderpool, R. W., Peters, T. M., McElroy, F. F., Gemmill, D. B., & Wiener, R. W. (2001). Federal reference and equivalent methods for measuring fine particulate matter. *Aerosol Science & Technology, 34*(5), 457–464. https://doi.org/10.1080/02786820121582

Okuda, T. (2013). Measurement of the specific surface area and particle size distribution of atmospheric aerosol reference materials. *Atmospheric Environment, 75*, 1–5. https://doi.org/10.1016/j.atmosenv.2013.04.033

Palm, B. B., Campuzano-Jost, P., Day, D. A., Ortega, A. M., Fry, J. L., Brown, S. S., Zarzana, K. J., Dube, W., Wagner, N. L., Draper, D. C., Kaser, L., Jud, W., Karl, T., Hansel, A., Gutiérrez-Montes, C., & Jimenez, J. L. (2017). Secondary organic aerosol formation from in situ OH, O_3, and NO_3 oxidation of ambient forest air in an oxidation flow reactor. *Atmospheric Chemistry and Physics, 17*, 5331–5354. https://doi.org/10.5194/acp-17-5331-2017

Parvez, F., Lamancusa, C., & Wagstrom, K. (2017). Primary and secondary particulate matter intake fraction from different height emission sources. *Atmospheric Environment, 165*, 1–11. https://doi.org/10.1016/j.atmosenv.2017.06.011

Patel, P., & Aggarwal, S. G. (2022). On the techniques and standards of particulate matter sampling. *Journal of the Air & Waste Management Association, 72*(8), 791–814. https://doi.org/10.1080/10962247.2022.2048129

Perraud, V., Bruns, E. A., Ezell, M. J., Johnson, S. N., Yu, Y., Alexander, M. L., Zelenyuk, A., Imre, D., Chang, W. L., Dabdub, D., Pankow, J. F., & Finlayson-Pitts, B. J. (2012). Nonequilibrium atmospheric secondary organic aerosol formation and growth. *Proceedings of the National Academy of Sciences of the United States of America, 109*(8), 2836–2841. https://doi.org/10.1073/pnas.1119909109

Perrone, M. R., Romano, S., Genga, A., & Paladini, F. (2018). Integration of optical and chemical parameters to improve the particulate matter characteriazation. *Atmospheric Research, 205*, 93–106. https://doi.org/10.1016/j.atmosres.2018.02.015

Popek, R., Fornal-Pieniak, B., Chyliński, F., Pawełkowicz, M., Bobrowicz, J., Chrzanowska, D., Piechota, N., & Przybysz, A. (2022). Not only trees matter—Traffic-related PM accumulation by vegetation of urban forests. *Sustainability, 14*(5). https://doi.org/10.3390/su14052973

Popek, R., Łukowski, A., & Karolewski, P. (2017). Particulate matter accumulation—Further differences between native *Prunus padus* and non-native *P. serotina*. *Dendrobiology, 78*, 85–95. https://doi.org/10.12657/denbio.078.009

Popek, R., Łukowski, A., Roy, A., Mandal, M., Przybysz, A., Zając, Z., & Sarkar, A. (2025). Combined efficiency of tree and shrub vegetation barriers in mitigating PM, TEs, and PAHs along urban roadways. *International Journal of Phytoremediation*, 1–9. https://doi.org/10.1080/15226514.2025.2555595

Popek, R., Roy, A., Mandal, M., Przybysz, A., Drążkiewicz, K., Romanowska, P., & Sarkar, A. (2024). Enhancing urban sustainability: How spatial and height variability of roadside plants improves pollution capture for greener cities. *Sustainability, 16*(24), 11131. https://doi.org/10.3390/su162411131

Przybysz, A., Sæbø, A., Hanslin, H. M., & Gawroński, S. W. (2014). Accumulation of particulate matter and trace elements on vegetation as affected by pollution level, rainfall and the passage of time. *The Science of the Total Environment, 481*, 360–369. https://doi.org/10.1016/j.scitotenv.2014.02.072

Querol, X., Viana, M., Alastuey, A., Amato, F., Moreno, T., Castillo, S., Pey, J., De La Rosa, J., Sánchez De La Campa, A., Artíñano, B., Salvador, P., García Dos Santos, S., Fernández-Patier, R., Moreno-Grau, S., Negral, L., Minguillón, M. C., Monfort, E., Gil, J. I., Inza, A., Ortega, L. A., et al. (2007). Source origin of trace elements in PM from regional background, urban and industrial sites of Spain. *Atmospheric Environment, 41*(34), 7219–7231. https://doi.org/10.1016/j.atmosenv.2007.05.022

Raja, S., Chandrasekaran, S. R., Lin, L., Xia, X., Hopke, P. K., & Valsaraj, K. T. (2017). Analysis of beta attenuation monitor filter rolls for particulate matter speciation. *Aerosol and Air Quality Research, 17*, 14–23. https://doi.org/10.4209/aaqr.2016.03.0122

Reijnders, J., Boot, M., & Goey, P. (2018). Particle nucleation-accumulation mode trade-off: A second diesel dilemma? *Journal of Aerosol Science, 124*, 95–111. https://doi.org/10.1016/j.jaerosci.2018.06.013

Roy, A., Das, S., Singh, P., Mandal, M., Kumar, M., Rajlaxmi, A., Sarkar, A., et al. (2024a). Summertime monitoring and source apportionment study of both coarse, fine, and ultra-fine particulate pollution in Eastern Himalayan Darjeeling: A hint to health risk during peak tourist season. *Mapan, 39*(4), 995–1009. https://doi.org/10.1007/s12647-024-00776-9

Roy, A., Mandal, M., Das, S., Kumar, M., Popek, R., Awasthi, A., Sarkar, A., et al. (2024b). Non-exhaust particulate pollution in Asian countries: A comprehensive review of sources, composition, and health effects. *Environmental Engineering Research, 29*(3), 44–68. https://doi.org/10.4491/eer.2023.384

Schraufnagel, D. E. (2020). The health effects of ultrafine particles. *Experimental & Molecular Medicine, 52*, 311–317. https://doi.org/10.1038/s12276-020-0403-3

Shahsavani, A., Tobias, A., Querol, X., Stafoggia, M., Abdolshanhnejad, M., Mayvaneg, F., Guo, Y., Hadei, M., Hashemi, S. S., Khosravi, A., Namvar, Z., Yarahmadi, M., & Emam, B. (2020). Short-term effects of particulate matter during desert and non-desert dust days on mortality in Iran. *Environment International*, 134. https://doi.org/10.1016/j.envint.2019.105299

Sicard, P., Agathokleous, E., Anenberg, S. C., De Marco, A., Paoletti, E., & Calatayud, V. (2023). Trends in urban air pollution over the last two decades: A global perspective. *Science of Total Environment, 858*, Article 160064. https://doi.org/10.1016/j.scitotenv.2022.160064

Song, Y., Maher, B. A., Li, F., Wang, X., Sun, X., & Zhang, H. (2015). Particulate matter deposited on leaf of five evergreen species in Beijing, China: Source identification and size distributrion. *Atmospheric Environment, 105*, 53–60. https://doi.org/10.1016/j.atmosenv.2015.01.032

Southerland, V. A., Brauer, M., Mohegh, A., Hammer, M. S., van Donkelaar, A., Martin, R. V., Apte, J. S., & Anenberg, S. C. (2022). Global urban temporal trends in fine particulate matter ($PM_{2.5}$) and attributable health burdens: Estimates from global datasets. *The Lancet. Planetary Health, 6*(2), e139–e146. https://doi.org/10.1016/S2542-5196(21)00350-8

U.S. Environmental Protection Agency. (2024). Reference method for the determination of particulate matter as PM_{10} in the atmosphere. In *Code of federal regulations (Title 40, Part 50, Appendix J)*. Government Printing Office.

Wiseman, C. L. S., Levesque, C., & Rasmussen, P. E. (2021). Characterizing the sources, concentrations and resuspension potential of metals and metalloids in the thoracic fraction of urban road dust. *Science of the Total Environment, 786*, Article 147467. https://doi.org/10.1016/j.scitotenv.2021.147467

Won, W. S., Oh, R., Lee, W., Ku, S., Su, P. C., & Yoon, Y. J. (2021). Hygroscopic properties of particulate matter and effects of their interactions with weather on visibility. *Scientific Reports, 11*, 16401. https://doi.org/10.1038/s41598-021-95834-6

World Health Organization. (2021). *WHO global air quality guidelines: Particulate matter ($PM_{2.5}$ and PM_{10}), ozone, nitrogen dioxide, sulfur dioxide and carbon monoxide.* https://www.who.int/publications/i/item/9789240034228/

Yang, L., Zhang, H., Zhang, X., Xing, W., Wang, Y., Bai, P., Zhang, L., Hayakawa, K., Toriba, A., & Tang, N. (2021). Exposure to atmospheric particulate matter-bound polycyclic aromatic hydrocarbons and their health effects: A review. *International Journal of Environmental Research and Public Health, 18*(4), 2177. https://doi.org/10.3390/ijerph18042177

Zeb, B., Alam, K., Sorooshian, A., Blaschke, T., Ahmad, I., & Shahid, I. (2018). On the morphology and composition of particulate matter in an urban environment. *Aerosol and Air Quality Research, 18*, 1431–1447. https://doi.org/10.4209/aaqr.2017.09.0340

Zhang, Y., Liu, P., Liu, X.-H., Jacobson, M. Z., McMurry, P. H., Yu, F., Yu, S., & Schere, K. L. (2010). A comparative study of nucleation parameterizations: 2. Three-dimensional model application and evaluation. *Journal of Geophysical Reaserch, 115*, D20. https://doi.org/10.1029/2010JD014151

Zhu, Y., Wan, Q., Hou, Q., Feng, Y., Yu, J., Shi, J., & Xia, C. (2024). Analysis of diffusion characteristics and influencing factors of particulate matter in ship exhaust plume in arctic environment based on CFD. *Atmosphere, 15*(5), 580. https://doi.org/10.3390/atmos15050580

Chapter 2
Role of Meteorological Parameters on Particulate Matter Distribution and Concentration

Avantika Pandey, Akanksha Adhikari, Sita Ram Mishra, and Vinod Kumar Dubey

Abstract Atmospheric pollution from increased particulate matter concentrations has emerged as a significant global concern, driven by rapid industrialization and urbanization. Particulate matter (PM) is a complex mixture of both liquid and solid particles of different natures, including minerals, ions, metals, organic matter, soot, dust, and polycyclic aromatic hydrocarbons. Based on aerodynamic size, these PM are classified into PM_{10} (with an aerodynamic diameter $\leq$ 10 μm), $PM_{2.5}$ (with an aerodynamic diameter $\leq$ 2.5 μm) and $PM_{0.1}$ (ultrafine, $\leq$ 0.1 μm). The concentration and dispersion of PM in the atmosphere are influenced by various factors, including emission sources, meteorological parameters and topography. Among these factors, meteorological parameters such as temperature, wind speed and direction, humidity, precipitation and solar radiation played a very important role in dispersion and concentration of PM in the atmosphere. A single meteorological parameter alone cannot adequately explain particulate matter concentration and dispersion; the combined influence of multiple meteorological parameters must be studied simultaneously. Consequently, this chapter addresses in depth the influence of several meteorological parameters on the dispersion and concentration of PM in the atmosphere.

A. Pandey (✉)
Department of Botany, University of Lucknow, Lucknow, Uttar Pradesh, India
e-mail: avantikabhu023@gmail.com

A. Adhikari
Department of Zoology, Radhey Hari Government Post Graduate College, Kashipur, Udham Singh Nagar, Uttarakhand, India

S. R. Mishra
Department of Agricultural Meteorology, College of Agriculture, Acharya Narendra Deva University of Agriculture and Technology, Ayodhya, Uttar Pradesh, India

V. K. Dubey
Department of Genetics and Plant Breeding, College of Agriculture, Acharya Narendra Deva University of Agriculture and Technology, Ayodhya, Uttar Pradesh, India

A. Sarkar et al. (eds.), *From Sky to Soil*, https://doi.org/10.1007/978-3-032-31991-3_2

Keywords Meteorological parameters · Particulate matter · Precipitation · Temperature · Wind speed · Wind direction

2.1 Introduction

Particulate matter (PM) is a mixture of many substances, including minerals, ions, metals, organic matter, soot, polycyclic aromatic hydrocarbons (PAH) and microorganisms, which differ in shape, size, morphology, surface area, origin, solubility, physical, chemical and biological properties (Yadav et al., 2015). PM is classified according to aerodynamic size as PM_{10} (particulate matter with an aerodynamic diameter $\leq$ 10 μm) and $PM_{2.5}$ (particulate matter with an aerodynamic diameter $\leq$ 2.5 μm), $PM_{0.1}$ (ultrafine, $\leq$ 0.1 μm). PM originates from both natural sources, such as windblown dust, wildfires, volcanic eruptions, and sea-salt aerosols, or from anthropogenic activities. Anthropogenic sources of PM encompass household combustion, road traffic, particularly emissions from diesel and petrol engines; pavement erosion due to vehicular movement; and tire and brake abrasion, alongside emissions from manufacturing and energy sectors, including construction, metal processing, smelting and mining operations, and cement and brick production (Kirešová & Guzan, 2022) (Fig. 2.1). The detrimental impacts of particulate matter on health and the environment have been extensively studied and documented (Kirešová & Guzan, 2022; Singh & Verma, 2023). It is established that particulate matter adversely impacts the respiratory and cardiovascular systems. Smaller particles have a greater effect on human health: coarse particles accumulate in the upper respiratory tract, fine particles penetrate the alveoli of the lungs, and ultrafine particles may enter the bloodstream (Kirešová & Guzan, 2022).

Nevertheless, particulate matter is not simply a health concern. Rather, it also has important environmental impacts that include reduced visibility, acid rain, destruction of ecosystems, and climate change. Understanding the effects of meteorological parameters on particulate matter concentration and dispersion is vital for developing effective policies to manage air quality and ensure the health and safety of the general population (Roy et al., 2025; Zender-Świercz et al. 2024).

It has been well known that the concentration and dispersion of these PM are strongly influenced by meteorological conditions (Xing & Sun, 2022) (Fig. 2.1).

For example, various researchers have described the constituents of $PM_{2.5}$, determining that precipitation mainly enhances removal of $PM_{2.5}$, whereas wind direction, wind speeds, and topography mainly influence the transport and dispersion of $PM_{2.5}$. In addition, Xing and Sun (2022) indicated that meteorological parameters primarily influenced variations in particulate matter pollution levels in Beijing. It has also been reported that an increase in temperature enhances chemical reactions in the atmosphere, which may, in turn, increase levels of secondary pollutants. At the same time, an increase in temperature elevates the height of the atmospheric boundary layer, creating space for dilution. Therefore, an increase in temperature may either elevate or reduce the levels of PM (Das et al., 2023).

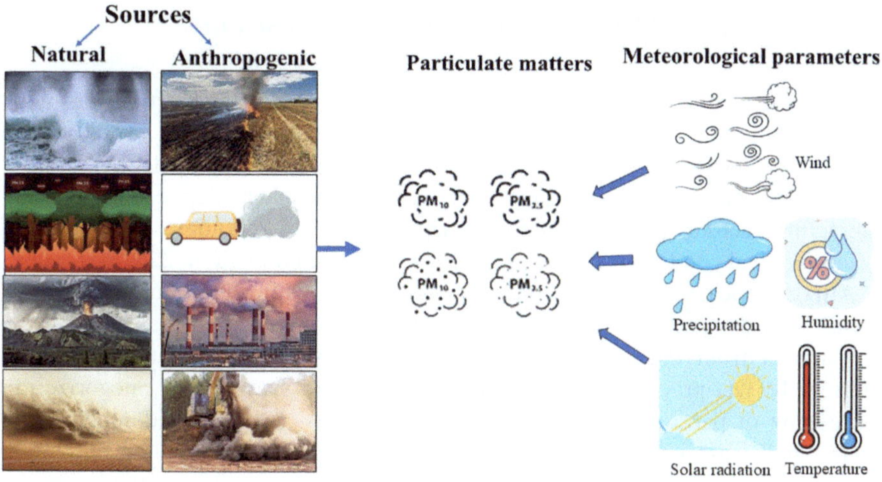

Fig. 2.1 Diagrammatic representation of various natural and anthropogenic sources of particulate matter and the influence of meteorological parameters on particulate matter

Lower wind speeds may cause atmospheric stagnation, thereby confining contaminants near the surface and elevating exposure levels. Rise in relative humidity levels enables the hygroscopic growth of atmospheric particulate matter, thus accelerating gravitational settling and consequently lowering $PM_{2.5}$ levels. Changes in air pressure influence the levels of $PM_{2.5}$, which may accumulate in some places and be diluted elsewhere (Basemera et al., 2025).

A study conducted by Basemera et al. (2025) at Mbarara, Uganda reported that temperature, wind speed, wind direction, and relative humidity were the variables that strongly linked with the diurnal change of $PM_{2.5}$ concentration. Polycyclic aromatic hydrocarbons (PAHs) are pervasive organic pollutants primarily produced by the pyrolysis or incomplete combustion of organic matter. PAHs can exist in both particulate and gas forms in the atmosphere. In particulate form, they may exist as PM_{10}-PAH or $PM_{2.5}$ PAH. Khan et al. (2018) reported that PAH concentrations are predominantly influenced by two factors: meteorological parameters and the intensity of emission sources. PAHs exhibited an inverse correlation with solar radiation, temperature, and wind velocity. This study indicates that biomass combustion for domestic heating and cooking, along with the adsorption of gaseous PAHs onto particles due to low air temperatures, were the primary factors contributing to elevated PAH concentrations during winter in the Veneto region (Khan et al., 2018). In the subsequent section, we discuss in detail the fundamentals of particulate matter and the role of various meteorological parameters in influencing particulate matter transport, deposition, and concentrations.

2.2 Particulate Matter Dynamics and Deposition

PM varies in size, source, chemical composition, and physical properties. It is one of the most complex and important elements of the atmosphere (Liu et al., 2020). The major characteristics of PM and their implications are shown in Table 2.1.

The time particles remain suspended in the atmosphere depends on multiple interacting factors, including particle size, density, morphology, and prevailing meteorological conditions (Yang et al., 2024) as depicted in Table 2.2.

Atmospheric deposition is a crucial method for mitigating air pollution. It encompasses both dry and wet processes. Dry deposition involves the transfer of gases or particles from the atmosphere directly to the surface without precipitation. Conversely, wet processes, often termed rain or snow scavenging, categorize rain scavenging of particulate matter as 'rainout particles' (act as cloud-condensation nuclei) and 'washout' (the removal of below-cloud particles by falling raindrops). Research has demonstrated that dry deposition effectively removes particulate matter and is influenced by surface type, temporal changes, spatial variations, diurnal fluctuations and meteorological conditions. Wet deposition also plays a significant role in decreasing air pollution through the removal of particulate matter and thereby directly influencing particulate matter concentrations. A study by Wu et al. (2018) in Wetland evaluated the effects of wet and dry deposition on particulate matter concentration. This study established the data underpinning measures that could enhance air quality.

Table 2.1 Characteristics of various particulate matter

Particulate matter	Aerodynamic diameter (μm)	Characteristics		Medical implications
PM_{10} (coarse particles)	≤ 10	Higher settling velocity	Shorter atmospheric residence time	Deposits mainly in the upper respiratory tract
$PM_{2.5}$ (fine particles)	≤ 2.5	Lower settling velocity	Remains airborne longer	Penetrates deep into the lungs
$PM_{0.1}$ (ultrafine particles)	≤ 0.1	Very weak gravitational settling; high diffusion	Can remain suspended for extended periods	Can reach alveoli and the bloodstream

Table 2.2 Prevalence of particle type in the atmosphere

Particle type	Residence time	Reasons
Coarse particles	Hours to a few days	Heavy, settles quickly by gravity
Fine particles	Several days	Smaller, slower settling
Ultrafine particles	Days to weeks	Very small, remains suspended easily

Smaller particles stay in the atmosphere longer and can travel long distances (on regional or continental scales)

The efficacy of $PM_{2.5}$ removal via wet deposition was significant, underscoring the need to regulate air pollutant emissions under dry weather. The effectiveness of $PM_{2.5}$ removal through this mechanism of wet deposition was substantial, which highlights the importance of regulating air pollutant emissions under dry weather conditions (Mandal et al., 2023).

2.3 Role of Meteorological Parameters in Affecting Particulate Matter

The weather plays a major role in the emission, transportation, and removal of particulate matter into the atmosphere. This is because the dispersion of PM is affected by the weather. In this context, stability, mixing height, and chemical reactions are controlled by meteorological parameters, which, in turn, affect the concentration and distribution of PM. Among the important meteorological parameters that affect particulate matter are wind speed and direction, temperature, precipitation, humidity, and solar radiation, as shown in Table 2.3.

Research has shown that an in-depth analysis of the relationship between different meteorological factors and the level of pollutants is a requirement for effective management. The ratios of particulate matter are drastically affected, such that pollutants can either accumulate or dissipate, depending on the velocity and direction of wind. For instance, research found that while stagnant air quality can increase the pollution levels, especially in places that experience the effects of automobiles, strong gusts of wind can reduce pollution by blowing pollutants out of big cities. However, the pattern of particulate matter is further compounded by the effects of climatic cycles. For instance, humid conditions associated with rainy seasons can reduce pollutant levels through sedimentation (Yadav et al., 2015).

In addition, the distribution and accumulation of such pollutants can vary significantly over time and space in the same geographical area. Furthermore, several meteorological parameters may influence particulate matter concentrations in a given region across different seasons of the year (Scudeler et al., 2025). Yang et al. (2017) studied the correlation between $PM_{2.5}$ concentrations and various meteorological conditions in China from February 2013 to November 2014. They observed that relative humidity significantly affected particulate matter concentrations in spring and summer, while wind speed and temperature were the major influencing factors in winter. It is worth noting that a single meteorological parameter alone cannot adequately explain particulate matter concentrations, deposition, and distribution; the combined influence of multiple meteorological parameters must be studied simultaneously (Scudeler et al., 2025).

Table 2.3 Various quantifications of meteorological parameters for the methodological evaluation

Quantification parameter	Methodological evaluation	Observations	Feedback assessment	References
Temperature and humidity favour particle accumulation during heat waves	Statistical correlation and heavy-tailed probability analysis	Urban basin geomorphology with summer focus	Heat waves and drought accelerate particle persistence	Pacheco Hernández et al. (2025)
Meteorological factors significantly influence pollutant levels with lagged effects	Time series analysis with lag models	Seasonal and diurnal variations in urban	Climate shapes pollutant transport and photochemical processes	Soleimanpour et al. (2023)
Chaotic dynamics link meteorology and pollutants in different morphologies	Chaos theory and fractal dimension analysis	Basin, mountain, and coastal urban locations	Pollutants dominate entropic forces affecting diffusion	Pacheco Hernández et al. (2025)
Wind speed and temperature strongly influence secondary pollutants	Gradient boosting regression and spatial analysis	Fine-scale urban meteorology	Urban heat and morphology interact with the weather for pollution	Guo et al. (2022)
Geopotential height, surface temperature, and wind speed impact aerosol variability	Autoregressive distributed lag (ARDL)	Nigeria's urban and rural areas are affected	Wind speed notably influences aerosol dynamics	Ibe et al. (2025)
Temperature and humidity show significant lagged correlations with PM_{10} and $PM_{2.5}$	Uni and multivariate linear regression	Seasonal urban aerosol trends	Meteorology crucial for seasonal urban pollution formation	Zender-Świercz et al. (2024)
Temperature and pollution interactions increase aerosols and precipitation	Regional numerical framework	Urban heat and pollution islands	Aerosols' feedback on urban heat island and precipitation	Wang et al. (2022)
Temperature, humidity, and traffic emissions	GIS-based imaging	Local-scale urban areas	Urban form and climate jointly affect air quality	Yaacob et al. (2024)
Seasonal meteorological changes	Predictive modeling	Urban Indonesia with seasonal pollutant focus	Weather and geography jointly influence pollution patterns	Amin et al. (2024)

(continued)

Table 2.3 (continued)

Quantification parameter	Methodological evaluation	Observations	Feedback assessment	References
Temperature, humidity, rainfall, wind speed	Multiple linear regression and HYSPLIT trajectory analysis	Bangkok urban area with biomass burning influence	Meteorology and biomass burning jointly affect PM levels	Tesfaldet and Chanpiwat (2023)
Wind behaviour in the urban basin	Chaos theory and entropy-based indicators	Urban basin micrometeorology with height sensitivity	Urbanisation reduces micrometeorological resilience	Li et al. (2026)
Urban heterogeneity impacts meteorology and air quality	Observational data analysis with local climate zones	Nanjing downtown with urban heterogeneity focus	Urban form influences thermal comfort and pollution resistance	Zhong et al. (2023)
Meteorological components influence pollutant spatial patterns	Physical modelling with chemical transport models	Southern Lake Michigan urban–rural gradients	Seasonal O_3 regimes and urban pollution patterns	Montgomery et al. (2023)
Cross-correlations between pollutants and temperature, humidity, and wind	Detrended fluctuation and cross-correlation coefficient methods	Coastal urban region in Brazil with multi-station data	Meteorology-pollution relationships with temporal scales	Taheri et al. (2025)
Negative correlations between PM and temperature, wind speed, and precipitation	Linear correlation	Metropolitan region with basin geography	Suspended particles affect PM transport, accumulation, and visibility	Li et al. (2026)
Urban densification alters meteorological and pollutant chaotic parameters	Nonlinear time series and chaos theory	Urban densification over two periods	Urbanisation increases complexity and reduces meteorological persistence	Montgomery et al. (2023)
Positive correlation of PM_{10} with wind speed and temperature during dust storms	Empirical modelling and correlation analysis	Middle Eastern desert dust storm events	Wind speed and pressure modulate dust storm PM_{10} peaks	Maleki et al. (2022)
Positive correlations of $PM_{2.5}$ with temperature, humidity, and wind speed	Spatial interpolation and regression modelling	Cartagena urban district with limited station data	Meteorology shapes spatial $PM_{2.5}$ distribution	Aldegunde et al. (2022)

(continued)

Table 2.3 (continued)

Quantification parameter	Methodological evaluation	Observations	Feedback assessment	References
Significant correlations between PM and precipitation, temperature, and humidity	Statistical, seasonal, and correlation analyses	Southeast Brazil urban centres with seasonal focus	Nonlinear meteorology-PM relationships vary by season	Moreira et al. (2024)
Wind speed, humidity, and temperature control the $PM_{2.5}$ spatial distribution	Multiple regression and machine learning models	Atlanta metropolitan area with urban form integration	Meteorology and urban form jointly affect $PM_{2.5}$ patterns	Tian et al. (2024)
Low wind, high humidity, and temperature inversion are linked to pollution episodes	Observational and remote sensing data analysis	Beijing urban boundary layer during pollution episodes	Meteorology modulates boundary layer and pollutant accumulation	Miao et al. (2022)
Temperature positively correlates with $PM_{2.5}$ and PM_{10}, while humidity negatively correlates with them	Correlation and regression	Urban and rural haze periods	Climate factors influence particles with spatial and temporal variability	Wang et al. (2024)
Strongly correlates with pollutant concentrations	Data mining methods	Urban area with multipollutant focus	Pollutant levels via complex interactions	
Pollutant decay and dispersion	Thermodynamic principles	Urban area with time series data	Meteorology-pollutant entropic interactions affect dispersion	

2.3.1 *Wind Speed and Direction*

Wind speed is a crucial determinant of air pollutant concentrations; specifically, pollutant concentrations generally decrease as wind velocity increases. Khan et al. (2018) observed a negative relationship between wind speed and PAH concentrations. In another study, carried out by Basemera et al. (2025) at Mbarara, Uganda, reported that wind speed exhibits a positive correlation with $PM_{2.5}$, indicating that local $PM_{2.5}$ sources are dispersed by the wind more slowly than the wind generates new airborne $PM_{2.5}$ through particle resuspension. Kliengchuay et al. (2018) conducted a study to find out the relationship between meteorological parameters and levels of particulate matter. Authors reported that wind direction is a significant factor influencing PM. Winds originating from various directions transport various amounts of pollutants.

Throughout this investigation, elevated PM_{10} concentrations were produced when the wind originated from the north or northeast. This suggests that winds from the north or northeast during the northeastern monsoon season affect PM_{10} concentrations.

Airflow regimes can also influence the dispersion of particulate matter. Feng et al. (2023) explain that a normal dispersion value less than one m s^{-1} is likely to produce poor dispersion and pollutant buildup. Similarly, Swa et al. (2025) stated that low-value dispersion, one to three m s^{-1}, limits mixing, resulting in increased PM near sources. The moderate-dispersion value of three to six m s^{-1} improves turbulent mixing and reduces surface concentrations. Finally, Zhang et al. (2020) reported that the strong pre-treatment of six to ten m s^{-1} allows the long-range transport of PM. The primary determinant of molecules' dry deposition velocity is wind speed, though degree is the next pressing ecological factor.

2.3.2 *Temperature*

Temperature also plays a vital role in deposition, dispersion and the concentration of PM in the atmosphere. A study carried out by Scudeler et al. (2025) in a small town in Southern Brazil over a period of two years observed that in winter, when lower temperatures occurred, the concentration of PM_{10} was found to be higher; however, in summer (when higher temperatures occurred), lower PM_{10} concentrations were found. The study suggested that the rise in anthropogenic emissions during winter was mainly attributable to the utilization of fossil fuels, predominantly coal, for residential heating. In addition to the emission source i.e., wood burning, elevated PM concentrations at the surface during winter may also be ascribed to the adverse circumstances for vertical dispersion of the pollutant experienced in this season. The enhanced thermal stability conditions prevalent in winter impede vertical air-mass movement (convection), a phenomenon that is more pronounced during summer (Scudeler et al., 2025). In another study, Khan et al. (2018) reported that atmospheric temperature was negatively associated with particulate PAH concentrations. In summer, from April to August, as ambient temperatures increased, a declining trend in PAH levels was observed. A comparable correlation was noted throughout the winter months, indicating that elevated temperatures facilitate the evaporation of particulate PAHs and lower concentrations. In a study performed by Kayes et al. (2019) in Bangladesh, PM_{10} also exhibited a negative correlation with air temperature during the pre-monsoon (March–May) post-monsoon period (October–November). A substantial negative association between PM and temperature during these periods suggests that increased PM concentration in the atmosphere may lead to a reduction in atmospheric temperature due to net negative radiative forcing. However, in the monsoon, there was a positive association between temperature and PM_{10} levels, which can be elucidated by the meteorological aspects of the monsoon season and their interactions with airborne particulate matter (Kayes et al., 2019).

2.3.3 Precipitation and Humidity

Precipitation is a critical factor in removing PM from the atmosphere through its "washing effect." Most researchers have noted that increased rainfall intensity is associated with lower PM concentrations. However, some studies also highlighted that particulate concentrations are not consistently reduced during rainfall events; this phenomenon primarily results from external sources (long-range transport) that can augment the particulate concentration at the monitored location (Scudeler et al., 2025). Additional research suggested that the impact of precipitation on particulate matter removal may also vary with the time-of-day rainfall occurs, particle size, and atmospheric stability (Scudeler et al., 2025). Xing and Sun (2022) reported that precipitation levels are associated with particulate matter levels in a consistent negative manner. The authors suggested that precipitation's efficacy in removing contaminants depends on its duration, intensity, and overall volume. The physical impacts of scouring, dilution, and collision from substantial rains also diminish atmospheric particulate matter.

Relative humidity is a crucial factor in the hygroscopic expansion of particulate matter and aerosols. As they decrease in size, the particles are deposited via Brownian diffusion. The larger particles demonstrate inertial effects, sedimentation, and interception. The hygroscopicity of particles is influenced by their temperature, composition, size, and humidity levels. The chemical composition of PM is contingent upon regional and environmental conditions (Molnár et al., (2020). The rise in sedimentation rate is attributed to the enhanced density of particles that have amalgamated with atmospheric moisture. Consequently, humidity influences not only visibility and radiation but also particulate matter concentration (Zender-Świercz et al. 2024). Furthermore, Scudeler et al. (2025) identified a negative association between relative humidity and PM_{10} concentration, attributable to the coalescence of particles with air humidity, leading to their deposition in the atmosphere.

2.4 Conclusions and Future Perspectives

The concentration of particulate matter (PM) is significantly influenced by weather conditions.

In addition to local sources, meteorological factors also affect particulate pollution in any given area by determining its concentration, deposition, and atmospheric dispersion. Understanding local and regional meteorology is essential to understanding the mechanisms that cause the temporal and spatial dispersion of PM concentrations. With the ultimate goal of lowering PM concentrations to a level not dangerous to people's health, wind speed, wind direction, and long-distance transport will help the government and other relevant authorities make accurate, sufficient decisions to reduce PM levels. To comprehend the pollution load, seasonal variations, climatology, source of apportionment, and trajectory pathway of these PMs were continuously measured in conjunction with meteorological factors.

References

Aldegunde, J. A. Á., Sánchez, A. F., Saba, M., Bolaños, E. Q., & Palenque, J. Ú. (2022). Analysis of $PM_{2.5}$ and meteorological variables using enhanced geospatial techniques in developing countries: A case study of Cartagena de Indias City (Colombia). *Atmosphere, 13*(4), 506.

Amin, M., Ariefianto, T., Kaula, D., Husni, N., Serlina, Y., Suryati, I., & Bachtiar, V. S. (2024). Seasonal anomaly of particulate matter concentration in an equatorial climate: Evaluating the transboundary impact from neighboring provinces on Padang City. *Indonesia. Environmental Monitoring and Assessment, 196*(11), 1013.

Basemera, S., Onyango, S., Tumwesigyire, J., Mukama, M., Santorino, D., North, C. M., & Parks, B. (2025). Impact of meteorological factors on seasonal and diurnal variation of $PM_{2.5}$ at a site in Mbarara, Uganda. *Air, 3*(1), 1.

Das, S., Roy, A., Masiwal, R., Mandal, M., Popek, R., Chakraborty, M., Sarkar, A., et al. (2023). Comprehensive analysis of PM_1 composition in the Eastern Indo-Gangetic Basin: A three-year urban study. *Sustainability, 15*(20), 14894.

Feng, X., Shao, L., Jones, T., Li, Y., Zhang, M., Ge, S., Zhang, D., et al. (2023). The changing sulphur content of a northern Chinese dust storm: Initiation, attenuation and culmination. *Atmospheric Environment, 297*, 119606.

Guo, Q., Wu, D., Yu, C., Wang, T., Ji, M., & Wang, X. (2022). Impacts of meteorological parameters on the occurrence of air pollution episodes in the Sichuan basin. *Journal of Environmental Sciences, 114*, 308–321.

Ibe, O. C., Nwofor, O. K., & Okoro, U. K. (2025). Evaluation of aerosol loading and its relationship with geopotential height in Nigeria's Guinea Coast. *Bulletin of the National Research Centre, 49*(1), 11.

Kayes, I., Shahriar, S. A., Hasan, K., Akhter, M., Kabir, M. M., & Salam, M. A. (2019). *The relationships between meteorological parameters and air pollutants in an urban environment.*

Khan, M. B., Masiol, M., Bruno, C., Pasqualetto, A., Formenton, G. M., Agostinelli, C., & Pavoni, B. (2018). Potential sources and meteorological factors affecting $PM_{2.5}$-bound polycyclic aromatic hydrocarbon levels in six main cities of northeastern Italy: an assessment of the related carcinogenic and mutagenic risks. *Environmental Science and Pollution Research, 25*(32), 31987–32000.

Kirešová, S., & Guzan, M. (2022). Determining the correlation between particulate matter PM_{10} and meteorological factors. *Eng, 3*(3), 343–363.

Kliengchuay, W., Cooper Meeyai, A., Worakhunpiset, S., & Tantrakarnapa, K. (2018). Relationships between meteorological parameters and particulate matter in Mae Hong Son province, Thailand. *International Journal of Environmental Research and Public Health, 15*(12), 2801.

Li, H., Shen, S., Gao, M., Wu, Y., Wang, Z., & Shen, H. (2026). Spatiotemporal temperature variations induced by urban extreme heat events at the micro-scale. *Urban Climate, 66*, 102831.

Liu, D., He, C., Schwarz, J. P., & Wang, X. (2020). Lifecycle of light-absorbing carbonaceous aerosols in the atmosphere. *NPJ Climate and Atmospheric Science, 3*(1), 40.

Maleki, H., Sorooshian, A., Alam, K., Fathi, A., Weckwerth, T., Moazed, H., Goudarzi, G., et al. (2022). The impact of meteorological parameters on PM_{10} and visibility during the Middle Eastern dust storms. *Journal of Environmental Health Science and Engineering, 20*(1), 495–507.

Mandal, M., Popek, R., Przybysz, A., Roy, A., Das, S., & Sarkar, A. (2023). Breathing fresh air in the city: Implementing avenue trees as a sustainable solution to reduce particulate pollution in urban agglomerations. *Plants, 12*(7), 1545.

Miao, Y., Che, H., Liu, S., & Zhang, X. (2022). Heat stress in Beijing and its relationship with boundary layer structure and air pollution. *Atmospheric Environment, 282*, 119159.

Montgomery, A., Schnell, J. L., Adelman, Z., Janssen, M., & Horton, D. E. (2023). Simulation of neighborhood-scale air quality with two-way coupled WRF-CMAQ over Southern Lake Michigan-Chicago Region. *Journal of Geophysical Research: Atmospheres, 128*(6), e2002JD037942.

Molnar, A., Imre, K., Ferenczi, Z., Kiss, G., & Gelencser, A. (2020). Aerosol hygroscopicity: Hygroscopic growth proxy based on visibility for low-cost PM monitoring. *Atmospheric Research, 236*, 104815.

Moreira, A. B., Wanderley, L. S. D. A., Duarte, C. C., & Matzarakis, A. (2024). Atmospheric conditions related to extreme heat and human comfort in the City of Rio de Janeiro (Brazil) during the first quarter of the year 2024. *Atmosphere, 15*(8), 973.

Pacheco Hernández, P., Mera Garrido, E., Navarro Ahumada, G., Wachter Chamblas, J., & Polo Pizan, S. (2025). Submicron particles and micrometeorology in highly densified urban environments: Heavy-tailed probability study. *Atmosphere, 16*(9), 1044.

Roy, A., Mandal, M., Binha, S. K., Prasad, D., Popek, R., Przybysz, A., & Sarkar, A. (2025). Festive pollution: A global concern—A comparative study of Diwali in India and new year's eve in Poland. *Atmosphere, 16*(4), 442.

Scudeler, A. F., Lui, E. S., de Souza, D. Z., Martins, K. G., & Schirmer, W. N. (2025). Impact of the meteorological variables and of restrictions to urban mobility associated with the COVID-19 pandemic on the concentrations of particulate matter (PM_{10}). *Revista Brasileira de Geografia Física, 18*(7), 5323–5338.

Singh, D., & Verma, S. K. (2023). Impacts of particulate matter pollution on plants. In *New Frontiers in Plant-Environment Interactions: Innovative Technologies and Developments* (pp. 483–500). Cham: Springer Nature Switzerland.

Soleimanpour, M., Alizadeh, O., & Sabetghadam, S. (2023). Analysis of diurnal to seasonal variations and trends in air pollution potential in an urban area. *Scientific Reports, 13*(1), 21065.

Swa, M. H., Hayat, T., Rahman, M. Z., Hossain, M. S., & Nower, N. (2025). *Spatio-temporal dynamics of tropospheric NO_2 over Dhaka division (2019–2024)*. Bangladesh.

Taheri, M., Bigdeli, M., Imanian, H., & Mohammadian, A. (2025). An overview of evapotranspiration estimation models utilizing artificial intelligence. *Water, 17*(9), 1384.

Tesfaldet, Y. T., & Chanpiwat, P. (2023). The effects of meteorology and biomass burning on urban air quality: The case of Bangkok. *Urban Climate, 49*, 101441.

Tian, Y., Yao, X. A., Madden, M., & Grundstein, A. (2024). Synergic effects of meteorological factors on urban form-outdoor exercise relationship: A study with crowdsourced data. *Journal of Geographical Systems, 26*(1), 47–72.

Wang, J., Cortes-Ramirez, J., Gan, T., Davies, J. M., & Hu, W. (2024). Effects of climate and environmental factors on childhood and adolescent asthma: A systematic review based on spatial and temporal analysis evidence. *Science of the Total Environment, 951*, 175863.

Wang, Z., Xue, L., Liu, J., Ding, K., Lou, S., Ding, A., Huang, X., et al. (2022). Roles of atmospheric aerosols in extreme meteorological events: A systematic review. *Current Pollution Reports, 8*(2), 177–188.

Wu, Y., Liu, J., Zhai, J., Cong, L., Wang, Y., Ma, W., Li, C., et al. (2018). Comparison of dry and wet deposition of particulate matter in near-surface waters during summer. *PLoS ONE, 13*(6), e0199241.

Xing, Q., & Sun, M. (2022). Characteristics of $PM_{2.5}$ and PM_{10} spatio-temporal distribution and influencing meteorological conditions in Beijing. *Atmosphere, 13*(7), 1120.

Yaacob, N. F. F., Yazid, M. R. M., Abdul Maulud, K. N., Khahro, S. H., & Javed, Y. (2024). Spatio-temporal analysis of CO_2 emissions from vehicles in urban areas: A satellite imagery approach. *Sustainability, 16*(23), 10765.

Yadav, S., Praveen, O. D., & Satsangi, P. G. (2015). The effect of climate and meteorological changes on particulate matter in Pune, India. *Environmental Monitoring and Assessment, 187*(7), 402.

Yang, F., Cui, P. Y., Chen, C., Zhao, Q. X., & Huang, Y. D. (2024). Effects of atmospheric stability on air flow and pollutant dispersion characteristics within different scale urban areas. *Physics of Fluids, 36*(12).

Yang, Q., Yuan, Q., Li, T., Shen, H., & Zhang, L. (2017). The relationships between PM2. 5 and meteorological factors in China: seasonal and regional variations. *International journal of environmental research and public health, 14*(12), 1510.

Zender-Świercz, E., Galiszewska, B., Telejko, M., & Starzomska, M. (2024). The effect of temperature and humidity of air on the concentration of particulate matter-$PM_{2.5}$ and PM_{10}. *Atmospheric Research, 312*, 107733.

Zhang, X., Lyu, J., Han, Y., Sun, N., Sun, W., Li, J., Liu, C., & Yin, S. (2020). Effects of the leaf functional traits of coniferous and broadleaved trees in subtropical monsoon regions on $PM_{2.5}$ dry deposition velocities. *Environmental Pollution, 265*, 114845.

Zhong, T., Zhang, N., Wang, B., Ma, X., Wang, Y., & Chen, Y. (2023). Observed neighborhood-scale meteorology and air quality characteristics in downtown area of Nanjing. *Urban Climate, 51*, 101604.

Chapter 3
Urban Expansion and Air Quality Dynamics in the Himalayan Foothills: Insights from a Regional Case Studies

Anamika Roy, Mamun Mandal, Prerna Singh, Manoj Kumar, and Abhijit Sarkar

Abstract Urbanization has become a major contributor to global environmental alteration in recent years, altering mountain regions. The Himalayan foothills have undergone rapid urbanization, economic growth, and unregulated anthropogenic activity, all of which have been associated with socio-ecological concerns such as depletion of natural resources, a higher likelihood of natural disasters, and a changing climate pattern. The decline in air quality and air pollution are significant challenges confronting the Himalayas in recent decades. Among the air pollutants, particulate matter (PM) has become a significant concern, particularly in hilly regions, often exceeding the National Ambient Air Quality Standard (NAAQS). Therefore, this chapter summarizes the association between urbanization and poor air quality in order to provide insight to city planners on future development and the trade-off between urbanization and air pollution. This chapter also intends to find gaps and limitations in the available literature for understanding the air quality and pollutant levels in the Himalayan foothills, which is critical for the climatic change trends and associated health consequences. Dedicated initiatives at the regional and national scales are essential for addressing in-situ and transboundary pollution in this region.

Keywords Himalaya · Air quality · Air pollution · Urbanization · Particulate matter (PM)

A. Roy · M. Mandal · A. Sarkar (✉)
Laboratory of Applied Stress Biology, Department of Botany, University of Gour Banga, Malda, West Bengal, India
e-mail: abbhijitbot@ugb.ac.in; abhijitbhu@gmail.com

P. Singh · M. Kumar
Environmental Science and Biomedical Metrology Division, CSIR-National Physical Laboratory, New Delhi, India

Academy of Scientific and Innovative Research, Ghaziabad, Uttar Pradesh, India

M. Kumar
e-mail: manojkumar.nplindia@csir.res.in

A. Sarkar et al. (eds.), *From Sky to Soil*, https://doi.org/10.1007/978-3-032-31991-3_3

3.1 Himalayan Foothills: The Tourism Heaven

The word "Himalaya" comes from Sanskrit, signifying "house of snow," where 'Him' refers to snow and 'Alaya' denotes house (Negi, 2009). This region is home to the highest peaks worldwide (Biswas & Kumar, 2024). The Himalayas are part of the Hindu-Kush Himalaya, spanning approximately 3 km and covering about 750 km^2 (Mittermeier, 2005). The Indian Himalayan Region (IHR) captivates many pilgrims and visitors due to its rich biodiversity and vibrant cultural heritage (Aayog, 2018). The hilly region attracts tourists due to majestic landscapes, scenic views, and clean environment, which will serve as a refreshing break from city life (Dkhar, 2024).

Tourism is a key factor in the economic growth of this region, that facilitates the expansion of various tourism-related services and infrastructure. It generates substantial revenue and constitutes a major proportion of the gross domestic product (GDP), thereby benefiting both state governments and private enterprises (Mohanty et al., 2024). Consequently, the advancement of tourism profoundly influences the national economy (Datta & Banerji, 2015; Nguyen, 2021). In recent decades, tourism is growing rapidly in the IHR, establishing one of the fastest-growing economic sectors of India, contributing approximately 10% to the total income of this region (Datta & Banerji, 2015; Pande & Bhardwaj, 2024). For instance, Ladakh, located in remote and elevated areas of IHR, attracted nearly 500 visitors when tourism was first introduced in this region in 1974. By 2010, this number had surged to 55,685 Indian tourists, and further heightened to approximately 1.5 lakh in 2011; 2024 represents the 50th anniversary of tourism here, attracting around 5 lakh visitors (Aayog, 2018). While the rapid growth of tourism has brought economic benefits, it has also caused environmental destruction and exploited cultural and social heritage (Green, 2005). The tourism pattern in the IHR has evolved over the past decade. The peak tourist season was predominantly observed in hill stations during the summer months from May to July (Pande & Bhardwaj, 2024). However, the notable surge in tourism activities during peak tourist season in the IHR region has been accompanied by adverse impacts, including environmental degradation and the erosion of socio-cultural heritage (Green, 2005). The traits of mass tourism, characterized by extensive sightseeing and visits to renowned tourism hubs, necessitate the continuous expansion of infrastructure, such as road networks and accommodation facilities. This expansion stresses the region's fragile natural resources, delicate ecosystems, and local social structures (Aayog, 2018). Furthermore, the unregulated tourism growth has led to construction in vulnerable areas, such as river valleys, landslide-prone regions, and floodplains, thereby increasing disaster risks (Tiwari & Rani, 2024).

3.1.1 The Biodiversity Hub

The Himalayan foothills are the youngest and most intricate mountain range, exhibiting diverse environments, from extremely wet to arid areas; this region is recognised as one of the 36 global biodiversity hotspots (Biswas & Kumar, 2024). The variation in climate, coupled with pronounced altitudinal gradients, enhanced the biological diversity, including an extraordinary array of flora and fauna; these factors work synergistically, more evident along both regional orientations (Kumari, 2024). The Himalayan region serves a crucial role as a transitional zone bridging the Palearctic and Indo-Malayan realms. This transitional significance facilitates the coexistence of species native to both regions, resulting in a rich biological hotspot. The geological compositions, variability in meteorological conditions, and differing altitudinal level profoundly influence the biological diversity within this hotspot (Päckert et al., 2012). Biodiversity-rich Himalayan foothills provide crucial ecosystem services to billions of people living in proximity to these mountains, highlighting their importance as both natural resources and economic assets (Sharma et al., 2023a). The IHR faces considerable challenges stemming from climate change. The impacts of rising temperatures and altered precipitation patterns pose significant risks to the delicate ecosystems, affecting both endemic and migratory species (Awasthi et al., 2023; Shrestha et al., 2012). This region hosts a distinct assemblage of endemic species, including medicinal plants and various economically valuable flora and fauna, many of which are rare or endangered (Kindlmann, 2011). Reports from Conservation International indicate that species diversity in the Himalayan foothills is impressive, comprising approximately 10,000 plant species, with 32% classified as endemic, along with 300 mammals (4% endemic), 977 birds (2% endemic), 176 reptiles (27% endemic), 105 amphibians (40% endemic), and 269 freshwater fish (12.3% endemic) (http://www.biodiversityhotspots.org). Furthermore, in the Eastern Himalayas, around 3,500 endemic plant species exist among approximately 9,000 species identified (Pullaiah, 2024). This region is considered to be a place of origin for flowering plants, and yet it still harbours some of the most ancestral families among the angiosperms, such as Degeneriaceae and Eupomatiaceae, Himantandraceae, Magnoliaceae, Trochodendraceae, and Winteraceae (Rawal et al., 2013). The Orchidaceae family, representing the largest family of flowering plants in this hotspot, comprises many orchids, most of them are endemic (Raju et al., 2010). Among five major flagship species, the Himalayan foothills are home to three of the species, following: the elephant, tiger, and rhinoceros (Rawal et al., 2013). The Central Himalayas are rich in biodiversity, containing 7000 floral species and 500 faunal species (Sundriyal & Sharma, 2016). The Trans-Himalayan region contains many wetlands and lakes, highly rich in biodiversity, featuring around 68 threatened plant species (Rawal et al., 2013). The Western Himalayan region, covering the Indian states of Himachal Pradesh, Jammu and Kashmir, and Uttarakhand, creates a natural barrier limiting species migration, leading to increased biodiversity richness of this area (Tewari et al., 2017).

3.2 Tourism and Urbanization: The Inseparable Duo

Mountains cover around 24% of the land surface worldwide and are home to 12% of total population (Tiwari et al., 2018). The captivating beauty and majestic landscapes of the Himalayan region have attracted visitors from across the globe (Aayog, 2018). However, the increasing number of tourists and seasonal visitors has surpassed the carrying capacity of Himalayan urbanized areas (Tiwari et al., 2018). Tourism acts as a major catalyst for urbanization in the Himalayas (Anees et al., 2022). Significant portions of arable land have been converted into built-up spaces each year due to rapid urbanization, infrastructure expansion, and increased anthropogenic and commercial activities (Tiwari & Joshi, 2016). Recently, even previously inaccessible areas have undergone rapid advancement, particularly due to road connectivity improvement that has increased new tourist spots, influencing significant growth in urbanization and tourism (Tiwari et al., 2018).

3.2.1 Increasing Population Load

The remote areas of the Himalayas exhibit a sparse population distribution largely due to their challenging topography and varied climatic conditions. Despite these challenges, this mountainous locale's population growth trend remains evident. The Census of India reported that the population has risen from 19.9 million in 1961 to 53.8 million in 2011, representing an impressive increase of 250% over the past five decades. If this growth rate of 3.3% annually continues, the population could exceed 260 million by 2061, representing a 13-fold increase, indicating expansion of urban areas (Apollo, 2017). Urban regions serve as vital catalysts for economic progress and development due to the diverse opportunities, infrastructure, amenities, living standards, and economic growth. These aspects attract individuals and businesses, resulting in increased population density that promotes both horizontal and vertical expansion of urban environments (Duranton, 2008). Significant urban expansion has been recorded in Kathmandu, which experienced an increase of 81.20 km^2 (714%) between 1972 and 2015. Similarly, Dehradun witnessed an increase of 45.99 km^2, marking a growth of 1123% during the same period (Kumar, 2017). The current population growth rates have strained the state's ability to invest in human resource development and improve living standards (Sati, 2006).

3.2.2 Increasing Concrete Construction

Urbanization has become a significant driver of global environmental change in recent years, particularly affecting mountain regions (Tiwari et al., 2018). The Himalayan foothills have witnessed rapid urban growth and unregulated human activities, resulting in socio-ecological challenges such as the depletion of natural resources and an increase in concrete construction, including roads, dams, and buildings (Bansal et al., 2023; Kumar et al., 2024; Tiwari et al., 2018). In urban areas, the scarcity of space leads to constructing higher buildings made of concrete rather than locally produced brick (Wester et al., 2019). Increasing concrete concentration in Himalayas not only alters visual aesthetic but also has significant ecological consequences, including urban heat island effect leading to warming of local climate (Bhochhibhoya et al., 2017; Pandit, 2013). In addition, rapid urbanization along with overpopulation increases slum areas, leading to poor infrastructure and overcrowding. This trend transformed hilly Himalayan region into concrete-dominated urban areas (Bhochhibhoya et al., 2017; Tiwari et al., 2018).

3.2.3 Nature and Natural Resources Destruction

The unplanned and rapid urbanization in the hilly mountain range significantly affected ecosystems and depleted natural resources, including biodiversity and ecological lands (Tiwari et al., 2018). Rapid urbanization has resulted in the degradation of forests, ranging from 2 to 10%, leading to alteration of the natural landscapes. Moreover, the percentage of irrigated land also reduced from 21 to 4%, mainly due to alteration in hydrological cycles in urban areas (Bhattacharyya et al., 2015). Hydrological cycles had been profoundly affected in different urban Himalayan region, reducing the ground water level, reducing the availability of drinking water, sanitation, and agricultural activities (Scott et al., 2019). The likelihood of natural hazards is more intensified in urban and sub-urban Himalayan areas, making the inhabitants more vulnerable concerning water, food, health, and livelihood (Dodman et al., 2022). Most populated and fast-growing urban areas in Uttarakhand Himalaya are expanding over the arable land in their surrounding rural areas. This expansion has resulted in significant changes in productive land use patterns in urban and sub-urban zones primarily due to the urban land use. As a consequence, surrounding rural areas have experienced significant reduction in productive agricultural lands, ranging from 5 to 13% over the past decades (Tiwari et al., 2018). The continuing depletion of essential agricultural resources—including arable land, biomass manure, and water for irrigation—coupled with deforestation prompted by urban development has led to a decline in agricultural productivity ranging from 19 to 55% across ten urban zones. Consequently, rural settlements on the periphery of these urban complexes now face alarming food deficits, estimated between 65 and 95% (Tiwari et al., 2018).

3.2.4 Increasing Rate of Transportation

The vehicular sector significantly contributes to air pollution, considerably impacting human health (Teri, 2015). In the remote and stunning regions of the Himalayas, trails and footpaths were the primary means of transportation. However, the landscape has evolved considerably as road transport has expanded, becoming a critical component of local infrastructure (Beazley, 2013). For instance, the Kathmandu Valley has witnessed a substantial increase in vehicle numbers over the past 15 years, reflecting a trend of rapid urbanization and development. Data indicates that registered vehicles surged from 24,003 in 2000–2001 to an impressive 779,822 by 2015–2016 (Saud & Paudel, 2018). Similarly, in Shimla, the scenic town, the vehicle population more than doubled from 240,000 in 2001 to 730,000 in 2012. This remarkable growth has been primarily driven by the increasing popularity of cars and two-wheelers, thereby enhancing mobility and accessibility for both residents and visitors (Teri, 2015). According to the Federation of Automobile Dealers Associations (FADA), the transportation landscape is experiencing rapid changes; the two-wheeler segment recorded a growth of 38%, while the three-wheeler category expanded by 27% in Jammu and Kashmir from 2023 to 2024 (http://himalayanmail.com/fada-releases-april-2024-vehicle-retail-data-for-the-ut-of-jammu-and-kashmir/). This upward trend is further exacerbated during peak tourist seasons when the influx of visitors considerably increases the demand for transport. Tourism in the Himalayan region exhibits significant variability, resulting in diverse patterns across different states. Northern Himalayan states such as Jammu and Kashmir, Himachal Pradesh, Uttarakhand, Assam, Meghalaya, Sikkim, and Tripura attract substantial tourists. Conversely, states such as Arunachal Pradesh, Manipur, Mizoram, and Nagaland experience relatively lower levels of tourist inflow (Kuniyal, 2023). Likewise, in Darjeeling, an eastern Himalayan state, the number of vehicles during peak tourist season can rise dramatically from 6000 to 20,000 (Roy et al., 2024a, 2024b).

3.2.5 Increasing Use of Fossil Fuel

The escalating demand for energy, particularly derived from fossil fuels, has adversely affected both the environment and public health. Fossil fuel combustion is a significant contributor to air pollution in this region, as it emits around 3.5 times more black carbon (BC) than biomass burning (Srivastava et al., 2023). Fossil fuel combustion not only contributes to rising pollution levels, but also accelerates the melting of glaciers (Awasthi et al., 2022). As population continues to grow and commercial activities tend to expand, electricity demands also increase, resulting in an expansion of electrical grids (Ramachandra & Krishnadas, 2011). In addition, extensive use of technologies and appliances in urbanizing areas requires higher fossil fuel combustion, further increasing pollution levels (Wester et al., 2019). Solar system implementation might be a feasible alternative that can reduce the dependence on traditional fuels, including firewood and fossil fuels (Wester et al., 2019).

3.3 Particulate Matter (PM) Pollution in the Himalayan Foothills

Anthropogenic activities along with natural factors are potential contributors to declining air quality in this region (Hassan et al., 2023). Among the pollutants, particulate matter (PM) has become a pressing concern in the Himalayan region. The unique topography of the Indo-Gangetic Plain (IGP) further exacerbates the pollution. The hilly mountain region with irregular basin and steep hills reduced the dispersion of airborne pollutants, facilitated PM accumulation, and thus increased atmospheric pollutants concentration (Ojha et al., 2012). In addition, density gradient pollutant concentration is observed across the Himalayas, following 3000 m elevation, highlighting the influence of topography on the dispersal of the pollutants throughout the region (Moore & Semple, 2021).

3.3.1 Sources

PM can be emitted by both anthropogenic and natural sources. Natural sources of PM are volcanic ash, volatile organic compounds (VOCs), sea-salt aerosols, byproducts of forest fires, and dust storms. Anthropogenic sources of PM are construction emission, fossil fuel combustion, agricultural waste management, commercial, industrial, and vehicular emission (Mandal et al., 2023a). Anthropogenic sources are further categorized into two major groups: stationary and mobile sources. Stationary sources are primarily non-movable sources, including construction and demolition activities, residential heating, fuel combustion, processing of woods, industrial emission, road dust emission from paved and unpaved roads, waste disposal and recycling (Cass et al., 2000). Mobile sources are movable sources including, vehicular emissions. Source apportionment studies reported that air quality in this region is primarily affected by air movements originating from different urbanizing areas of IGP, resulting in decline in air quality. Long-range transports also contribute to air pollution level over the area (Hassan et al., 2023; Roy et al., 2024a, 2024b). The primary sources of air pollution in the Himalayan region and IGP are conventional cooking practices, fossil fuel combustion, biomass burning, vehicular emission, biomass burning, agricultural waste burning, and byproducts of forest fires (Hassan et al., 2023; Rajput et al., 2013). Fossil fuel combustion reduced oxygen levels, which in turn, influencing air quality in this hilly terrain (Wester et al., 2019). Air pollution dynamics in this region requires in-depth analysis of various air pollution sources along with other factors such as topographical and meteorological conditions (Hassan et al., 2023).

3.3.2 Types

PM is a heterogeneous composition of liquid–solid particles which can differ in size, sources and composition (Adams et al., 2015). The toxicity of PM particles can be affected by their physical and chemical properties (Mandal et al., 2023b). PM can be primarily classified into three major groups: coarse PM (PM_{10} with a diameter of 10 μm or less), fine PM ($PM_{2.5}$ with a diameter of 2.5 μm or less), and ultrafine PM (PM_1 with a diameter 1 μm or less) (Roy et al., 2024b). Ultrafine PM is a smaller proportion of total PM, but it can penetrate the bloodstream and brain, causing significant health risks (Brugge, 2018).

The chemical composition of PM can be complex and affected by many factors such as long-range transport, meteorological parameters, and atmospheric chemistry. The chemical components are carbonaceous particles [organic carbon (OC) and black carbon (BC)], ions like sulphate, ammonium, and nitrate, elements such as lead, copper, nickel, silicon, etc., and biological components (fungal spores, pollen, fungal spores, endotoxins, and biological constituents) (Roy et al., 2024b). Carbonaceous particles and secondary inorganic aerosols like sulphate, nitrate, and ammonium are particularly more concerning, as these components can significantly influence the acidic nature and toxicity of PM (Morakinyo et al., 2016). Understanding the diverse nature of PM is crucial for evaluating the composition and potential health impacts of PM.

3.4 Eastern, Central, and Western Himalayan Foothills Case Study

The case studies from eastern, central, and western Himalayan foothills were documented in Table 3.1. The eastern Himalayan region comprises Bhutan, north-eastern India, southern, central, and eastern part of Nepal, including Assam, Arunachal Pradesh, Darjeeling, Manipur, Meghalaya, Mizoram, Nagaland, Sikkim, and Tripura (Profile, 2005). Previous studies reported that this region experienced poor air quality and associated health issues, primarily attributed to higher levels of PM_{10} and $PM_{2.5}$ (Roy et al., 2024a; Sharma et al., 2020). Darjeeling, known for its majestic hills and scenic views, has experienced a significant decline in air quality over the past decades (Nag et al., 2018). A study conducted in Darjeeling from January to December, 2005, found that the average concentration was 19.6 ± 11.1 μg/m^3 and 29.5 ± 20.8 μg/m^3 for coarse and fine PM, respectively (Chatterjee et al. (2010)). Another recent study reported that the average concentration of coarse, fine, and ultrafine PM in Darjeeling was 233, 199, and 192 μg/m^3, respectively during summer; the findings of the study highlighted a significant surge in level of PM pollutants (Roy et al., 2024a). Kathmandu valley, known as 'city of temples', has also experienced a significantly increasing level of dust pollution, which transforms it into the 'city of pollution' (Saud & Paudel, 2018). A study conducted in Kathmandu revealed that the average concentration of coarse PM was 132 μg/m^3 during 2012–2013, which was higher than the standards (Kim et al., 2015). Thimphu, the capital of Bhutan, has also experienced a significantly increasing level of pollution, mainly during the winter months. In addition, conventional cooking using firewood, biomass burning, forest fires further exacerbate the pollution level, affecting the environment and health of local people (Sharma et al., 2020). Sharma et al. (2020) reported that the concentration of $PM_{2.5}$ was around 32 μg/m^3, which was higher than the recommended limit set by the World Health Organization (WHO). Meghalaya, once known for its scenic beauty, currently grapples with poor air quality (Majaw, 2014). Rajput et al. (2013) found that average fine PM concentration was ranging between 39 and 348 μg/m^3, indicating the poor air quality in this region. Another study conducted by Lamare and Chaturvedi (2014) in Shillong, the capital of Meghalaya, reported that the average concentration of respirable suspended particulate matter (RSPM) was varied between 81 and 261 μg/m^3, highlighting the poor quality of this area.

The central Himalayan region comprises Uttarakhand, known as Uttarakhand Himalaya or Uttaranchal, and features a hilly terrain influencing aerosol dynamics of this region, making it challenging to accurately forecast air quality by predictive models (Nair et al., 2013; Rao et al., 2016). Nainital is a popular tourist destination located in the central Himalayan region, known for scenic beauty, and also grapples with escalating levels of pollution (Gupta et al., 2024). Choudhury et al. (2022) reported that the concentration of coarse PM exceeded the National Ambient Air Quality Standards (NAAQS) in this region particularly due to vehicular emissions, coal/fuel combustion, and biomass burning. A study conducted in Dehradun, the capital city of Uttarakhand, disclosed that the concentration of coarse and fine PM exceeded the NAAQS, particularly during the winter months (Soni et al., 2020). The decline in air quality can be attributed to long-range transport of pollutants and different anthropogenic activities, such as biomass burning, vehicular and industrial emissions, and waste management practices. Likewise, Sinha et al. (2021) also observed similar findings, identifying vehicular and industrial emissions, waste burning, and resuspended road dust as major contributors to higher levels of pollution in Dehradun. Furthermore, air quality significantly worsens during Diwali due to the use of firecrackers and fireworks. Prabhu et al. (2019) reported that the average PM concentration increased by 1.8–2.1 times during Diwali than the pre-Diwali period.

The western Himalayan region, the western part of the Himalaya, comprises Himachal Pradesh, Uttar Pradesh, Jammu and Kashmir, and the western part of Nepal (Tewari et al., 2017). This region experienced escalating level of air pollution, mainly due to particulate matter (PM) influencing local air quality and visibility. Himachal Pradesh, a hilly state with scenic views, has witnessed a decline in air quality in recent years, mainly attributed to primary contributors such as heavy traffic, industrial emissions, and secondary sources, including long-range transports of pollutants (Sharma et al., 2014; Kumar et al., 2020). Kumar et al. (2020) monitored coarse PM concentration over three distinct areas, residential, industrial, and ecologically sensitive areas in Himachal Pradesh. They found that the PM_{10} level exceeded the NAAQS standard in residential and industrial zones, while it remained within the limit in ecologically sensitive areas; the concentration was 1.5 times more

Table 3.1 Particulate pollution scenario in the Himalayan region

State/Country	Place of study	Study period	Sources	Concentration of PM	Elemental composition	References
Nepal	Kathmandu	2012–2013	Vehicular emission, brick kilns, and biomass/garbage burning	PM_{10}: 132 μg/m^3	–	Kim et al. (2015)
	Chaurikharka	2013–2015 (winter)	Biomass burning	PM_{10}: 113 ± 64 μg/m^3 $PM_{2.5}$: 35 ± 24 μg/m^3	–	Pratali et al. (2019)
West Bengal	Darjeeling	2018–2019	Vehicular emission, coal/fuel combustion and biomass burning	PM_{10}: 54 ± 17 μg/m^3	–	Choudhary et al. (2022)
		2021	Biomass burning, fossil fuel combustion, crustal/soil dust, and industrial emissions	PM_{10}: 54 ± 17 μg/m^3 PM_{10}: 54 ± 17 μg/m^3 PM_{10}: 54 ± 17 μg/m^3	Si, N, Na, B, Ba, Al, and K	Roy et al. (2024a, 2024b)
Bhutan	Thimpu	2018–2020	–	$PM_{2.5}$: 32.47 μg/m^3	–	Sharma et al. (2020)
Meghalaya	Barapanu	2009	Vehicular emission, coal/fuel combustion and biomass burning	$PM_{2.5}$: 39–348 μg/m^3	–	Rajput et al. (2013)
	Shillong	2010	Vehicular emission	RSPM: 81.24–261.43 μg/m^3	–	Lamare and Chaturvedi (2014)
Uttarakhand	Bhagwanpur	2014–2015	Industrial, vehicular exhaust and non-exhaust emission, agricultural activity	RSPM: 259.6 ± 52.6 μg/m^3	Al, Ca, Cd, Co, Cr, Cu, Fe, K, Mg, Mn, Ni, Pb, Zn	Panwar et al. (2020)
	Dehradun	2015 - 2016	Vehicular emission, biomass burning, resuspended road dust	PM_{10}: 170.6 ± 23.7 μg/m^3	Al, Ca, Cd, Co, Cu, Fe, Mg, Mn, Na, Ni, Cr, Pb, Zn	Shridhar (2019)

(continued)

Table 3.1 (continued)

State/Country	Place of study	Study period	Sources	Concentration of PM	Elemental composition	References
		2014–2015	Vehicular emission, industrial emission, and biomass burning	PM_{10}: 87 ± 40–90 ± 32 μg/m^3 $PM_{2.5}$: 39 ± 19–63 ± 27 μg/m^3	–	Soni et al. (2020)
	Nainital	2018–2019	Vehicular emission, coal/fuel combustion and biomass burning	PM_{10}: 65 ± 41 μg/m^3	–	Choudhary et al. (2022)
		2021	Vehicular, industrial emission, coal/fuel combustion, non-exhaust emission, and biomass burning	PM_{10}: 64 ± 6 μg/m^3	Ca, Al, Fe, S, Mg, Pd, B, Mo, Zn, Ag,	Gupta et al. (2024)
		2018–2019	Dust, biomass burning and industrial emissions	PM_{10}: 44 ± 13 μg/m^3	Al, Fe, Ca, Mg and Ti	Sharma et al. (2020)
	Srinagar	2017	Vehicular emission and anthropogenic activities	–	–	Sandeep et al. (2020)
Himachal Pradesh	–	2011–2015	Local–regional sources	PM_{10}: 115.9 ± 47.8 μg/m^3 (industrial) PM_{10}: 87.16 ± 35.8 μg/m^3 (residential) PM_{10}: 59.02 ± 34.8 μg/m^3 (ecologically sensitive)	–	Kumar et al. (2020)
	Mandi	2009	Vehicular emission, coal/fuel combustion and biomass burning	PM_{10}: 111.3 μg/m^3	–	Thakur and Kuniyal (2016)
	Mohal-kullu	2018–2019	Vehicular emission, coal/fuel combustion and biomass burning	PM_{10}: 57 ± 32 μg/m^3	–	Choudhary et al. (2022)

(continued)

Table 3.1 (continued)

State/Country	Place of study	Study period	Sources	Concentration of PM	Elemental composition	References
Jammu and Kashmir	Sonamarg	2009	Vehicular emission, coal/fuel combustion, and non-exhaust emission	PM_{10}: 216 μg/m^3	–	Jehangir et al. (2011)
	Thajwas			PM_{10}: 155 μg/m^3		
	Baltal			PM_{10}: 172 μg/m^3		
	Jammu	2018	Vehicular emission, agricultural activities, and non-exhaust emission	$PM_{2.5}$: 24.8–88.3 μg/m^3	–	Gupta et al. (2021)

than the NAAQS limit even in residential zones. Another study conducted in the urban areas in the western Himalayas, including Kothi and Mohal, disclosed that emission sources were significantly affected by seasonal variation (Kuniyal et al., 2015). For instance, local emissions like biomass burning were more pronounced during winter months, while long-range transport was a major contributor during summer. Escalated level of air pollution was also observed in Jammu and Kashmir, western Himalayan region over the past decades, primarily due to transportation and industrial emissions (Jehangir et al., 2011). Sonamarg, the hilly region of Kashmir, known as 'meadow of gold' has also experienced escalating level of pollution. Long-range transport of contaminants facilitated by the unique topography of this region was a major contributor to pollution in this region (Bhat & Gaga, 2022). Gupta et al. (2021) reported that rising level of PM often exceeded the NAAQS limit and poses a significant concern for public health.

3.5 Particulate Matter Pollution Impact on the Biodiversity of the Himalayan Foothills

The biodiversity of a region has direct and indirect advantages. Direct advantages are the consumable products such as food fuel, fodder, and medicinal herbs, while indirect advantages were the ecosystem services, breakdown of pollutants, pollination, and micro-climatic management (Sundriyal and Shaarma, 2016). The Himalayas are a major global hotspot for biodiversity and endemic species, providing substantial benefit to local population and communities (Payne et al., 2020). The Himalayan foothills contain one-tenth of high-altitudinal floral and faunal species worldwide and almost half of the Indian native plants (Tewari et al., 2017). Unique topography, higher elevation gradient, climatic differences influence the biodiversity of the mountain region. However, this biodiversity-rich region currently experiences significant challenges, resulting in rapid decline in biodiversity (Payne et al., 2020). This decline is attributed to anthropogenic activities. For instance, cultivation shifting disrupts traditional agricultural farming, while expansion of agricultural areas over natural lands alters ecological dynamics. Other anthropogenic factors like urban growth, overpopulation, and unregulated human activities collectively affected biodiversity loss and ecosystem services (Sheikh et al., 2021). PM pollution has been more pronounced in recent years and may adversely affect biodiversity, particularly within urban forest ecosystems. PM deposition can serve as a nutrient source for forests. However, it also alters leaf surface properties, prolongs leaf moisture retention, modifies epiphytic organisms' habitats, and raises the risk of pathogen exposure (Rai, 2016). Furthermore, the deposition of PM can directly affect photosynthesis by altering the optical properties of leaf surfaces and obstructing stomata, ultimately shortening the lifespan of plants (Roy et al., 2024b), and can lead to exposed soil, thereby increasing the risk of floods and heavy rainfall (Masroor et al., 2021). Indirect effects of PM may involve changes to both the chemical and physical properties of soil. The germination and early growth stages are particularly vulnerable in a plant's life cycle; thus, environmental stress, combined with species sensitivity, can hinder successful establishment (Grantz et al., 2003).

Furthermore, PM plays a pivotal role in climatic changes, exacerbating existing ecological challenges through habitat loss and fragmentation, the decline of forest biodiversity, and the degradation of wetland and riverine ecosystems. This not only leads to a loss of forage and fodder resources but also causes a reduction in agro-biodiversity (Tse-Ring et al., 2010). The impact of PM is not restricted to terrestrial environments; it also infiltrates aquatic ecosystems, causing acidification and result in adverse biodiversity (Hartono et al., 2017). The aquatic ecosystem is more susceptible to fine PM, as it can magnify its concentration and resulting in adverse impacts (Zhao et al., 2014). PM can negatively impact fish health by reducing respiratory function, increasing susceptibility to diseases, and accumulating in fish tissues, which in the long term poses severe risks to human health (Gokul et al., 2023). Moreover, PM also affects faunal biodiversity, by increasing respiratory diseases, resulting in mortality and morbidity (Losacco & Perillo, 2018).

3.6 Particulate Matter Pollution Impact on Human Health Over the Himalayan Region

Exposure to particulate matter (PM) can increase health risks among people. PM pollution is associated with acute and chronic health effects (Chow et al., 2006). PM particles can cause a range of diseases, including respiratory diseases, more particularly chronic respiratory syndromes, morbidity, and even mortality (Mehmood et al., 2019). Health issues PM exposures also increase anxiety, mental health-related disorders, including, schizophrenia, autism, alter behaviour pattern, decrease overall happiness and life satisfaction, and elevate suicidal tendencies, particularly in men compared to women (Abed Al Ahad, 2024; Braithwaite et al., 2019; Lu, 2020). Fine and ultrafine PM are more hazardous as they can infiltrate the respiratory system (Panwar et al., 2020). Inhalable particles can reach up to alveoli and other organs, even in blood and accelerated cancer incidence, even among non-smokers (Swanton et al., 2022; Yu et al., 2021). PM particles can also cause oxidative stress, endocrine disruption, thus, increasing the risks of premature birth and lower birth weight (Giri et al., 2023). Air pollution and its associated health impacts constitute significant challenges, particularly in the eastern Himalayan region. Research indicates that indoor air pollution can occasionally surpass outdoor pollution levels, severely threatening public health (Tran et al., 2020). A recent study has classified urban air quality in Nepal as among the worst in the world. Among the essential rights outlined in Nepal's 2015 constitution (such as air, food, shelter, and health care), access to clean air is frequently compromised (Giri et al., 2023). In response to alarming air quality levels, the Government of Nepal ordered the closure of all schools for four days, from March 30 to April 2, 2021 (Post, 2021). A case study in Kopawa village, Nepal, has revealed that children are more susceptible to dust exposure than adults (Yadav et al., 2019b). Kathmandu has been designated the most polluted city in the world (Giri et al., 2023). Additionally, a study in the Kathmandu Valley has shown health issues associated with elevated indoor pollution levels (Bhandari et al., 2021). In Bhutan, concerns related to pollution have primarily focused on indoor environments, as many households used conventional cooking methods, which increased their exposure to indoor air pollution and related health problems, including bronchitis, asthma, and miscarriages (BMCI, 2016; Sharma et al., 2022). Wangchuk et al. (2015) investigated the effects of indoor air pollution on school children during the dry season and found that the highest contribution to daily ultrafine PM exposure occurred during cooking in rural villages of Bhutan. According to the Central Bureau of Health Intelligence (CBHI) reported that around 20,000 acute respiratory infection (ARI) cases and 200 deaths in Assan, 2016, which increased to around 23,000 ARI cases and 225 deaths in 2017 (Bhunia & Ding, 2020).

Nainital, a town in the central Himalayas, also experienced various health problems due to rising PM levels. Gupta et al. (2024) reported increased carcinogenic risks for adults and non-carcinogenic risks for children due to exposure to PM associated with heavy metals primarily from vehicular emissions, industrial activity, coal and fuel combustion, non-exhaust emissions, and biomass burning. A study in Dehradun, the capital city of Uttarakhand, also indicated that inhalable PM particles pose several health risks, including cancer risk for all age groups (Soni et al., 2020). Prabhu and Shridhar (2019) found that PM exposure associated with heavy metals could cause 137 cancer cases per one million people. Pant et al. (2022) noted that higher levels of PM_{10} in the Kashipur and Rudrapur regions of Uttarakhand led to increased hospitalization cases due to air pollution, particularly among individuals with respiratory issues. The study reported that residents might experience more health effects due to moderately polluted air quality; however, conditions tend to improve during the monsoon season.

Jammu and Kashmir, known as a natural paradise, faces a significant prevalence of lung diseases. Srinagar, the capital, showed the highest lung cancer incidences among males in India (Koul et al., 2010). A recent study reported that Jammu and Kahmir have one of the highest rates of chronic obstructive pulmonary disease (COPD), affecting 4750 individuals per 10,000 population primarily due to air pollution (Salvi et al., 2018). Around 10,000 people died every year can be attributed to air pollution, highlighting serious health risks due to pollutants exposure (Balakrishnan et al., 2019). Similarly, rising pollution level impacts residents' health and their work efficiency in Himachal Pradesh (Teri, 2015). A study in the Parwanoo and Baddi, industrial areas of Himachal Pradesh reported that children were more vulnerable to respiratory diseases might be due to their underdeveloped metabolic and pulmonary capacities than the adults (Barall et al., 2020). Uttar Pradesh, the largest Indian state, also experiences rising pollution level and associated health-related problems. The state has classified as the state with second-highest number of non-attainment cities (Bera et al., 2024). Sharma et al. (2023b) reported that health risks increased due to exposure to pollutants, particularly PM_{10} in 11 non-attainment cities in Uttar Pradesh. The health risks increased up to 13% in Ghaziabad (Sharma et al., 2023b).

The health impacts of PM over eastern, central, and western Himalayan foothills were presented in Table 3.2.

3.7 Impact of Himalayan Pollution on the Global Climate Change

Climate change in the Himalayan region has become a major concern, altering water availability, influencing ecosystem services, and agricultural production. Himalayan region characterized by unique ecosystems and climatic conditions was a significant indicator of climate change (Shrestha et al., 1999). The rising level of particulate matter (PM) can change the climate in the Himalayas; coarse PM is a major concern for environments, while fine PM is a major concern for climate and human health (Hassan et al., 2023; Romshoo et al., 2021). Carbonaceous aerosols also influence climatic changes (Li et al., 2021).

Air pollution influences temperature pattern within the Himalayan range; increasing temperature is more pronounced at higher elevation than lower altitudes (Romshoo et al., 2021; Shrestha et al., 1999). Current reduction in snow and glacier in the Himalayan region may rapidly increase warming trends in the higher elevation (Shrestha et al., 1999). Qian et al. (2011) reported that BC-covered snow can raise the air temperature by 1 °C. The projection trend suggested that the average mountain temperature may increase by 2 °C by 2050, leading to warmer winter and altering seasonal change pattern (Shrestha et al., 2015; Wester et al., 2019). BC is categorized as the second-largest anthropogenic contributor to global warming, after carbon dioxide (CO_2) (Gertler et al., 2016). Glacial melting is another significant climate change-related concern in the Himalayas (Tewari et al., 2017). Glacier have retreated over the past two decades, as much as 20% over this region (Wester et al., 2019). Snow-deposited PM and BC can create a darkening effect, reducing surface albedo, increasing solar radiation absorption, and thus, finally, increasing the melting rates (Gertler et al., 2016). The expansion of glacial lake mainly due to glacial retreat could result in devastating floods.

Aerosols and their components significantly impact precipitation pattern and rainfall dynamics, which are further accelerating global warming in this region (Wester et al., 2019). Rainfall is likely increased by around four to twelve percent in the nearest future, while four to twenty five percent in the long term (Krishnan et al., 2019). Global warming along with altered precipitation pattern significantly influence glacial retreat in the Himalayas (Shrestha et al., 1999). Glacier melting further disrupts hydrological cycle (Gertler et al., 2016). However, glacial changes of the Khumbu Himal were more closely associated with air temporal changes than the altered precipitation pattern (Seko, 1991; Shrestha et al., 1999).

Table 3.2 Health impacts of particulate pollution in the Himalayan region

State/Country	Place of study	Sources	Age group	Exposure dose/ model	Health impacts	References
Uttarakhand	Bhagwanpur	Industrial, vehicular exhaust and non-exhaust emission, agricultural activity	Both juvenile and adult	USEPA model	Increased carcinogenic risks	Panwar et al. (2020)
	Dehradun	Vehicular emission, biomass burning, resuspended road dust	Both juvenile and adult	USEPA model	Increased carcinogenic risks	Shridhar (2019)
		Vehicular emission, industrial emission, and biomass burning	Both juvenile and adult	USEPA model	Carcinogenic risks among all age groups	Soni et al. (2020)
	Nainital	Vehicular, industrial emission, coal/fuel combustion, non-exhaust emission, and biomass burning	Both juvenile and adult	USEPA model	Carcinogenic risks among adult and non-carcinogenic risks among children	Gupta et al. (2024)
	Kashipur and Rudrapur	–	Both juvenile and adult	–	People with asthma experienced more health problem	Pant et al. (2022)
Nepal	Chaurikharka	Biomass burning	Both juvenile and adult	Spirometry measurement	Pulmonary and cardiovascular diseases	Pratali et al. (2019)
	Kathmandu, Pokhara, Birgunj, and Biratnagar	Natural and anthropogenic activities	Both juvenile and adult	USEPA model	Non-significant risks, children were more prone to risks than adult	Yadav et al. (2019a)

(continued)

Table 3.2 (continued)

State/Country	Place of study	Sources	Age group	Exposure dose/ model	Health impacts	References
	Kopawa	Anthropogenic activities	Both juvenile and adult	USEPA model	Non-significant risks, children were more prone to risks than adult	Yadav et al. (2019b)
	Kathmandu	Anthropogenic activities	Both juvenile and adult	USEPA model	Non-significant carcinogenic and non-carcinogenic risks	Bhandari et al. (2021)
West Bengal	Darjeeling	Anthropogenic activities	Both juvenile and adult	USEPA model	Moderate carcinogenic risk	Rani et al. (2024)
Jammu and Kashmir	Srinagar and Pulwama	Anthropogenic activities	Both juvenile and adult	USEPA model	Potential carcinogenic risks	Shah et al. (2023)
Himachal Pradesh	Parwanoo and Baddi	Industrial emission	Juvenile	Spirometry measurement	Respiratory diseases	Barall et al. (2020)
Bhutan	Kanglung block	Indoor air pollution	Juvenile	–	Health problems	Wangchuk et al. (2015)

3.8 Future Prospects

Air pollution has become a significant environmental challenge in the Himalayan foothills, significantly affecting flora and fauna. The Himalayan ranges, known for their scenic beauty, various floral and faunal species, have been subjected to rapid and unplanned urbanization and anthropogenic activities. This has resulted in depletion of natural resources, more prone to natural calamities, change in climatic pattern, and increasing pollution level. As expansion of urban areas is unavoidable, urbanization planning and urban–rural land use can be sustainable alternatives. In addition, regulatory measures can be implemented to reduce industrial and transportation-related emissions. Implementation of land use policies is essential to effectively protect the Himalayan biodiversity, water resources, and agricultural land. Future studies should not only monitor pollutant concentrations but also emphasize examining the characteristics of pollutants and their sources to guide the designing of mitigation plans. In addition, there is a critical need for comprehensive epidemiological studies to understand better the specific health impacts associated with PM in each age group.

Acknowledgements AR acknowledges the Council of Scientific Research (CSIR), New Delhi, for providing fellowship in form of Junior Research Fellowship (CSIR-HRDG Ref No: Sept/06/22(i)EU-V). MM acknowledges the University Grants Commission (UGC, New Delhi, Govt. of India) as the author receives fellowship in the form of Senior Research Fellowship (CSIR-UGC SRF; Award No. 16-6(DEC.2018)/2019(NET/CSIR); UGC-Ref. No.: 900/(CSIR-UGC NET DEC.2018); dated: April 02, 2019). AS, MM, and AR also acknowledge the infrastructural support in the form of the DBT-BOOST program, Department of Science and Technology and Biotechnology, GoWB, GoI (vide Ref. No. 1089/BT(Estt)/1P-07/2018; dated: January 24, 2019) to the Department of Botany, University of Gour Banga, Malda, West Bengal, India.

References

Aayog, N. I. T. I. (2018). *Report of working group II sustainable tourism in the Indian Himalayan Region*. https://www.niti.gov.in/writereaddata/files/document_publication.Doc2.pdf. Last access: January 25, 2022.

Abed Al Ahad, M. (2024). Air pollution reduces the individuals' life satisfaction through health impairment. *Applied Research in Quality of Life*, 1–25.

Adams, K., Greenbaum, D. S., Shaikh, R., van Erp, A. M., & Russell, A. G. (2015). Particulate matter components, sources, and health: Systematic approaches to testing effects. *Journal of the Air & Waste Management Association, 65*(5), 544–558.

Anees, M. M., Sharma, R., & Joshi, P. K. (2022). Urbanization in Himalaya—An interregional perspective to land use and urban growth dynamics. *Mountain Landscapes in Transition: Effects of Land Use and Climate Change*, 517–538.

Apollo, M. (2017). The population of Himalayan regions—By the numbers: Past, present and future. *Contemporary studies in environment and tourism*, 145–160.

Awasthi, A., Dewan, R., & Singh, G. (2022). Glacier melting: Drastic future. *Advances in Behavioral Based Safety: Proceedings of HSFEA, 2020*, 255–266.

Awasthi, A., Pattnayak, K. C., Tandon, A., Sarkar, A., & Chakraborty, M. (2023). Implications of climate change on surface temperature in North Indian states: Evidence from CMIP6 model ensembles. *Frontiers in Environmental Science, 11*, 1264757.

Balakrishnan, K., Dey, S., Gupta, T., Dhaliwal, R. S., Brauer, M., Cohen, A. J., Dandona, L., et al. (2019). The impact of air pollution on deaths, disease burden, and life expectancy across the states of India: The global burden of disease study 2017. *Lancet Planet Heal, 3*, e26–e39.

Bansal, V., Jagadisan, S., & Sen, J. (2023). Understanding the urbanization induced issues in mountainous ecosystems of India: a comparative study between Nilgiris (Tamil Nadu), and lower Himalayas (Uttarakhand), India. In *Understanding Soils of Mountainous Landscapes* (pp. 303–323). Elsevier.

Barall, D., Mazta, S. R., Kashyap, R., & Goyel, P. (2020). Effects of industrial air pollution on lung functions of primary school children in Himachal Pradesh. *International Journal of Community Medicine and Public Health, 7*(7), 2607.

Beazley, R. (2013). *Impacts of expanding rural road networks on communities in the Annapurna conservation area.* Nepal.

Bera, O. P., Venkatesh, U., Pal, G. K., Shastri, S., Chakraborty, S., Grover, A., & Joshi, H. S. (2024). Assessing the impact of the national clean air programme in Uttar Pradesh's non-attainment cities: a prophet model time series analysis. *The Lancet Regional Health-Southeast Asia, 30.*

Bhandari, S., Shakya, S., Adhikari, B., Shrestha, M., Shakya, B. D., Pradhananga, A. R., Shakya, P. R., et al. (2021). Non-carcinogenic and carcinogenic risk assessment of heavy metals exposure to indoor dust in Kathmandu, Nepal. *Journal of Nepal Chemical Society, 42*(1), 16–28.

Bhat, M. A., & Gaga, E. O. (2022). Air pollutant emissions in the pristine Kashmir valley from the Brick Kilns. *Biodiversity, Conservation and Sustainability in Asia: Volume 2: Prospects and Challenges in South and Middle Asia* (pp. 959–979). Springer International Publishing.

Bhattacharyya, R., Ghosh, B. N., Mishra, P. K., Mandal, B., Rao, C. S., Sarkar, D., Franzluebbers, A. J., et al. (2015). Soil degradation in India: Challenges and potential solutions. *Sustainability, 7*(4), 3528–3570.

Bhochhibhoya, S., Zanetti, M., Pierobon, F., Gatto, P., Maskey, R. K., & Cavalli, R. (2017). The global warming potential of building materials: An application of life cycle analysis in Nepal. *Mountain Research and Development, 37*(1), 47–55.

Bhunia, G. S., & Ding, D. (2020). Temporal and spatial statistical analysis of ambient air quality of Assam (India). *Journal of the Air & Waste Management Association, 70*(8), 775–794.

Biswas, S. N., & Kumar, P. (Eds.). (2024). *Mountain tourism and ecological impacts: Himalayan region and beyond: Himalayan region and beyond*. IGI Global.

BMCI. (2016). *Bhutan Climate + Change Bhutan Media and Communications Institute.*

Braithwaite, I., Zhang, S., Kirkbride, J. B., Osborn, D. P. J., & Pollution, J. H. A. (2019). *Exposure and associations with depression, anxiety, bipolar, psychosis and suicide risk: A systematic review and meta-analysis* (Vol. 127, p. 126002). https://doi.org/10.1289/EHP4595

Brugge, D. (2018). *Particles in the air: The deadliest pollutant is one you breathe every day.* Springer.

Cass, G. R., Hughes, L. A., Bhave, P., Kleeman, M. J., Allen, J. O., & Salmon, L. G. (2000). The chemical composition of atmospheric ultrafine particles. *Philosophical Transactions of the Royal Society of London. Series A: Mathematical, Physical and Engineering Sciences, 358*(1775), 2581–2592.

Chatterjee, A., Adak, A., Singh, A. K., Srivastava, M. K., Ghosh, S. K., Tiwari, S., Raha, S., et al. (2010). Aerosol chemistry over a high altitude station at northeastern Himalayas, India. *PLoS ONE, 5*(6), e11122.

Choudhary, N., Srivastava, P., Dutta, M., Mukherjee, S., Rai, A., Kuniyal, J. C., Sharma, S. K., et al. (2022). Seasonal characteristics, sources and pollution pathways of PM_{10} at high altitudes Himalayas of India. *Aerosol and Air Quality Research, 22* (7).

Chow, J. C., Watson, J. G., Mauderly, J. L., Costa, D. L., Wyzga, R. E., Vedal, S., Dockery, D. W., et al. (2006). Health effects of fine particulate air pollution: Lines that connect. *Journal of the Air & Waste Management Association, 56*(10), 1368–1380.

Datta, D., & Banerji, S. (2015). Local tourism initiative in an eastern Himalayan village: Sustainable ecotourism or small-scale nature exploitation? *Bulletin of Geography. Socio-Economic Series, 27*, 33–49.

Dkhar, M. (2024). Tourism development in the West Jaintia Hills District, Meghalaya, in the Foothills of the Eastern Himalayas: Prospects and challenges. *Mountain Tourism and Ecological Impacts: Himalayan Region and Beyond*, 220–238.

Dodman, D., Hayward, B., Pelling, M., Castán Broto, V., & Chow, W. T. (2022). *Cities, settlements and key infrastructure.*

Duranton, G. (2008). From cities to productivity and growth in developing countries. *Canadian Journal of Economics/revue Canadienne D'économique, 41*(3), 689–736.

Gertler, C. G., Puppala, S. P., Panday, A., Stumm, D., & Shea, J. (2016). Black carbon and the Himalayan cryosphere: A review. *Atmospheric Environment, 125*, 404–417.

Giri, J., Raut, S., Rimal, B., Adhikari, R., Joshi, T. P., & Shah, G. (2023). Impact of air pollution on human health in different geographical locations of Nepal. *Environmental Research, 226*, 115669.

Gokul, T., Kumar, K. R., Prema, P., Arun, A., Balaji, P., & Faggio, C. (2023). Particulate pollution and its toxicity to fish: An overview. *Comparative Biochemistry and Physiology Part c: Toxicology & Pharmacology, 270*, 109646.

Grantz, D. A., Garner, J. H. B., & Johnson, D. W. (2003). Ecological effects of particulate matter. *Environment International, 29*(2–3), 213–239.

Green, R. (2005). Community perceptions of environmental and social change and tourism development on the island of Koh Samui, Thailand. *Journal of Environmental Psychology, 25*(1), 37–56.

Gupta, S., Rajput, A., & Kumar, R. (2021). Spatial variability of finer particulate matter ($PM_{2.5}$) mass concentrations over the Jammu urban area. *Applied Ecology and Environmental Sciences, 9*(4), 465–469.

Gupta, S., Tiwari, P., Srivastava, P., Naja, M., Choudhary, N., Vijayan, N., Kumar, M., & Sharma, S. K. (2024). Source apportionment and health risk assessment of PM10 bound carbonaceous and elemental species in the central Himalayas. *Current Natural Sciences & Engineering Journal, 1*(2), 115–126.

Hartono, D., Lioe, B., Zhang, Y., Li, B., & Yu, J. (2017). Impacts of particulate matter ($PM_{2.5}$) on the behavior of freshwater snail Parafossarulus striatulus. *Scientific Reports, 7*(1), 644.

Hassan, M. A., Mehmood, T., Liu, J., Luo, X., Li, X., Tanveer, M., Abid, M., et al. (2023). A review of particulate pollution over Himalaya region: Characteristics and salient factors contributing ambient PM pollution. *Atmospheric Environment, 294*, 119472.

Jehangir, A., Dar, N. A., Yousuf, A. R., & Sofi, A. H. (2011). Air quality at Sonamarg-a tourist hill station in Kashmir valley, India. *Journal of Experimental Sciences, 2*(6).

Kim, B. M., Park, J. S., Kim, S. W., Kim, H., Jeon, H., Cho, C., Yoon, S. C., et al. (2015). Source apportionment of PM_{10} mass and particulate carbon in the Kathmandu Valley, Nepal. *Atmospheric Environment, 123*, 190–199.

Kindlmann, P. (Ed.). (2011). *Himalayan biodiversity in the changing world.* Springer Science & Business Media.

Koul, P. A., Kaul, S. K., Sheikh, M. M., Tasleem, R. A., & Shah, A. (2010). Lung cancer in the Kashmir valley. *Lung India, 27*(3), 131–137.

Krishnan, R., Shrestha, A. B., Ren, G., Rajbhandari, R., Saeed, S., Sanjay, J., Ren, Y., et al. (2019). Unravelling climate change in the Hindu Kush Himalaya: rapid warming in the mountains and increasing extremes. In *The Hindu Kush Himalaya assessment: Mountains, climate change, sustainability and people* (pp. 57–97).

Kumar, A. (2017). Analysing urban sprawl and land consumption patterns in major capital cities in the Himalayan region using geoinformatics. *Applied Geography, 89*, 112–123.

Kumar, R., Goel, R., Singh, T., Priyadarshi, N., & Twala, B. (2024). Incentivizing green building technology: A financial perspective on sustainable development in India. *F1000Research, 13*, 924. https://doi.org/10.12688/f1000research.154056.2

Kumar, R., Lee, J. H., Shie, R. H., Chio, C. P., & Chan, C. C. (2020). Patterns and sources of PM10 in the ecologically sensitive Himalayan Region in Himachal Pradesh, India. *Aerosol and Air Quality Research, 20*(3), 410–418.

Kumari, B. (2024). Himalayan harmony: Balancing biodiversity conservation and development in great Himalayan National Park, Western Himalayas. *International Journal of Zoology Studies, 9*(1), 43–50.

Kuniyal, J. C. (2023). *Safety from numbers: Tourism in the Himalayas is not uniform; its impacts though are detrimental.* https://www.downtoearth.org.in/environment/safety-from-numbers-tourism-in-the-himalayas-is-not-uniform-its-impacts-though-are-detrimental-87949. Accessed on October 20, 2024.

Kuniyal, J. C., Sharma, M., Chand, K., & Mathela, C. S. (2015). Water soluble ionic components in particulate matter (PM_{10}) during high pollution episode days at Mohal and Kothi in the North-Western Himalaya, India. *Aerosol and Air Quality Research, 15*(2), 529–543.

Lamare, R. E., & Chaturvedi, S. S. (2014). Suspended particulate matter in ambient air of Shillong City, Meghalaya, India. *Indian Journal of Science and Technology, 2*, 37–41.

Li, C., Yan, F., Kang, S., Yan, C., Hu, Z., Chen, P., Stubbins, A., et al. (2021). Carbonaceous matter in the atmosphere and glaciers of the Himalayas and the Tibetan plateau: An investigative review. *Environment International, 146*, 106281.

Losacco, C., & Perillo, A. (2018). Particulate matter air pollution and respiratory impact on humans and animals. *Environmental Science and Pollution Research, 25*(34), 33901–33910.

Lu, J. G. (2020). Air pollution: A systematic review of its psychological, economic, and social effects. *Current Opinion in Psychology, 32*, 52–65.

Majaw, B. (2014). Meghalaya-small but not so beautiful. *World Affairs: THe Journal of International Issues, 18*(4), 126–139.

Mandal, M., Das, S., Roy, A., Rakwal, R., Jones, O. A., Popek, R., Sarkar, A., et al. (2023a). Interactive relations between plants, the phyllosphere microbial community, and particulate matter pollution. *Science of the Total Environment, 890*, 164352.

Mandal, M., Popek, R., Przybysz, A., Roy, A., Das, S., & Sarkar, A. (2023b). Breathing fresh air in the city: Implementing avenue trees as a sustainable solution to reduce particulate pollution in urban agglomerations. *Plants, 12*(7), 1545.

Masroor, M., Sajjad, H., Rehman, S., Singh, R., Hibjur Rahaman, M., Sahana, M., Ahmed, R., & Avtar, R. (2021). Analysing the relationship between drought and soil erosion using vegetation health index and RUSLE models in Godavari Middle Sub-basin, India. *Geoscience Frontiers, 13*(2), 101312. https://doi.org/10.1016/j.gsf.2021.101312

Mehmood, T., Tianle, Z., Ahmad, I., & Li, X. (2019). Integration of AirQ+ and particulate matter mass concentration to calculate health and ecological constraints in Islamabad, Pakistan. In *2019 16th International Bhurban Conference on Applied Sciences and Technology (IBCAST)* (pp. 248–254). IEEE.

Mittermeier, R. (2005). *Hotspots revisited: Earth's biologically richest and most endangered terrestrial ecoregions.*

Mohanty, R., Samal, A., & Kumar, R. (2024). Green finance supports sustainable environment: Empirical approach towards applicability in tourism sector. *Theoretical and Practical Research in Economic Fields, 15*(1), 114–126.

Moore, G. W. K., & Semple, J. L. (2021). Himalaya air quality impacts from the COVID-19 lockdown across the Indo-Gangetic Plain. *GeoHealth, 5*(6), e2020GH000351.

Morakinyo, O. M., Mokgobu, M. I., Mukhola, M. S., & Hunter, R. P. (2016). Health outcomes of exposure to biological and chemical components of inhalable and respirable particulate matter. *International Journal of Environmental Research and Public Health, 13*(6), 592.

Nag, D., Paul, S. K., Saha, S., & Goswami, A. K. (2018). Sustainability assessment for the transportation environment of Darjeeling, India. *Journal of Environmental Management, 213*, 489–502.

Nair, V. S., Babu, S. S., Moorthy, K. K., Sharma, A. K., Marinoni, A., & Ajai. (2013). Black carbon aerosols over the Himalayas: Direct and surface albedo forcing. *Tellus B: Chemical and Physical Meteorology, 65*(1), 19738.

Negi, S. P. (2009). Forest cover in Indian Himalayan states—An overview. *Indian Journal of Forestry, 32*(1), 1–5.

Nguyen, Q. H. (2021). Impact of investment in tourism infrastructure development on attracting international visitors: A nonlinear panel ARDL approach using Vietnam's data. *Economies, 9*(3), 131.

Ojha, N., Naja, M., Singh, K. P., Sarangi, T., Kumar, R., Lal, S., & Chandola, H. C., et al. (2012). Variabilities in ozone at a semi-urban site in the Indo-Gangetic Plain region: Association with the meteorology and regional processes. *Journal of Geophysical Research: Atmospheres, 117*(D20).

Päckert, M., Martens, J., Sun, Y. H., Severinghaus, L. L., Nazarenko, A. A., Ting, J. I., Tietze, D. T., et al. (2012). Horizontal and elevational phylogeographic patterns of Himalayan and Southeast Asian forest passerines (Aves: Passeriformes). *Journal of Biogeography, 39*(3), 556–573.

Pande, S., & Bhardwaj, M. (2024). Changing tourism trends and vulnerability assessment of built environment in hill stations of Indian Himalayan Region. *Current World Environment, 19*(1), 237.

Pandit, M. K. (2013). The Himalayas must be protected. *Nature, 501*(7467), 283–283.

Pant, A., Sharma, S., & Joshi, R. C. (2022). Air quality modeling for effective environmental management in Uttarakhand, India: A comparison of logistic regression and naive bayes. *Journal of Air Pollution and Health, 7*(3), 287–298.

Panwar, P., Prabhu, V., Soni, A., Punetha, D., & Shridhar, V. (2020). Sources and health risks of atmospheric particulate matter at Bhagwanpur, an industrial site along the Himalayan foothills. *SN Applied Sciences, 2*(4), 632.

Payne, D., Spehn, E. M., Prescott, G. W., Geschke, J., Snethlage, M. A., & Fischer, M. (2020). Mountain biodiversity is central to sustainable development in mountains and beyond. *One Earth, 3*(5), 530–533.

Post, T. K. (2021). *Government Decides to shut schools for four days as air pollution reaches hazardous levels*. https://kathmandupost.com/national/2021/03/29/government-decides-to-shut-schools-for-four-days-as-air-pollution-reaches-hazardous-levels. Accessed on October 10, 2024.

Prabhu, V., Prakash, J., Soni, A., Madhwal, S., & Shridhar, V. (2019). Atmospheric aerosols and inhalable particle number count during Diwali in Dehradun. *City and Environment Interactions, 2*, 100006.

Prabhu, V., & Shridhar, V. (2019). Investigation of potential sources, transport pathway, and health risks associated with respirable suspended particulate matter in Dehradun city, situated in the foothills of the Himalayas. *Atmospheric Pollution Research, 10*(1), 187–196.

Pratali, L., Marinoni, A., Cogo, A., Ujka, K., Gilardoni, S., Bernardi, E., Fuzzi, S., et al. (2019). Indoor air pollution exposure effects on lung and cardiovascular health in the High Himalayas, Nepal: An observational study. *European Journal of Internal Medicine, 61*, 81–87.

Profile, C. E. (2005). Eastern Himalayas Region. *WWF-US, Asia Program.*

Pullaiah, T. (Ed.). (2024). *Biodiversity Hotspot of the Himalaya* (1st ed.). Apple Academic Press. https://doi.org/10.1201/9781003455202

Qian, Y., Flanner, M. G., Leung, L. R., & Wang, W. (2011). Sensitivity studies on the impacts of Tibetan Plateau snowpack pollution on the Asian hydrological cycle and monsoon climate. *Atmospheric Chemistry and Physics, 11*(5), 1929–1948.

Rai, P. K. (2016). Impacts of particulate matter pollution on plants: Implications for environmental biomonitoring. *Ecotoxicology and Environmental Safety, 129*, 120–136.

Rajput, P., Sarin, M., & Kundu, S. S. (2013). Atmospheric particulate matter ($PM_{2.5}$), EC, OC, WSOC and PAHs from NE–Himalaya: abundances and chemical characteristics. *Atmospheric Pollution Research, 4*(2), 214–221.

Raju, V. S., Reddy, C. S., & Suthari, S. (2010). Flowering plant diversity and endemism in India: An overview. *Algae, 2*(40000), 6–25.

Ramachandra, T. V., & Krishnadas, G. (2011, November). Decentralized renewable energy options for Himalayan States in India. In *7th National Conference on Indian Energy Sector Synergy with Energy* (pp. 80–86).

Rani, M., Ahlawat, S., Vijayan, N., Yadav, L., Banerjee, T., Chatterjee, A., Mandal, T. K., et al. (2024). Spatial heterogeneity in health risk assessment of heavy metals during North-East Monsoon and South-West Monsoon over India. *Aerosol Science and Engineering*, 1–16.

Rao, K. S., Saxena, K. G., & Tiwari, B. K. (2016). *Biodiversity, climate change and socio-economic development in the Indian Himalaya: An overview*. Bishen Singh Mahendra Pal Singh.

Rawal, R. S., Bhatt, I. D., Sekar, K. C., & Nandi, S. K. (2013). *The Himalayan biodiversity: Richness, representativeness, uniqueness and life-support values* (p. 84). GB Pant Institute of Himalayan Environment and Development (GBPIHED), Kosi-Katarmal, Almora, Uttarakhand, India.

Romshoo, S. A., Bhat, M. A., & Beig, G. (2021). Particulate pollution over an urban Himalayan site: Temporal variability, impact of meteorology and potential source regions. *Science of the Total Environment, 799*, 149364.

Roy, A., Das, S., Singh, P., Mandal, M., Kumar, M., Rajlaxmi, A., & Sarkar, A., et al. (2024a). Summer-time monitoring and source apportionment study of both coarse, fine, and ultra-fine particulate pollution in Eastern Himalayan Darjeeling: A hint to health risk during peak tourist season. *MAPAN*, 1–15.

Roy, A., Mandal, M., Das, S., Popek, R., Rakwal, R., Agrawal, G. K., & Sarkar, A., et al. (2024b). The cellular consequences of particulate matter pollutants in plants: Safeguarding the harmonious integration of structure and function. *Science of the Total Environment*, 169763.

Salvi, S., Kumar, G. A., Dhaliwal, R. S., Paulson, K., Agrawal, A., Koul, P. A., Dandona, L., et al. (2018). The burden of chronic respiratory diseases and their heterogeneity across the states of India: The Global Burden of Disease Study 1990–2016. *The Lancet Global Health, 6*(12), e1363–e1374.

Sandeep, K., Negi, R. S., Panicker, A. S., Gautam, A. S., Bhist, D. S., Beig, G., Das, S., et al. (2020). Characteristics and variability of carbonaceous aerosols over a semi urban location in Garhwal Himalayas. *Asia-Pacific Journal of Atmospheric Sciences, 56*, 455–465.

Sati, V. P. (2006). Population and sustainability issues in mountains: A case for the Uttaranchal Himalaya. *Himalayan Ecology, 14*(2), 29.

Saud, B., & Paudel, G. (2018). The threat of ambient air pollution in Kathmandu, Nepal. *Journal of Environmental and Public Health, 2018*(1), 1504591.

Scott, C. A., Zhang, F., Mukherji, A., Immerzeel, W., Mustafa, D., & Bharati, L. (2019). Water in the Hindu kush himalaya. In *The Hindu Kush Himalaya assessment: Mountains, climate change, sustainability and people* (pp. 257–299).

Seko, K. (1991). Characteristics of winter precipitation and its effect on glaciers in the Nepal Himalayas. *Bulletin of Glacier Research, 9*, 9–16.

Shah, R. A., Shafi, A., Andrabi, S. M. A., Bhat, S. U., Hamid, A., & Mondal, N. C. (2023). Spatial distribution and health risk assessment of heavy metals in the groundwater of selected watersheds of Srinagar and Pulwama districts of Kashmir Himalaya. *Water, Air, & Soil Pollution, 234*(11), 665.

Sharma, K., Garg, A., Joshi, V., & Kumar, A. (2023a). Assessment of health risks for criteria air pollutants present in 11 non-attainment cities of Uttar Pradesh, India. *Human and Ecological Risk Assessment: An International Journal, 29*(1), 103–122.

Sharma, S., Hussain, S., & Singh, A. N. (2023a). Ecosystem services of the Indian Himalayan Region in a changing environment: an overview. *Climate Change in the Himalayas*, 185–200.

Sharma, S., Sharma, R., Sahu, S. K., & Kota, S. H. (2022). Transboundary sources dominated $PM_{2.5}$ in Thimphu, Bhutan. *International Journal of Environmental Science and Technology, 19*(6), 5649–5658.

Sharma, S. K., Choudhary, N., Srivastava, P., Naja, M., Vijayan, N., Kotnala, G., & Mandal, T. K. (2020). Variation of carbonaceous species and trace elements in PM_{10} at a mountain site in the central Himalayan region of India. *Journal of Atmospheric Chemistry, 77*(3), 49–62. https://doi.org/10.1007/s10874-020-09402-9

Sharma, S. K., Mandal, T. K., Sharma, C., Kuniyal, J. C., Joshi, R., Dhyani, P. P., Rohtash, Sen, A., Ghayas, H., Gupta, N.C., & Kumar, A. (2014). Measurements of particulate (PM2. 5), BC and trace gases over the northwestern Himalayan region of India. *Mapan, 29*(4), 243–253.

Sheikh, M. A., Kumar, M., Todaria, N. P., Bhat, J. A., Kumar, A., & Pandey, R. (2021). Contribution of Cedrus deodara forests for climate mitigation along altitudinal gradient in Garhwal Himalaya, India. *Mitigation and Adaptation Strategies for Global Change, 26*, 1–19.

Shrestha, A. B., Wake, C. P., Mayewski, P. A., & Dibb, J. E. (1999). Maximum temperature trends in the Himalaya and its vicinity: An analysis based on temperature records from Nepal for the period 1971–94. *Journal of Climate, 12*(9), 2775–2786.

Shrestha, U. B., Gautam, S., & Bawa, K. S. (2012). Widespread climate change in the Himalayas and associated changes in local ecosystems. *PLoS ONE, 7*(5), e36741.

Shrestha, A. B., Agrawal, N. K., Alfthan, B., Bajracharya, S. R., Maréchal, J., & Oort, B. V. (2015). *The Himalayan climate and water atlas: Impact of climate change on water resources in five of Asia's major river basins.* International Centre for Integrated Mountain Development (ICIMOD), Kathmandu, Nepal.

Shridhar, V. (2019). Investigation of potential sources, transport pathway, and health risks associated with respirable suspended particulate matter in Dehradun city, situated in the foothills of the Himalayas. *Atmospheric Pollution Research, 10*(1), 187–196.

Sinha, S., Kumar, R., Kumar, A. R., Prabhu, V., Kumar, R. P., & Tripathi, A. K. (2021). Elemental characteristics of respirable particulate matter in the urban atmosphere of Dehradun, Uttrakhand, India. *Current World Environment: an International Research Journal of Environmental Sciences, 16*(1), 319–328.

Soni, A., Kumar, U., Prabhu, V., & Shridhar, V. (2020). Characterization, source apportionment and carcinogenic risk assessment of atmospheric particulate matter at Dehradun, situated in the Foothills of Himalayas. *Journal of Atmospheric and Solar-Terrestrial Physics, 199*, 105205.

Srivastava, P., Naja, M., & Seshadri, T. R. (2023). Sources and radiative impact of carbonaceous aerosols using four years ground-based measurements over the central Himalayas. *Aerosol and Air Quality Research, 23*(10), 220381.

Sundriyal, M., & Sharma, B. (2016a). Status of Biodiversity in Central Himalaya. *Applied Ecology and Environmental Sciences, 4*(2), 37–43.

Swanton, C., Hill, W., Lim, E., Lee, C., Weeden, C. E., Augustine, M., Jamal-Hanjani, M., et al. (2022). LBA1 Mechanism of action and an actionable inflammatory axis for air pollution induced non-small cell lung cancer: Towards molecular cancer prevention. *Annals of Oncology, 33*, S1413.

Teri. (2015). *Air pollution in Himachal Pradesh* (p. 10). The Energy and Resources Institute.

Tewari, V. P., Verma, R. K., & Von Gadow, K. (2017). Climate change effects in the Western Himalayan ecosystems of India: Evidence and strategies. *Forest Ecosystems, 4*(1), 1–9.

Thakur, H. K., & Kuniyal, J. C. (2016). Ambient air quality status and its sources in urban and semi-urban locations of Himachal Pradesh, India. *Globalization and Marginalization in Mountain Regions: Assets and Challenges in Marginal Regions*, 173–189.

Tiwari, C., & Rani, M. (2024). Trouble in paradise: Evaluating the effects of unorganized tourism on the Himalayan ecology. *The Himalayas in the Anthropocene: Environment and Development*, 165–183.

Tiwari, P. C., & Joshi, B. (2016). *Rapid urban growth in mountainous regions: The case of Nainital, India.* Urbanization and Global Environment Change (UGEC) Viewpoints, Global Institute of Sustainability, Arizona State University, Tempe.

Tiwari, P. C., Tiwari, A., & Joshi, B. (2018). Urban growth in Himalaya: Understanding the process and options for sustainable development. *Journal of Urban and Regional Studies on Contemporary India, 4*(2), 15–27.

Tran, V. V., Park, D., & Lee, Y. C. (2020). Indoor air pollution, related human diseases, and recent trends in the control and improvement of indoor air quality. *International Journal of Environmental Research and Public Health, 17*(8), 2927.

Tse-Ring, K., Sharma, E., Chettri, N., & Shrestha, A. B. (2010). *Climate change vulnerability of mountain ecosystems in the Eastern Himalayas*. International Centre for Integrated Mountain Development (ICIMOD).

Wangchuk, T., Mazaheri, M., Clifford, S., Dudzinska, M. R., He, C., Buonanno, G., & Morawska, L. (2015). Children's personal exposure to air pollution in rural villages in Bhutan. *Environmental Research, 140*, 691–698.

Wester, P., Mishra, A., Mukherji, A., & Shrestha, A. B. (2019). *The Hindu Kush Himalaya assessment: Mountains, climate change, sustainability and people* (p. 627). Springer Nature.

Yadav, I. C., Devi, N. L., Singh, V. K., Li, J., & Zhang, G. (2019a). Spatial distribution, source analysis, and health risk assessment of heavy metals contamination in house dust and surface soil from four major cities of Nepal. *Chemosphere, 218*, 1100–1113.

Yadav, I. C., Devi, N. L., Singh, V. K., Li, J., & Zhang, G. (2019b). Measurement of legacy and emerging flame retardants in indoor dust from a rural village (Kopawa) in Nepal: Implication for source apportionment and health risk assessment. *Ecotoxicology and Environmental Safety, 168*, 304–314.

Yu, P., Guo, S., Xu, R., Ye, T., Li, S., Sim, M. R., Guo, Y., et al. (2021). Cohort studies of long-term exposure to outdoor particulate matter and risks of cancer: A systematic review and meta-analysis. *The Innovation, 2*(3).

Zhao, Y., Lin, Z., Jia, R., Li, G., Xi, Z., & Wang, D. (2014). Transgenerational effects of traffic-related fine particulate matter ($PM_{2.5}$) on nematode *Caenorhabditis elegans*. *Journal of Hazardous Materials, 274*, 106–114.

Chapter 4
Fate and Long-Range Movement of Volatile Pesticides as Particulate Organic Pollutants: Mechanisms and Possible Solutions

Sukanya Acharyya, Sumedha Saha, and Malay Bhattacharya

Abstract Particulate organic pollutants or "POPs" derived from volatile pesticides have been recognized as an immediate and formidable threat to the environment. Their long-range transport and deposition due to the "grasshopper effect," presents a significant challenge in environmental management and protection of human health. Certain pollutants, upon evaporation from their source, are transported over long distances, undergoing condensation and deposition onto soil, vegetation and water bodies. Thus, these pollutants can "hop" from one location to another, even across continents, via atmospheric currents leading to widespread contamination of ecosystems far from the original source. Moreover, its impact is increasing with increase in global temperatures. This is especially relevant in areas having altitudinal gradients as present in montane environments. An indication of its influence was observed when tea exports from India were recently returned due to concerns over pesticide residue levels, prompting heightened scrutiny and stricter quality control measures. Studies have uncovered copious levels of pesticide residue in remote areas, harming indigenous biodiversity and soil fertility through the destruction of nutrient cycling microbes. Crops unmonitored due to unawareness are carrying unsafe pesticide levels, posing significant health-risks. This chapter aims to provide a comprehensive overview on volatile pesticides, tracing their journey as staple pest management solutions to a silent threat wreaking havoc on soil, water, biodiversity and human health. It begins with an account of the various sources, emphasizing mismanaged agricultural and industrial practises as primary contributors and goes on to explore their transport and environmental fate. Potential proposed solutions are explored including regulatory measures and sustainable agricultural practises aimed at minimizing pesticide usage, enhanced degradation kinetics or tolerance abilities in biostimulant microbes, decreased tolerance in pests and reduction in atmospheric emission/dispersion. Moreover, the importance of international cooperation and policy frameworks in tackling transboundary pollution is highlighted.

S. Acharyya · S. Saha · M. Bhattacharya (✉)
Department of Tea Science, University of North Bengal, Raja Rammohunpur, Siliguri, Darjeeling, West Bengal, India
e-mail: malaytsnbu@gmail.com

A. Sarkar et al. (eds.), *From Sky to Soil*, https://doi.org/10.1007/978-3-032-31991-3_4

Keywords Persistent Organic Pollutants (POPs) · Grasshopper effect · Volatile pesticides · Long-range transport · Pesticide residues

4.1 Introduction

Agriculture stands as a cornerstone for human civilization. It is the single most essential factor in sustaining an ever-increasing global population. As such, governments around the world focus a large part of their resources towards the betterment of the agricultural sector. This equates to increased yield and quality of crops that provide food, fibre and other essential commodities. Scientists and policy makers unite in their endeavour to formulate and implement plans utilizing state-of-the-art technologies and the latest scientific discoveries that can enhance outputs generated per unit of input. Agriculture has long acted as a catalyst for the advancement of science. Historically, numerous successful societies have been based solely around the mastery of food production and management, leading to thriving economies and cultural revolutions. It has served as a kind of barometer for measuring the capabilities of a nation or government. Innovations such as crop domestication, irrigation systems, automation, and the advent of biotechnology have allowed the dominance of the human race on this planet in the years since. Yet, this pursuit is not without its challenges and complexities.

At its root, food production is driven by farmers. They lend the labour for cultivation and harvest various raw materials for other downstream usages. One of the biggest challenges faced by this community is the invasion of pests. These unwelcome guests feed on plant tissues, reducing crop quantity, as well as quality, and can devastate entire fields if left unchecked. Various plant diseases are spread by pests that can act as vectors for the pathogens. Infestations lead to destruction of crops and economic hardship for the farmers. Significant advancements have been made in different pest control methods. However, farmers have continued to rely heavily on pesticides for pest control since their introduction due to high efficacy and ease of application. In 2019, the worldwide pesticide market achieved a total value of $36.4 billion. Unsurprisingly, China emerged as the leading exporter, generating $6.11 billion, while Brazil, the primary importer, topped the list with imports valued at $3.69 billion (OEC, 2021). This really highlights the magnitude of dependence of the agricultural sector on pesticides.

Pesticides are synthetically produced compounds designed to target either a specific pest or pest groups. Each has a unique mode of action depending on the type (insecticides, herbicides, fungicides, rodenticides, bactericides) and chemical nature. Application methods are tailored to the crop and the target pest. The high efficacy displayed by these chemical agents has encouraged heavy usage among farmers and increased exploration into more such compounds. Unfortunately, the toxicity of pesticides extends beyond their effect on pests. It was well understood early on that their usage needed to be strictly regulated. The toxicity of pesticides exerted on unintended organisms was first brought to attention in the publication

Silent Spring (Carson, 1962). Since then, various regulatory agencies (EFSA, EPA, MoEFCC, FSSAI, etc.) have begun evaluating the efficacy and safety of pesticides through standardized testing. As the market for pesticides has expanded with an ever-increasing population demanding more food production, their use has become ubiquitous among farmers and spawned a new concern. Although effective, with continued use, target pests adapt and become resistant to assigned doses for these pesticides. In order to cope, farmers are forced to use higher and higher doses. This has resulted in the abuse of pesticides in agriculture.

Although for long serving as an indispensable tool to agriculture, recent studies have uncovered an overlooked consequence of overuse of pesticides that may outweigh its benefits. Pesticide residue has been detected as contaminants in various sections of the environment, all over the world, as it has the potential to spread unintentionally beyond its intended target zones (Aston & Seiber, 1997; Coronado et al., 2011; Weber et al., 2010; Zhao et al., 2013). This can occur through various mechanisms, such as drifting from nearby agricultural crops or even evaporation. Each application of pesticides results in drift, leading to a loss of chemicals that can range anywhere between 2 and 25% (Tudi et al., 2021). The reach of the drift can extend from short distances to hundreds of miles (Glotfelty & Schomburg, 1989). Volatilization is another common occurrence, whereby pesticides from soil and foliage evaporate and disperse into the surrounding environment (Farha et al., 2016). A substantial proportion of pesticides, approximately 80–90%, undergo volatilization within a short period following application (Majewski & Capel, 1995). These volatile pesticides can be classified under the umbrella of persistent organic pollutants or POPs.

4.2 Pesticides as POPs

Persistent organic pollutants are a vast group of chemical compounds that can, as the name suggests, "persist" for a considerable amount of time in the environment. In their persistence they also continually exert toxicity (Mandal et al., 2023; Pariatamby & Kee, 2016). The exact period of time that grants such a compound the attribute persistent is debated. However, a longevity of several years to decades in soil and several days in the atmosphere is generally agreed upon (Bonmatin et al., 2015) (Table 4.1). The following are properties that are commonly found in POPs.

4.2.1 Properties of POPs

i. Persistence: All POPs exhibit an ability to persist, unchanged in soil, air, water and even living organisms. They have extended half-lives and are resistant to environmental degradation either by chemicals or microbiota that would normally break down pollutants into simpler or less toxic substances.

Table 4.1 List of some volatile pesticide pollutants

Pesticide	Chemical class	Vapor pressure (mm Hg)	Solubility in water	Toxicity
Aldrin*	Organochlorine	1.2×10^{-4} at 25 °C	Insoluble	Acute
Azinfos-methyl	Organophosphorus	2.0×10^{-7} at 20 °C	Poorly soluble	Acute
Atrazine	Triazine	2.89×10^{-7} at 25 °C	Soluble	Moderate
Alachlor	Organochlorine	1.6×10^{-5} at 25 °C	Poorly soluble	Low
Chlorpyrifos	Organophosphorus	2.02×10^{-5} at 25 °C	Insoluble	Moderate
Chlordane*	Organochlorine	9.75×10^{-6} at 25 °C	Insoluble	Moderate
Carbaryl	Carbamate ester	1.36×10^{-6} at 25 °C	Insoluble	Low
Carbendazim	Benzimidazole	7.5×10^{-10} at 20 °C	Poorly soluble	Low acute
Chlorothalonil	Organochlorine	5.7×10^{-7} at 25 °C	Slightly soluble	Moderate
Difenoconazole	Triazole fungicide	2.5×10^{-10} at 25 °C	Slightly soluble	Low acute
Diuron	Dimethylurea	6.9×10^{-8} at 25 °C	Poorly soluble	Slight
Dimethoate	Organophosphate	1.875×10^{-5} at 25 °C	Soluble	Moderate
Dieldrin*	Organochlorine	3.1×10^{-6} at 20 °C	Insoluble	High
Diazinon	Organophosphate	9.01×10^{-5} at 25 °C	Insoluble	Low
DDT*	Organochlorine	1.89×10^{-7} at 20 °C	Insoluble	Moderate
Endrin*	Organochlorine	3×10^{-6} at 20 °C	Insoluble	Acute
Endosulfan	Organochlorine	1.73×10^{-7} at 25 °C	Insoluble	Acute
Ethoprophos	Organophosphate	3.8×10^{-4} at 20–25 °C	May be soluble	High
Folpet	Phthalimide	1.57×10^{-7} at 25 °C	Insoluble	Low
Heptachlor*	Organochlorine	4×10^{-4} at 25 °C	Insoluble	High
Hexachlorobenzene*	Organochlorine	1.72×10^{-5} at 25 °C	Insoluble	Acute

(continued)

Table 4.1 (continued)

Pesticide	Chemical class	Vapor pressure (mm Hg)	Solubility in water	Toxicity
Glyphosate	Organophosphate	9.8×10^{-8} at 25 °C	Soluble	Low
Metolachlor	Organochlorine	3.14×10^{-5} at 25 °C	Slightly soluble	Acute
Metazachlor	Organochlorine	6.75×10^{-7} at 20 °C	Poorly soluble	Moderate
Metribuzin	Triazine	4.35×10^{-7} at 20 °C	Poorly soluble	Low
Malathion	Organophosphate	3.97×10^{-5} at 30 °C	Poorly soluble	Low
Mirex*	Organochlorine	8×10^{-7} at 25 °C	Insoluble	Moderate
Oxamyl	Carbamate	2.3×10^{-4} at 20–25 °C	Highly soluble	High
Propiconazole	Triazole	1×10^{-6} at 25 °C	Soluble	Low to moderate
Phorate	Organophosphate	6.38×10^{-4} at 25 °C	Insoluble	Extremely
Monocrotophos	Organophosphate	7×10^{-6} at 20 °C	Soluble	Acute
Trifluralin	Organofluorine	4.58×10^{-5} at 25 °C	Insoluble	Low

Pesticides with * are reported to be persistent organic pollutants (POPs), according to the Stockholm Convention. *Sources* PubChem; https://enb.iisd.org/articles/Stockholm-convention

ii. Toxicity: All POPs are a hazard to the environment owing to their toxic nature. The level and mechanism of toxicity depend on the chemical nature of the particular POP. Many have been reported to be highly carcinogenic or contributors to disorders of the reproductive, nervous, cardiac, endocrine and immune systems. Moreover, POPs can disrupt natural ecological balance, and cause declines in local biodiversity.
iii. Lipophilicity: Most POPs are found to be lipophilic in nature. This promotes strong bonds with components like soil organic matter, allowing them to persist further in the environment. The low solubility in water also restricts them from dispersing in aqueous environments.
iv. Bioaccumulation and biomagnification: POPs are resistant to metabolic processes in living systems. As such, they have a tendency to accumulate in the fatty tissue of animals over time with repeated exposure. The POP accumulated becomes increasingly concentrated, i.e., it is biomagnified, higher up in the food chain. This leads to the presence of high levels of POPs in apex predators.

v. Volatility: Some POPs have the ability to volatilize at ambient temperatures, converting to a gaseous phase. This ability enables the travel of these pollutants through the air over considerable distances, further spreading pollution.

POPs can come from a variety of chemical families. While some arise as inadvertent by-products of combustion or industrial processes, others are deliberately synthesized for industrial or agricultural applications, as is the case with POPs arising from pesticides. Human exposure to POPs occurs primarily via contaminated water or foods like fish, livestock or crops, in which they readily bioaccumulate. Studies have consistently found pesticide residues among the most prevalent food contaminants (Khan & Rahman, 2017). Some examples of pesticides that are already recognized as POPs are endosulfan, mirex, toxaphene, dichlorodiphenyltrichloroethane (DDT), aldrin, chlordane, dieldrin, endrin, hexachlorobenzene (HCB), heptachlor, and diazinon. Several new ones are being added to the list regularly.

The consequence of these pollutants can be multifaceted. However, the focus of this chapter will be on volatilization/deposition as a mechanism of long-range transport. This particular phenomenon has captured the attention of scientists worldwide as it allows for the spread of pesticide POPs and other such pollutants beyond the limitations of the agricultural field. The chapter starts off by discussing the various sources and practices associated with the spread of pesticide POPs and goes on to give an account on the current understanding of their transport mechanisms that allow contamination of varied environments. The following sections list the various ecological and human-health risks that are posed by this type of pollution and the solutions that have been proposed to mitigate the problem. The chapter finishes off with a summary of the concept and discussions on directions that should be taken in research and solutions that may be implemented in the future for a cleaner and safer environment.

4.3 Sources of Volatile Pesticides

Some major sources that contribute to the volatile pesticide reserves include agricultural practises, industrial activities and emissions, and domestic uses.

4.3.1 Agricultural Practices

Pesticides have long played a crucial role in keeping a check on agripests. Their extensive usage and direct dependency commenced in the nineteenth-century "Green Revolution." Almost 2.7 million tonnes of synthetic pesticides are manufactured and utilized globally with Asia being the leading exporter and the Americas being the largest consumers (FAO, 2022). It was discovered much later that some pesticides are able to drift or volatilize and migrate through the air, spreading contamination (Galon et al., 2021; Mandal et al., 2025).

It has been observed that the rate of emission of pesticides is much higher in the case of liquid than solid formulations (He et al., 2022). Pesticides with emulsifiable concentrate (EC) formulations and fumigants are top contributors, with the latter holding a 100% emission potential (Kumar et al., 2011). Synthesis of commercial pesticides often employs various solvents that aid the emission of volatile compounds. This was observed in the San Joaquin Valley, where pesticides were reported to have contributed almost 5.5% of the total volatile compound emissions (He et al., 2022). In the report of CDPR 2008, an EC-based chlorpyrifos pesticide (organophosphate) was deemed to be the highest contributor to emissions (Kumar et al., 2011). Application of weed oils that primarily comprise petroleum-based hydrocarbons such as diesel-type oils or mineral oils is another source from agricultural lands. Horticulture oils as pesticides were reported to be extensively in use in California during the year of 2016. Also, these have been observed to largely contribute to photochemical smog in the San Joaquin Valley. Moreover, higher molecular weight chemicals mixed with these solvents persist in the soil and undergo much slower dissipation, thus polluting the soil of the region (Woodrow et al., 2019). Surface runoffs or leaching of these chemicals to nearby waterbodies or groundwater poses further threat to all the lifeforms (Tudi et al., 2021).

4.3.2 Industrial Activities

When it comes to volatile organic pollutants, industrial emissions make up the majority of anthropogenic sources. Various organic solvents like benzene, chlorobenzene, formaldehyde, toluene, etc., used during the synthesis of pesticides, contribute to emissions (Yuan et al., 2022). Moreover, synthesis also involves the production of several intermediates that may be highly volatile and toxic in nature (aromatic hydrocarbons, aldehydes, ketones, ethers, etc.). The vapour forms of these toxic chemicals present in and around these manufacturing plants pose an immediate threat to the workers and the people around coming in direct contact through inhalation. A very well-known and reported instance of such exposure to methyl isocyanate gas in the production of carbamate pesticides was seen in the "Bhopal Gas Tragedy" of 1984 (Varma & Varma, 2005). Another such example was observed in an incident at a tunnel under construction linking Denmark and Sweden, where workers were inadvertently exposed to potent neurotoxin and carcinogen monomeric acrylamide (Kjuus, 2001). Hence, strict monitoring of exposure in workplace environments and maintaining a safe exposure limit becomes a quintessential criterion.

4.3.3 Domestic Use

Pesticides are usually sprayed to maintain lawns/gardens or maintain houses free from flies, mosquitoes, cockroaches, mice, etc. The volatile or semi-volatile components released from these chemicals persist in the air for an extended period. Considering the closed space, residential exposure to these pesticides tends to affect health more seriously if the exposure is long and frequent. Al-Alam et al. (2022), in their work found that both air samples and dust particles from places like bedrooms and living rooms carried pesticides all throughout the year. Also, direct exposure was highest towards pyrethroids due to their semi-volatile nature. The concentration of pesticides found would vary depending on factors such as contaminated dust or particulate matter, household surfaces that collect them, and the presence of pesticide containers. Various disinfectants used also add to the volatile compound load in indoor air.

4.4 Transport Mechanisms and Environmental Fate

Understanding the transport mechanisms of persistent pesticide pollutants is an area of active research that has gained more attention with passing time. The sheer number of factors and their relatedness that affect the distribution patterns of such pollutants has made it difficult to fully comprehend all the parts of the puzzle. It is imperative to elucidate the transport mechanisms to facilitate the development of proper monitoring technologies, guidelines and regulations to mitigate this problem. In the case of volatile pesticides that act as POPs, transport mainly takes place through the atmosphere in multiple volatilization-transport-deposition cycles (Bedos et al., 2002). This enables the pesticides to reach off-target sites, far away from the source of origin, wreaking havoc on the environment upon exposure. There are three main instances wherein pesticides are released into the atmosphere, i.e., direct volatilization, while field spraying and after field spraying (Farha et al., 2016).

By their work description, pesticides are required to persist on the crops they are sprayed on in order to be effective against the target pest. However, this in itself, promotes release and subsequent dispersion of these harmful chemicals into the air. During application, 30–50% of the sprayed pesticide is immediately lost through "spray drift" or simple evaporation. After application, loss can still occur through wind erosion or volatilization (Ucar & Hall, 2001). Once released, the pesticides exist in the atmosphere either as aerosols themselves or latch onto pre-existing aerosol particles or carriers like fog or snow. Wind can then sweep these forms of atmospheric pesticides away to off-target locations where they may be deposited along with precipitation events and accumulate in soil, water or on vegetation. This cycle can repeat itself depending on the re-emission rate, which in turn depends on local climate, solar radiation, wind pattern, temperature, etc.

4.4.1 Spray Drift

The very first opportunity for dispersion comes at the time of spraying of the pesticide (Maybank et al., 1978). Spray drift is the unintentional movement of sprayed pesticide away from the intended target area during application. It occurs due to wind or air currents and can result in the deposition of pesticides onto neighbouring crops, vegetation, water bodies, and residential areas, leading to off-target contamination. This poses significant risks to human health, the ecosystem and agricultural productivity of non-target sites. The volume of spray drift can be influenced by various factors. The drifting pesticide can occur as droplets, particles or vapour depending on climate conditions. When present as drops, they can be a variety of sizes based on the size of the nozzle of the spraying equipment (Hofman & Solseng, 2001). The larger drops have an increased chance of deposition within the target site. Alternatively, when present as particles, drift volume is dependent on quality of the applicator, especially in the case of pneumatic seeding devices (Biocca et al., 2019). In these devices, the pesticide takes the form of a powder supplemented with the active ingredient. At the time of application, the powder can disperse unevenly leading to inefficient deposition of the pesticide at the point of application. In this way, most of the pesticide is lost to drift instead.

Furthermore, the characteristics of the pesticide being sprayed like viscosity or volatility are important in assessing how much of it can be lost to spray drift. Meteorological conditions (relative humidity, temperature, air stability, wind speed, and direction) at the area of application, as well as the site of precipitation are all involved in determining drift pattern. Higher temperature at the target site encourages evaporation or volatilization, while colder temperature at non-target sites encourages precipitation. Pesticides in the form of vapour, induced by relatively high temperature are more easily loaded onto the air and trafficked by wind currents. The speed and direction of blowing wind at the site of application decides how fast, how far, and where the impending precipitation takes place.

4.4.2 Wind Erosion

After application, the amount of pesticide remaining at the target site can still come under the influence of other dispersion factors. One of them is wind erosion. This is a process by which the pesticide adheres to small particles like granules that are picked up and carried away by wind in the traditional way of topsoil erosion. For example, soils with greater clay content or aggregates smaller than 0.01 mm promote erosion by wind while coarser particles are more resilient (Mina et al., 2022). When the particle size is larger, they can be carried horizontally by the wind, but not vertically. Hence, the adhered pesticide residue does not reach the atmosphere for dispersal. It is not the most efficient transfer mechanism for pesticides to enter the atmosphere but it contributes nonetheless. Research revealed a loss of only 1.4% for trifluralin and triallate (Cessna et al., 2006), but notable proportions for 2,4-D, bromoxynil, mecoprop, and diclofop (Larney et al., 1999). This mechanism is thus dependent on properties that also determine soil erodibility in general like characteristics of the soil, weathering, temperature, precipitation and cultivation practices.

4.4.3 Volatilization

Perhaps the most important mechanism of transport of pesticide POPs is through volatilization. This process occurs after the settling of the sprayed pesticide onto the agricultural field and results in its conversion into gaseous form. It is an amalgamation of several different physicochemical processes that take place all together or a few at a time, producing large plumes of gaseous pesticide that can be easily translocated by wind to far away locations (Bedos et al., 2002). Volatilization also facilitates the farthest distance covered in significant volumes. The mechanisms involved are varied. When the pesticide is in liquid form or dissolved in an aqueous phase, it can evaporate, whereas in the solid state, sublimation occurs. Another method called desorption can occur to release bound state pesticide from the soil matrix. Vapour pressure of the pesticide applied heavily influences release. In the absence of wind, pesticide-laden fog may form at the site of volatilization, directly exposing farmers to the toxic fumes. More volatilization also means less pesticide actively tackling pests in the targeted agricultural field, resulting in losses in yield and finances in the long term. Conversely, volatilization reduces the risk of soil and water contamination. The rate of volatilization is dictated by intermolecular forces, properties of the chemical compound, climate, vegetation cover, etc.

Studies on chemical agents like dicamba and 2,4-D have illustrated the consequences of such processes (Niu et al., 2023). Dicamba was used extensively in the agricultural field for over six decades as a commonplace herbicide. However, its high volatility made it susceptible to displacement, especially in conditions of high temperatures and low humidity (Egan & Mortensen, 2012). Despite attempts at new formulations (such as Xtendimax, FeXapan, and Engenia) and improved application equipment, incidents of off target were still reported in the USA in 2017 (Alves et al., 2017; Werle et al., 2018). Similarly, another auxinic herbicide, 2,4-D, demonstrated phytotoxic effects on off-target non-resistant plants in Brazil (da Silva Brochado et al., 2022).

In soil systems, deposited pesticide remains in a bound state to either organic or inorganic counterparts through intermolecular interactions. Binding may be a result of physical adsorption through van der Waals interactions, chemical adsorption involving coulomb forces or interactions dependant on hydrogen bonding. Adsorption is inversely related to the rate of volatilization as it reduces the chemical activity of the bound pesticide. Thus, more binding equals less volatilization. In the preparation of application, adjuvants may be added to the formulation which brings about changes in the chemical nature of the pesticide. This can affect volatilization rates, most often demonstrating a positive correlation. Such has been observed in the 3–6 times increased volatilization rates of atrazine when paired with adjuvant polytetrafluoroethylene (Green, 2000). The reverse has also been observed in the case of DDT which volatilizes 45–70% less when paired with the same adjuvant. Semi-volatile pesticides like chlorpyrifos, pyrimethanil or trifluralin have displayed a further dependency on temperature along with the adjuvant. Volatilization only increased when the temperature remained below 24.5 °C and decreased above it Das and Hageman (2020).

Volatilization typically increases with an increase in soil temperature but can decrease if the increase in temperature dries out the soil, thereby affecting vapour pressure of the pesticide residue. This is an example of the relatedness of the many factors influencing such processes. A study on the volatilization of carbofuran at an ambient temperature of 25 °C reported a more substantial loss in flooded soil compared to soil that was not soaked (Benicha et al., 2011). Higher rates of volatilization have also been demonstrated by diazinon, atrazine, 2,4-D and glyphosate at moisture content that is higher than field capacity and a temperature of 30 °C. Diazinon displayed the highest dissipation rates of a whopping 43% in a matter of 72 h when humidity was as high as 90–110% (Góngora-Echeverría et al., 2019). Microbial activity, i.e., enzymatic reactions conducted by soil-dwelling organisms are instrumental in processes such as conjugation and oligomerization (Chaplain et al., 2011). If the soil has high organic matter content and low degradability by local microbes, the cumulative effect of these two factors would allow extended residence of the pesticide.

The bioavailability of certain pesticides depends heavily on soil pH, as it can alter their ionic properties. pH also determines the distribution of charge of the soil, and dictates the pathway for uptake through plant roots. Research experiments have demonstrated that increasing soil pH can enhance the phytotoxic effects and persistence of atrazine and metribuzin (Holford et al., 1989; Peek & Appleby, 1989). The pH also affects associations and/or dissociations of pesticide residue with components in the soil. Thus, it is important to take all such aspects into consideration when trying to fully comprehend the flux dynamics of pesticide volatilization from soil.

Volatilization can also take place from plant surfaces. In fact, some studies suggest more volatilization takes place from plants than from soil (Bedos et al., 2002). Pesticides also accumulate more in plant leaves than the vegetables/crops themselves, as evidenced by studies on zucchini and pepper plants (Park et al., 2011). Upon deposition, some of the pesticide may penetrate the leaf surface and become part of chemical, photochemical or microbial degradation processes. A portion of it may also be washed away by rain, especially in the initial 24 h after application. These processes lead to a reduction in volatile emissions from the plant surface. The remaining pesticide comes under the influence of photochemical alterations under sunlight and volatilizes under the right temperature and humidity. Higher volatilization rates than soil may be due to higher adsorption of pesticide to soil than leaf surfaces. Pesticide loosely attached to the plant surface leaves it prone to volatilization by environmental factors. A direct positive relationship can be drawn between the number of leaves and the rate of volatilization. Air turbulence within the foliar cover may increase convection rates between the leaf surface and the air above it, further facilitating increased rates of volatilization.

Following transport, at the site of precipitation, the deposited pesticide can bind to various elements. Depending on the strength of the bond, they may accumulate in that environment. As is the case at the point of origin, the retention capacity of soil or vegetation cover for the deposited pesticide once again comes under the influence of the same factors that were at play at the source site. Relationships between soil, vegetation and pesticide, as well as new meteorological parameters of the deposition site dictate whether the pesticide will or how much of it will rejoin the emission-transport-deposition cycle. This cycle can repeat itself a number of times carrying the pesticide residues to locations thousands of kilometres away to be deposited at pristine remote regions like the Antarctic. This constitutes the global biogeochemical cycle of these pollutants. If the deposition site has had no previous exposure to such chemicals, the introduction of the pesticide residue can have grave consequences on the microbial biodiversity, plant life, wildlife and overtime, human health. It can cause a complete breakdown of the balance of the ecosystem.

Another point of interest is the deposition of the volatile pesticides upon agricultural fields that practise sustainable organic farming. Without knowledge or intention of the farmers, their fields are exposed to the toxic effects of pesticides sprayed elsewhere that have been carried over to their farmland via air travel. This can lead to crop failures, plant disease, loss of beneficial plant-growth-promoting or nutrient-cycling microbes, loss of soil fertility, adverse effects on farm animals and the farmhands. Pesticides that are water-soluble can dissolve into the water and be leached away along with water percolation. Whereas pesticides that are hydrophobic are adsorbed onto organic matter reducing their bioavailability. Organic matter has high affinity for such pollutants that essentially traps them in the soil, discouraging re-emission into the atmosphere. Although in this way the pesticides can persist in the soil for months or even years, the reduced bioavailability may translate into reduced risk to the environment.

4.5 Role of Environmental Factors on Dissipation of Pesticides

4.5.1 Photodegradation

Photodegradation is the action of light on pesticides that can cause degradation. It requires the absorption of energy, in this case photons or solar energy, for the photolysis of the pesticide. This is a well-known factor significantly influencing pesticide dissipation rates. A correlation has been established between the rate constant of the photolysis mechanism and the half-life of the pesticide (Zongmao & Haibin, 1988). The amount of incident light can vary from location to location, as can the quantity of photons absorbed by the medium (soil, water, plant) or the pesticide. The effect of photolysis has been documented in the case of pendimethalin in turf grass and rotenone from olive fruits (Cabizza et al., 2004; Cooper et al., 1990). Furthermore, UV irradiation has been shown to trigger three different photochemical reactions in ethylene vinyl acetate (EVA), known as Norrish I, II, and III (Beinert et al., 2014). Even the most persistent pesticides like DDT will degrade to some degree under radiation (Fig. 4.1).

The degradation of pesticides through photochemical methods can operate through direct or indirect mechanisms. Direct photodegradation is only significant for compounds that can absorb light within specific wavelengths (>290 nm) and occurs when the pesticide molecule absorbs photons, leading to the breaking or rearrangement of bonds and the formation of more stable byproducts. Indirect photodegradation involves a photosensitizer that absorbs light and generates reactive species that degrade the pesticide. Absorbed photons induce excitation that leads to various organic reactions (Katagi, 2004) like hydroxylation and decarboxylation, which constitute direct photolysis, or cause the generation of reactive oxygen species (OH, O_3, O_2) that oxidize functional groups in pesticides (Katagi, 2004). Portions

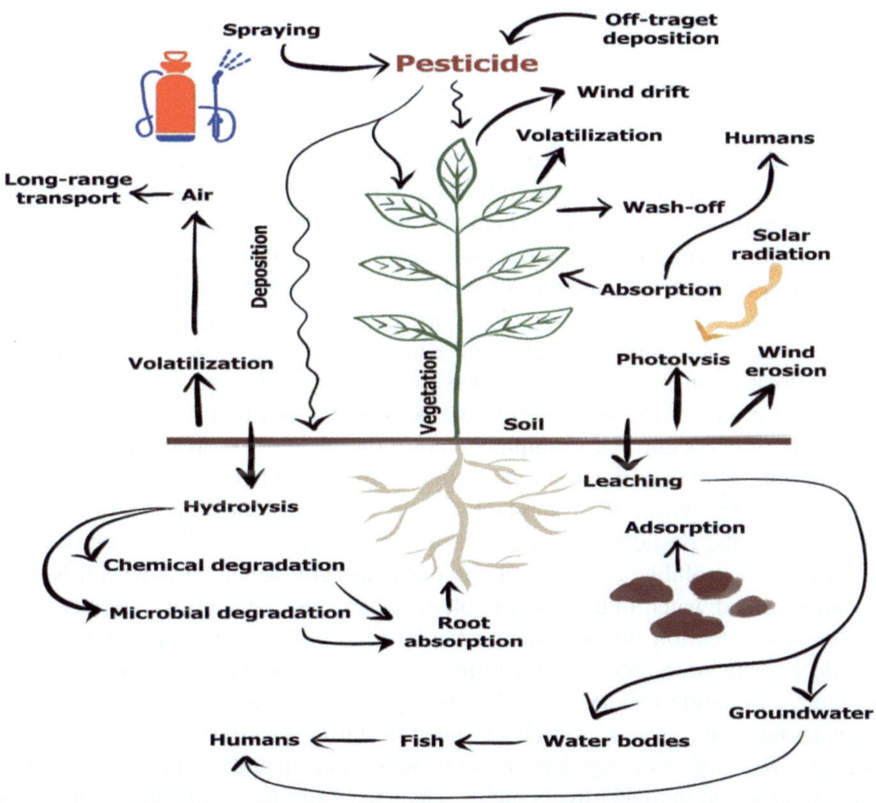

Fig. 4.1 Different interactions of volatile pesticides with the environment

of the pesticide volume that undergo these processes may no longer be available to take part in further transport outside of the target area, reducing the risk. However, it can also cause the volatilization of pesticide leading to the opposite end result.

4.5.2 Temperature

As volatility is under the direct influence of temperature, it automatically becomes an important factor in the dissipation of pesticides. Apart from volatilization rates, temperature also affects the rate of degradation in soil and on plants. Moreover, it determines the amount of pesticide that is absorbed by the plant through the roots or cuticle (Elango et al., 2022). Temperature can affect solubility and toxicity of the pesticide (Verheyen et al., 2022). Higher temperature supports evasion of pesticide as is observed in the case of greenhouses (Stenersen, 2004) and from the drying out of soil (Mao et al., 2019). In colder climates, photodecomposition and microbial degradation processes also become significantly reduced (Katagi, 2004).

4.5.3 Precipitation

Rainfall holds great significance in the deposition as well as elimination of pesticide residues. The intensity of the rainfall directly influences washing off of pesticide residue from plant surfaces. Duration of the rainfall further determines how thorough the washing is. Time elapsed since application, pesticide formulation and water solubility of the pesticide are some other factors that can be taken into consideration in this context (Elango et al., 2022). Moist plant surfaces and soil are known to enhance pesticide uptake through cuticle and roots. Alternatively, when there is excess moisture in plants, it is excreted by the leaves and roots which becomes an evasion route for the pesticide. Pesticide formulations that add solubility may also affect this process. For example, bendiocarb efficacy is reduced with long and low-intensity rainfall compared to short bursts of high-intensity rainfall (Taylor & Matthews, 1986). Wash-off from plant surface due to the mechanical action of rain can mobilize the pesticide into the soil and also influence leaching processes. In this way, water soluble and non-volatile pesticides can percolate downwards until reaching ground water. In the case of volatile pesticides, transport to soil may result in increased retention times as the pesticides bind to different components in the soil, decreasing emissions. Cloud and fog are also great scavengers of such volatile pollutants. Pesticide residue from the atmosphere can thus deposit onto land and water surfaces along with rainfall or snowfall during precipitation events at sink locations. Rain affects spray deposits in various ways, including direct effects such as dissolution and indirect effects such as changes in surface temperatures, humidity and physical damages.

4.5.4 Humidity

High humidity significantly aids in the absorption of pesticides by leaf surfaces (Hull, 1970). Post-spraying, plants exhibit heightened uptake under humid conditions, potentially due to mechanisms delaying droplet drying (Ramsey et al., 2002). High humidity also promotes cuticle hydration for enhanced absorption, expands stomatal opening, and increases the permeability of water-soluble pesticides into leaf surfaces. This causes reduction in emissions. Expectedly, water-soluble pesticides are more prominently affected by the influence of humidity (Varanasi et al., 2016).

4.6 Overview of the Grasshopper Effect

As discussed in the previous section, volatile pesticides can volatilize at relatively warmer regions due to high vapour pressure and travel long distances through the atmosphere. During their journey, the pesticides undergo processes such as evaporation, condensation, and deposition. On a global scale, this is referred to as the "grasshopper effect" (Gong et al., 2021). The name of the phenomenon comes from these chemicals being transported in "hops" from one place to another or between soil, water, air and living organisms, mimicking the movements of a grasshopper. This effect is particularly significant for volatile POPs because it allows these pollutants to spread far beyond their source of origin, affecting remote and pristine areas where they may not have been used or produced. For example, POPs released in industrialized regions can be transported to polar regions in this way (Weber et al., 2010). Once deposited in these remote areas, they can accumulate in the soil, water, vegetation and animals, posing risks to wildlife and indigenous communities who rely on traditional diets for sustenance.

In this process, the pollutants require a much longer time to reach each destination, possibly months. If the pollutant is able to reach the troposphere or stratosphere through emission, range and distribution are enhanced. On this scale, solar radiation induced temperature differences among equatorial zones and the poles, as well as global longitudinal wind patterns become the major directing factors setting distribution trends (Scheringer, 2009). Just as a warmer climate promotes volatilization and transport by air, a colder climate promotes precipitation. Thus, the direction of global spread of such pollutants is from the mid-latitudes towards the poles. Many pitstops are crossed along the way that constitute the hops. Disturbingly, at least seven new pesticides (2-methyl-4-chlorophenoxyacetic acid (MCPA), metribuzin, pendimethalin, phosalone, quizalofop-ethyl, tefluthrin, triallate), besides the already studied endosulfan have been discovered to be accumulating in the Arctic since 2010 (Balmer et al., 2019).

Geographical areas of high emission volume and rates contributing to the grasshopper effect are known as "global source areas." They are typically located at the mid-latitudes and heavily populated and industrialized. High DDT usage recorded in such zones has been found to correspond to residue in the atmosphere, volatilized by the warmer temperatures (Malusá et al., 2020). Unfortunately, DDT also still finds use in many countries in the tropical and temperate zones. Upon exposure to a colder climate at a higher latitude, the residues can undergo atmospheric depletion events and deposit onto the soil. Wet precipitation along with rain or snow may also occur. Re-emission also becomes more and more limited at colder climates, especially at the poles due to lack of volatilization causing enhanced accumulation with latitudinal increase. Thus, this is a method of transcontinental transport whereby the pollutant is fractionated along a latitudinal gradient.

A distinct bias in distribution can currently also be recognized based in regions that are part of the northern hemisphere (Lohmann et al., 2007). The northern hemisphere holds the largest proportion of the population of earth and is where most industries are based. Hence, unsurprisingly, it contributes far more POPs to the atmosphere. These regions either produce the pesticides or use them extensively to support the burden of feeding a large population. Predictions have been made that this dynamic will shift in the coming years towards the southern hemisphere. As the data suggests, industries are increasingly being shifted to Asian countries like India and China, slowly converting them into some of the largest producers of atmospheric POPs. Shifting monsoons will cause these atmospheric deposits to drift into the southern hemisphere (Semeena et al., 2006). Additionally, both India and China are agronomy-based countries that utilize a large number of pesticides in their practices (Pozo et al., 2006). Based on these it may be assumed that a shift in POP distribution towards the southern hemisphere may be seen in the foreseeable future.

The polar regions at the deposition end of the spectrum may be regarded as the global sink for such pollutants generated around the world. This has fuelled research on Arctic pollution with the objective of reducing exposure of the native populations to elevated levels of toxic POPs who rely on fat-heavy diets, making them more vulnerable to this pollution. In light of this, attention is now being allotted to the establishment of global inventories that can record and monitor emission trends. In the case of pesticide derived POPs, this presents with certain limitations. Changes in the future trends in POP distribution and toxicology can change depending on the nature of the sorption medium, properties of the migration site, food web dynamics, meteorological parameters, etc. There is a scarcity of data on pesticide emissions and distribution in most places. that does not allow for the proper evaluation of their presence in the atmosphere. The lack of time-series data on emission/deposition at different locations has resulted in very few changes being observed in POP distribution dynamics. Even the extent of influence the organic carbon cycle or flooding and drought events can have on the remobilization or persistence of POPs remains unclear to this day. Moreover, governments may be apprehensive in disclosing statistics on industrially produced POPs, when available, that makes studying such trends on the global scale difficult. Before realizing the bigger picture of global pesticide distribution and its impacts, data must be collected and analysed on the regional trends of distribution. The cumulation of this data can provide insights into the full scope of the problem (Fig. 4.2).

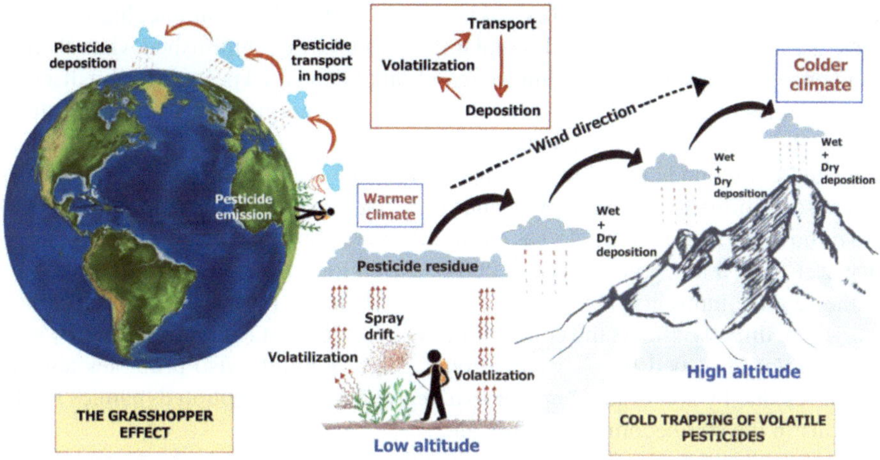

Fig. 4.2 The grasshopper effect (left) in global transport of volatile pesticides (from equator to both poles) and the regional long-range transport (right) of volatile pesticides

4.7 Cold Trapping of Volatile Pesticides

While the grasshopper effect operates on a global scale, cold trapping can refer to the regional dispersion of volatile pesticides based on an altitudinal instead of latitudinal gradient. The processes involved in transport remain the same but operate over a smaller distance. In this case, atmospheric pesticide residue is carried from agricultural fields in low-altitude regions with a warmer climate into the colder climate of high-altitude land masses like mountains to be deposited. The high-altitude sinks can be pitstops in the global transport chain of the pollutant while also taking a portion of the pesticide residue out of circulation though local accumulation (Daly et al., 2007). Much like in the case of spray drift, meteorological factors significantly influence the fate of pesticides that miss their intended targets. Temperature, for instance, plays a pivotal role in particle-vapour partitioning (Carlson & Hites, 2005). The combined effect of wind direction with temperature on the concentrations of endosulfan have also been determined (Burgoyne & Hites, 1993). Higher wind speed can promote more evaporation and influence the distance covered, while the direction of the wind current determines the deposition spot (Meire et al., 2012). Thermal inversion, occurring when cooler air near the surface gets trapped beneath warmer upper layers, hinders vertical movement and dispersion, thereby raising pesticide concentrations and toxicity in the affected area. However, as the surface temperature continues to rise with more incident solar radiation, this effect becomes insignificant and the restricted movements resume (Enz et al., 2017).

For the volatilized pesticide residue to remain intact and persist in the air for transport, it has to avoid degradation through photochemical reactions charged by solar radiation. Thus, the intensity of solar radiation, which varies with the time of day, cloud cover/fog, and altitude, matters greatly to the concentration of pesticide persisting in the atmosphere (Majewski & Capel, 1996). For example, high-altitude areas like mountains receive far less solar radiation, further diminished when there is cloud or fog cover. This also promotes retention of deposited pesticides and discourages re-emission. Rainfall facilitates wet deposition, purging them from the atmosphere. Rainfall volume and intensity also determine the maximum depth to which deposited compounds are able to penetrate into the soil surface, and consequently their susceptibility to degradation by microbes. Snow is porous and cold. This enables it to remove pollutants (particle/gas) from the atmosphere and, much like rain, influence wet deposition (Lei & Wania, 2004). Retention-emission dynamics also rely on the relative humidity. For example, high humidity informs the rate of evaporation which directly affects the concentration of residue lost from or retained by soil/ water and/or absorbed by air water vapour. High humidity also promotes absorption of residue by soil organic matter to a degree where absorption dynamics by the minerals in the soil becomes insignificant.

Studies have found ample evidence for the regional and global distribution of pesticides. In a series of articles, volatile pesticide transfer from agricultural fields of Central Valley of California to mountain ranges in the Sierra Nevada, was reported (LeNoir et al., 1999). Further studies were carried out on the impact this had on local frog populations (Bradford et al., 2011; Fellers et al., 2004). Low levels of HCB, HCH, and DDTs recorded in the Sichuan Province of China along the east-facing side of the Balang Mountain, were observed to increase exponentially with increase in elevation (Chen et al., 2008). This supported the occurrence of mountain cold-trapping mechanism. Plenty such reports have been made for pesticide deposition in high-altitude forests and lakes (Arellano et al., 2011; Davidson et al., 2003; Daly et al., 2007; Meire et al., 2012; Masclet, 1997; Shunthirasingham et al., 2013). The Tibetan Plateau, called the "third pole" of the world and the Himalayas have also emerged as a significant convergence zone for such pollutants (Acharyya et al., 2024; Bhattacharya et al., 2024; Gong et al., 2015; Huang et al., 2019; Li et al., 2006; Pokhrel et al., 2018; Wang et al., 2023; Yang et al., 2007, 2013). The sources of organochlorine pesticides and PCBs in the Arctic, their transport and distribution have been assessed in several studies showcasing how these pesticide pollutants contaminate even some of the most remote locations (Bailey et al., 2000; Cabrerizo et al., 2019). Data on the topic is severely limited. As such, regional accounts as the ones mentioned above, can provide valuable insights into these phenomena.

4.8 Risks and Impacts

Research in this field in the last few decades has been conducted on the impacts of pesticides and their constituents on the environment and divulged the associated risks. Some of the major impacts and risks have been discussed below.

4.8.1 Environment

Degradation of soil fertility due to pesticides is a serious threat to the soil ecosystem. Pesticides have adverse effects on the soil microbiota, responsible for maintaining soil fertility. Contamination leads to alteration in their enzymatic activities and reduction of biomass. Such changes decrease the rate of litter decomposition and fresh organic matter in the soil, which in turn affects the biogeochemical cycles in the contaminated area (Yarwood, 2018). Hydrophobic pesticides bind strongly to organic matter helping them accumulate and persist longer (Tudi et al., 2021). Entry and accumulation of the pesticides and their residues in non-target plants have led to contamination of the food that challenges food security. Exposure to certain pesticides often damages the plant itself, as is the relationship between potato plants and clopyralid (Colquhoun et al., 2017). In a study on the presence of pesticide residues in regular fruits and vegetables, 40 pesticides were detected (mean residue level of 7–951 μg/kg in vegetables and 8–775 μg/kg in fruits) (El-Sheikh et al., 2022). Moreover, herbicides like glyphosate have been reported to affect non-target crops by reducing the quality of seeds and increasing their vulnerability to diseases (Mishra et al., 2018).

Pesticides also affect beneficial organisms. Reports have suggested imidacloprid adversely affects the scavenging conduct of bees (Decourtye et al., 2003). DDT has been implicated in the decline of bald eagles' populations in the USA (Liroff, 2000). Usage of various bug sprays and fungicides has led to neural toxicity in beneficial worms (Mishra et al., 2018). Pesticide runoffs from fields or precipitation from the air has been reported to restrict aquatic plants responsible for dissolved oxygen content in the water, leading to suffocation of aerobic lifeforms (Helfrich et al., 2009). Atrazine and pyrethroids are highly toxic to fishes (Mishra et al., 2018; Yang et al., 2020), while glyphosate can kill tadpoles (Mishra et al., 2018). Some herbicides even decrease the number of zooplanktons, the primary food in the water, affecting the entire food chain (Hassan & El Nemr, 2020).

4.8.2 Human-Health Risks

Human body exposure to pesticides can occur through inhalation, drinking and/or consumption with inhalation being potentially the most harmful route. Workers in agricultural fields, industries, pest control and people residing near farms and industries, are statistically at a very high risk of developing respiratory issues. Organophosphates and carbamates were found to be involved in causing asthma attacks in India, Spain, Canada, and South Africa (Pathak et al., 2022). Some pesticides, such as permethrin and sumithrin were key contributors to tremors, convulsions, headaches, along with asthma and can even be lethal (Amaral, 2014). Pesticides are very potent carcinogens (Rajmohan et al., 2020), can disrupt the endocrine and immune systems (Zhang et al., 2016), central nervous system (CNS), redox activity and metabolism in the mitochondria (Pezzoli & Cereda, 2013; Qi et al., 2014), cardiac and reproductive health. An increased risk of coronary artery disease due to a reduction in paraoxonase activity has been reported as well (Kabir et al., 2015; Mandal et al., 2022). Some of these pesticides are even known to mimic hormones. Studies suggest that sperm damage and reduction in sperm quality is directly linked with the presence of pyrethroid metabolites (Meeker et al., 2008). Children are found to be at higher risk as higher amounts of dialkylphosphate (DAP) metabolites have been detected in urine samples of children living in agricultural communities than adults (Koch et al., 2002). Moreover, various studies have reported the lethal effects of pesticides on developing foetuses as well (Arbuckle & Sever, 1998).

4.9 Proposed Solutions

Addressing the issue of pesticide pollution will require a multistep approach with innovative solutions at various stages of pesticide management. All aspects of pesticide use, from production and application to regulation and monitoring, should be given equal attention. Strategies proposed aim to either minimize the release of volatile pesticides, enhance their degradation to non-toxic forms or effectively remove them from the environment. Pest management itself through more creative means will also promote a holistic and sustainable option. Additionally, as this problem and its consequences become more recognized, efforts are underway to improve international cooperation and develop policy frameworks to effectively manage pesticide pollution on the global scale.

4.9.1 Sustainable Agriculture

The first obvious approach is to reduce emission rates of pesticides in use. This can be achieved if usage is either stopped completely or regulated to a stricter degree. Thus, sustainable farming practises are often encouraged to mitigate the issue. Furthermore, the use of proper equipment, better technologies and formulations may be a method of controlling pesticide residue in the air. Organic farming may be adopted by farmers. This prohibits the use of synthetic pesticides entirely and promotes the use of natural methods of pest and disease control. However, shifting to organic farming can be quite expensive and laborious. Without any input of pesticides at all, the ease of pest eradication is lost. Initial yields may also suffer as it takes time for soil quality to adjust to the new practises. Furthermore, organic farms run the risk of being exposed to pesticides when they are in the vicinity of other inorganically practising farms due to spray drift or leaching or volatilization deposition cycles. Nevertheless, advocates of organic farming practises maintain that reduced applications of chemical pesticides will reduce pressure on pests to adapt and mutate as well.

Another familiar method that can cater to reduced use of pesticide, is Integrated Pest Management or IPM (Kogan, 1998). This constitutes several processes, the cumulative effect of which results in proficient and environment-friendly management of pests. The different aspects of IPM are as follows:

A. ***Biological Control***: Pests infesting agricultural fields also have abundant naturally existing enemies that may be utilized for controlling their populations. These organisms can include predators, parasitoids or microbial pathogens (bacteria, fungi, viruses) that attack, parasitize or infect pests, helping to suppress their numbers. Insect-eating invertebrates, arthropods, birds and herbivore mammals are also known to be employed. When properly utilized, the merit lies in the fact that these natural enemies have a narrow target range and a life cycle that increases and decreases parallelly with target pests. However, it is important to understand the relationship between the natural enemy and the pest and its impact on the surrounding environment to avoid the organism becoming an invasive species (Roy et al., 2024). For example, when cane toads were introduced to manage the population of cane beetles, they were found to feed more on other beneficial natural enemies instead and became a threat to local reptiles and mammal species (Shine et al., 2020).
B. ***Cultural Practices***: Crop rotation, in which the type of crops grown in a field is alternated over successive seasons, fall under the realm of cultural practises that aim to reduce pests as well as enhance production (Francis & Porter, 2011). It helps end plant disease cycle propagation and provides additional benefits such as the prevention of soil nutrient depletion and improvement of soil structure. Substituting crops with genetically modified resistant varieties like the Bt maize and cotton can chemically target and kill pests (Abbas, 2018). Permaculture (introduction of a permanently growing crop cover over the agricultural land to suppress weeds and soil erosion), or perennial polycultures (highly diverse combination of crops is planted that are not desirable as pest food), or strip cropping (different species of plants are planted in alternating strips) etc., together with crop rotation can help keep the number of pests under the economic threshold. Moreover, adjusting planting density and spacing of crops can influence light interception and airflow promoting reduced pest and disease pressure, and maximize yield potential. Tillage practises, application of mulch, maintaining clean farming equipment, tools and infrastructure, as well as removing crop residues and weeds regularly, help reduce pest and disease carryover between seasons.
C. ***Mechanical and Physical Methods***: Manual removal of pests is a labour-intensive but effective method, particularly for larger pests such as caterpillars, beetles, or snails. Sticky, pheromone, light, or pitfall traps can be used to capture and monitor smaller-sized pests. Simple shaking, brushing or spraying crops with mineral oils, surfactants and soaps or high-pressure water can dislodge aphids, mites and other soft-bodied pests from plants. Physical barriers, such as trenches, fences, row covers, screens, netting or particle films, can be installed. Captured pests can be exposed to steam sterilization or hot water immersion or solarization. Weeds can be mowed or hand-weeded. Planting trap crops or

decoy plants that attract pests away from main crops, combined with repellent intercrops or border plants, creates another physical barrier.

D. ***Judicious Use of Pesticides***: While the use of chemical pesticides cannot be fully eliminated, it should be minimized and the pesticides should be selected based on thorough examination of its efficacy, pest specificity and impact on the environment. Factors such as pest biology, pesticide application timing, dosage etc., can greatly affect the outcome. The concept of economic threshold advocates the use of pesticides only when a specific pest population reaches a point where the cost of control will be more than the returns in increased crop production. Precision agriculture, on the other hand, utilizes GPS monitoring, remote sensing and data analytics to maximize output and give an idea of when and how much pesticide application is necessary. The further development of such technologies can promote a more calculated application of chemical control agents. Use of calibrated equipment, proper nozzle selection, right formulations etc., can ensure uniform pesticide distribution and minimize drift. Maintaining accurate records of pesticide applications, including product usage, application rates, dates, and locations, to track compliance with regulations can help evaluate performance, and facilitate traceability in case of emergencies or regulatory inquiries.

4.9.2 Microbial Remediation

Fast and effective pesticide degradation can be accomplished using resistant microbes that can break them down into harmless metabolites. Pesticide bioremediation offers a way to sustainably reduce pollution via bacteria, fungi, algae or mixed consortia. While natural pesticide degradation does occur in agricultural settings, it is typically slow and insufficient when the site is heavily contaminated (Singh & Singh, 2017). To accelerate the process, pesticide degrading microorganisms can be deployed. Bacteria, fungi and microalgae secrete various enzymes that facilitate the breakdown. Various mechanisms like hydroxylation, demethylation, dechlorination and oxidation have been identified. Fungi may even utilize some pesticides as a carbon source (Vaksmaa et al., 2023). Enzymes involved in pesticide biodegradation include monooxygenases, oxidoreductases, dehalogenases, phosphotriesterases, esterases, dehydrogenases, cytochrome p450, dioxigenases, dioxygenases, etc. Only a few degradation pathways have so far been elucidated but it is understood that the mechanism involved is targeted to the pesticide and the enzymatic capabilities harboured by the remediating microorganism (Mandal et al., 2024).

Microbial remediation technologies are being increasingly employed for cleaning up polluted sites due to their flexibility, cost-effectiveness and ability to treat various contaminants in all parts of the environment. Microbes may also have additional characteristics that help soil regain fertility or produce plant growth promoting factors that boost crop growth and development. Following are examples of microbial remediation strategies:

A. ***Bacterial remediation***: Bacterial strains are exceptionally proficient and notoriously easy to culture and maintain. They may harbour specific genes responsible for the breakdown of specific pesticides in its environment (Sakamoto & Tsutsumi, 2004). The exact mechanisms in the pathways are still largely unknown. Research has illuminated degradation pathways for some, such as cyhalothrin (Chen et al., 2015). Many strains have been successful in degrading pesticides. *Novosphingobium* sp., *Pseudomonas* sp., and *Chryseobacterium* sp., have proven effective against 2,4-D, endosulfan and DDT respectively (Dai et al., 2015; Qu et al., 2015; Zaffar et al., 2018). *Bacillus* strains have degraded triazole fungicides and organochlorines like Dieldrin and Chlorpyrifos (Matsumura & Boush, 1967; Salunkhe et al., 2015; Zhu et al., 2010). *Acinetobacter* strains have shown degradation of diazinon (Amani et al., 2019). All these mentioned pesticides are volatile pollutants. Utilizing microbial consortia instead of a single strain has also been tested and resulted in enhanced degradation. A consortium of *Pseudomonas*, *Stenotrophomonas*, *Klebsiella*, and others, was reported to effectively remediate chlorpyrifos (Singh et al., 2016). Another combination of *Pseudomonas aeruginosa*, and *Bacillus firmus* achieved notable degradation of DDT (Powthong et al., 2016). It has been found that indigenous consortia are better suited for remediation as they are well-adjusted and adapted to particular geographical regions.

B. ***Fungal and algal remediation***: While bacteria have dominated the field of biological remediation for a long time, fungi and algae are also being increasingly utilized. Certain species like *Trametes versicolor* and *Phanerochaete* have demonstrated the ability to treat mixtures of imiprothrin, cypermethrin, alachlor and atrazine. The first three are volatile pesticides. Additionally, *Aspergillus sp.*, *Phlebia sp.*, *Trichoderma sp.* have displayed potential in degrading oxyfluorfen, heptachlor and DDT respectively (Matsumura & Boush, 1968; Xiao et al., 2011). Some fungi have also been used in consort with bacteria for enhanced removal capacity in the remediation of endosulfan contaminated soil (Mohanasrinivasan et al., 2013). Apart from fungi, microalgal species like *Chlorococcum* and *Scenedesmus*, as well as cyanobacterial strains like *Anabaena azotica*, have effectively degraded endosulfan, lindane and hexachlorocyclohexane (Sethunathan et al., 2004). *Chlamydomonas reinhardtii* has been reported to accumulate and degrade prometryne and fluroxypyr (Jin et al., 2012). Some research has even suggested that cyanobacteria and microalgae may be better at bioremediation of such pollutants than bacteria and fungi. Nevertheless, this is a comparatively newer facet of microbial bioremediation that needs further studies to draw concrete conclusions.

4.9.3 Monitoring and Control of Atmospheric Emission

Pesticide management practises and remediation techniques can only be successfully implemented with proper monitoring and gathering of data. As the long-range transport, deposition of pesticides and its impact becomes recognized, the need for monitoring systems has also become apparent. Even developing regulations for the control of local emissions requires sufficient data. Furthermore, sharing of this data among nations can be pivotal to alleviating the repercussions of "grasshopper effect." Initiatives such as the Stockholm Convention on POPs or the United Nations Economic Commission for Europe protocol (UNECE) have endorsed global discourse on the topic. Collaborating at the international level to address transboundary pesticide transport and transparency among governing bodies in sharing of data, best practises and technologies can help to develop coordinated strategies for monitoring and controlling atmospheric pesticide pollution.

Air quality monitoring stations, satellite imagery and aerial drones can identify pesticide dispersion patterns and high-risk areas. In addition, mathematical models like atmospheric dispersion models can simulate pesticide transfer into the atmosphere and subsequent deposition. These take into account meteorological conditions, emission sources, and pesticide chemistry for the prediction. However, most important is that regulators should set emission limits and introduce emission trading and pollution taxes to ensure manufacturers reduce their emissions to the acceptable level. Educational campaigns should also be launched to inform the farmers and the public of risks associated with atmospheric pesticide pollution and ways to avoid it.

4.10 Conclusion

4.10.1 Recap of Key Points

Pesticides today are an integral part of assuring agricultural productivity and product quality. In such a world, it is understandable to overlook the negative aspects of pesticides. There have been many instances where the broad-spectrum toxicity of a pesticide has resulted in ecological disasters. In such cases, the particular pesticides have been officially banned and efforts have been made to remove them from circulation. However, often, the concern is thought to have been resolved with curbing the use of that particular pesticide. In reality, there are many factors and phenomena that direct environmental fate.

One of the relatively recently discovered facets of pesticide pollution is its long-range transport through the atmosphere. Aptly named the "grasshopper effect," this phenomenon is controlled by the pesticide's chemical nature, vapour pressure, application technique, and various meteorological parameters of the application area (solar radiation, moisture, temperature, humidity, wind speed, direction, etc.). This greatly enhances the potential of certain pesticides as a pollutant. Re-emission from a site of deposition may also follow depending on similar factors, as well as soil properties or vegetation cover in case of terrestrial deposition. The process is repeated in volatilization-transport-deposition cycles until a final sink is reached that limits re-volatilization to the lowest level. This is where the pesticide residue then accumulates over time. However, it causes damage at each stop along the way.

4.10.2 Call to Action for Effective Management of Volatile Pesticides

Harm to the environment is increased significantly due to emission of volatile pesticides. Hence, the best way to control the pollution and prevent ecological risk would be to curb emissions. This can be achieved by either eliminating the use of pesticides entirely or by diminishing it through supplementation with other types of pest management strategies. Since the problem is a global one, it requires the large-scale collaboration of scientists and policy-makers alike from several world nations. Although this is a huge undertaking that can be difficult to bring to fruition due to several geopolitical reasons, efforts must still be made.

The EU POP regulation is enforced in companies in the EU that prohibit or restrict the production, marketing and use of POPs and direct the management of already present stockpiles. The signatory parties of the Stockholm Convention are committed to the reduction of POP emissions through global corporation. There are currently 186 countries party to it. Another important part of the job is the classification and listing of POPs. Such inventories are important to keep track of the target pollutants. The open-source availability of such data facilitates seamless exchange of ideas and also raises awareness among the public regarding such issues. When implemented as planned, proposed strategies can target and reduce emissions at the time of manufacturing or application. Farmers should also receive proper education on the cost of using such chemicals and of alternative routes they can take. Consumers should be made to understand the hazards of having pesticide residue in their food. Only with awareness and education can this problem be effectively addressed.

4.10.3 Future Directions for Research

There are quite a few areas in the research of POPs that are still lacking. The many factors that decide the trajectory, relationships and environmental fate of POPs are still being explored. Furthermore, the effect of environmental factors or pesticide formulations on the percentage of drift during and after application is another aspect that needs study. Better designs of application equipment and alternative biopesticides can minimize consequence to the environment. Monitoring stations should be established at the right locations to keep an accurate account of the amount of pesticide in the air. Ecological impact assessment should be conducted on different groups of living organisms and more microorganisms must be explored for remediation solutions that are eco-friendly and sustainable. Moreover, how the pesticide residues enter the food chain is important for understanding how to prevent them from causing harm to livestock and humans. The impact of pesticide residue on the residents of remote locations with no prior exposure to it can help us better comprehend the risks of the grasshopper effect. We may know a lot about the effects of acute exposure to high doses of pesticides but knowledge on long-term chronic exposure of low doses needs investigation. Thus, it is evident that much more study of the different aspects of this pollution problem is warranted to be able to mitigate the issue on the global scale and work towards sustainability in agriculture.

References

Abbas, M. S. T. (2018). Genetically engineered (modified) crops (Bacillus thuringiensis crops) and the world controversy on their safety. *Egyptian Journal of Biological Pest Control, 28*(1), 1–12.

Acharyya, S., Majumder, S., & Bhattacharya, M. (2024). Presence of pesticide-tolerant microorganisms in high-altitude pristine lakes within Singalila Ridge of the Himalayas. *Environmental Monitoring and Assessment, 196*(10), 925.

Al-Alam, J., Sonnette, A., Delhomme, O., Alleman, L. Y., Coddeville, P., & Millet, M. (2022). Pesticides in the indoor environment of residential houses: A case study in Strasbourg, France. *International Journal of Environmental Research and Public Health, 19*(21), 14049.

Alves, G. S., Kruger, G. R., da Cunha, J. P. A., de Santana, D. G., Pinto, L. A. T., Guimarães, F., & Zaric, M. (2017). Dicamba spray drift as influenced by wind speed and nozzle type. Weed Technology, 31(5), 724–731. https://doi.org/10.1017/wet.2017.61

Amani, F., Safari Sinegani, A. A., Ebrahimi, F., & Nazarian, S. (2018). Biodegradation of chlorpyrifos and diazinon organophosphates by two bacteria isolated from contaminated agricultural soils. *Biological Journal of Microorganism, 7*(28), 27–39.

Amani, F., Student, P., Akbar, A., Sinegani, S., Ebrahimi, F., & Nazarian, S. (2019). Biodegradation of chlorpyrifos and diazinon organophosphates by two bacteria isolated from contaminated agricultural soils. *Biological Journal of Microorganism, 7*(28), 27–39.

Amaral, A. F. (2014). Pesticides and asthma: Challenges for epidemiology. *Frontiers in Public Health, 2*, 6.

Arbuckle, T. E., & Sever, L. E. (1998). Pesticide exposures and fetal death: A review of the epidemiologic literature. *Critical Reviews in Toxicology, 28*(3), 229–270.

Arellano, L., Fernández, P., Tatosova, J., Stuchlik, E., & Grimalt, J. O. (2011). Long-range transported atmospheric pollutants in snowpacks accumulated at different altitudes in the Tatra Mountains (Slovakia). *Environmental Science and Technology, 45*(21), 9268–9275.

Aston, L. S., & Seiber, J. N. (1997). *Fate of summertime airborne organophosphate pesticide residues in the Sierra Nevada mountains* (Vol. 26, No. 6, pp. 1483–1492). American Society of Agronomy, Crop Science Society of America, and Soil Science Society of America.

Bailey, R., Barrie, L. A., Halsall, C. J., Fellin, P., & Muir, D. C. G. (2000). Atmospheric organochlorine pesticides in the western Canadian Arctic: Evidence of transpacific transport. *Journal of Geophysical Research: Atmospheres, 105*(D9), 11805–11811.

Balmer, J. E., Morris, A. D., Hung, H., Jantunen, L., Vorkamp, K., Rigét, F., et al. (2019). Levels and trends of current-use pesticides (CUPs) in the arctic: An updated review, 2010–2018. *Emerging Contaminants, 5*, 70–88.

Bedos, C., Cellier, P., Calvet, R., Barriuso, E., & Gabrielle, B. (2002). Mass transfer of pesticides into the atmosphere by volatilization from soils and plants: Overview. *Agronomie, 22*(1), 21–33.
Beinert, A., Peike, C., Dürr, I., Kempe, M. D., Reiter, G., & Weiß, K. A. (2014, September). The influence of the additive composition on degradation induced changes in poly (ethylene-co-vinyl acetate) during photochemical aging. In *29th European PV solar energy conference and exhibition.*
Benicha, M., Mrabet, R., & Azmani, A. (2011). Biodegradation and dissipation of 14C-carbofuran in clay soil from Loukkos perimeter, Northwestern Morocco. *Journal of Soil Science and Environmental Management, 2*(12), 404–410.
Bhattacharya, M., Acharyya, S., Saha, S., Majumder, S., Ghosh, A., & Subba, P. (2024). Prevalence of pesticide tolerance in soil bacteria from Darjeeling Himalayas: A growing concern of grasshopper effect. *Khulna University Studies, 21*(1), 207–216.
Biocca, M., Fanigliulo, R., Pochi, D., & Gallo, P. (2019). Dust drift mitigating devices applied on precision pneumatic seed drills: a mini-review.
Bonmatin, J. M., Giorio, C., Girolami, V., Goulson, D., Kreutzweiser, D. P., Krupke, C., et al. (2015). Environmental fate and exposure; neonicotinoids and fipronil. *Environmental Science and Pollution Research, 22*, 35–67.
Bradford, D. F., Knapp, R. A., Sparling, D. W., Nash, M. S., Stanley, K. A., Tallent-Halsell, N. G., et al. (2011). Pesticide distributions and population declines of California, USA, alpine frogs, Rana muscosa and Rana sierrae. *Environmental Toxicology and Chemistry, 30*(3), 682–691.
Burgoyne, T. W., & Hites, R. A. (1993). Effect of temperature and wind direction on the atmospheric concentrations of. alpha.-endosulfan. *Environmental Science and Technology, 27*(5), 910–914.
Cabizza, M., Angioni, A., Melis, M., Cabras, M., Tuberoso, C. V., & Cabras, P. (2004). Rotenone and rotenoids in cube resins, formulations, and residues on olives. *Journal of Agricultural and Food Chemistry, 52*(2), 288–293.
Cabrerizo, A., Muir, D. C., Teixeira, C., Lamoureux, S. F., & Lafreniere, M. J. (2019). Snow deposition and melting as drivers of polychlorinated biphenyls and organochlorine pesticides in Arctic rivers, lakes, and ocean. *Environmental Science and Technology, 53*(24), 14377–14386.
Carlson, D. L., & Hites, R. A. (2005). Temperature dependence of atmospheric PCB concentrations. *Environmental Science and Technology, 39*(3), 740–747.
Carson, R. (1962). Silent spring III. *New Yorker, 23.*
Cessna, A. J., Larney, F. J., Kerr, L. A., & Bullock, M. S. (2006). Transport of trifluralin on wind-eroded sediment. *Canadian Journal of Soil Science, 86*(3), 545–554.
Chaplain, V., Mamy, L., Vieublé, L., Mougin, C., Benoit, P., Barriuso, E., & Nélieu, S. (2011). Fate of pesticides in soils: Toward an integrated approach of influential factors. *Pesticides in the Modern World-Risks and Benefits, np.*
Chen, D., Liu, W., Liu, X., Westgate, J. N., & Wania, F. (2008). Cold-trapping of persistent organic pollutants in the mountain soils of Western Sichuan China. *Environmental Science and Technology, 42*(24), 9086–9091.
Chen, S., Deng, Y., Chang, C., Lee, J., Cheng, Y., Cui, Z., et al. (2015). Pathway and kinetics of cyhalothrin biodegradation by Bacillus thuringiensis strain ZS-19. *Scientific Reports, 5*(1), 8784.
Colquhoun, J. B., Heider, D. J., & Rittmeyer, R. A. (2017). Seed potato growth and yield as affected by mother plant exposure to herbicides. *Weed Technology, 31*(1), 136–147.
Cooper, R. J., Jenkins, J. J., & Curtis, A. S. (1990). *Pendimethalin volatility following application to turfgrass* (Vol. 19, No. 3, pp. 508–513). American Society of Agronomy, Crop Science Society of America, and Soil Science Society of America.
Coronado, G. D., Holte, S., Vigoren, E., Griffith, W. C., Barr, D. B., Faustman, E., & Thompson, B. (2011). Organophosphate pesticide exposure and residential proximity to nearby fields: Evidence for the drift pathway. *Journal of Occupational and Environmental Medicine, 53*(8), 884–891.
da Silva Brochado, M. G., Mielke, K. C., de Paula, D. F., Laube, A. F. S., Alcántara-de la Cruz, R., Gonzatto, M. P., & Mendes, K. F. (2022). Impacts of dicamba and 2, 4-D drift on

'Ponkan' mandarin seedlings, soil microbiota and Amaranthus retroflexus. *Journal of Hazardous Materials Advances, 6*, Article 100084.
Dai, Y., Li, N., Zhao, Q., & Xie, S. (2015). Bioremediation using Novosphingobium strain DY4 for 2, 4-dichlorophenoxyacetic acid-contaminated soil and impact on microbial community structure. *Biodegradation, 26*, 161–170.
Daly, G. L., Lei, Y. D., Teixeira, C., Muir, D. C., Castillo, L. E., & Wania, F. (2007). Accumulation of current-use pesticides in neotropical montane forests. *Environmental Science and Technology, 41*(4), 1118–1123.
Das, S., & Hageman, K. J. (2020). Influence of adjuvants on pesticide soil–air partition coefficients: Laboratory measurements and predicted effects on volatilization. *Environmental Science and Technology, 54*(12), 7302–7308.
Davidson, D. A., Wilkinson, A. C., Blais, J. M., Kimpe, L. E., McDonald, K. M., & Schindler, D. W. (2003). Orographic cold-trapping of persistent organic pollutants by vegetation in mountains of western Canada. *Environmental Science and Technology, 37*(2), 209–215.
Decourtye, A., Lacassie, E., & Pham-Delègue, M. H. (2003). Learning performances of honeybees (Apis mellifera L) are differentially affected by imidacloprid according to the season. *Pest Management Science: Formerly Pesticide Science, 59*(3), 269–278.
Egan, J. F., & Mortensen, D. A. (2012). Quantifying vapor drift of dicamba herbicides applied to soybean. *Environmental Toxicology and Chemistry, 31*(5), 1023–1031.
El-Sheikh, E. S. A., Ramadan, M. M., El-Sobki, A. E., Shalaby, A. A., McCoy, M. R., Hamed, I. A., et al. (2022). Pesticide residues in vegetables and fruits from farmer markets and associated dietary risks. *Molecules, 27*(22), 8072.
Elango, K., Arunkumar, P., & Vijayalakshmi, G. (2022). Factors affecting the toxicity of pesticides: An overview. *Advances in Agricultural and Horticultural Sciences, 1.*
Enz, J. W., Hofman, V. L., & Thostenson, A. (2017). *Air temperature inversions: causes, characteristics and potential effects on pesticide spray drift*. NDSU Extension Service.
Farha, W., Abd El-Aty, A. M., Rahman, M. M., Shin, H. C., & Shim, J. H. (2016). An overview on common aspects influencing the dissipation pattern of pesticides: A review. *Environmental Monitoring and Assessment, 188*, 1–21.
Fellers, G. M., McConnell, L. L., Pratt, D., & Datta, S. (2004). Pesticides in mountain yellow-legged frogs (Rana muscosa) from the Sierra Nevada Mountains of California, USA. *Environmental Toxicology and Chemistry: An International Journal, 23*(9), 2170–2177.
Francis, C. A., & Porter, P. (2011). Ecology in sustainable agriculture practices and systems. *Critical Reviews in Plant Sciences, 30*(1–2), 64–73.
Galon, L., Bragagnolo, L., Korf, E. P., Dos Santos, J. B., Barroso, G. M., & Ribeiro, V. H. V. (2021). Mobility and environmental monitoring of pesticides in the atmosphere—a review. *Environmental Science and Pollution Research, 28*(25), 32236–32255.
Glotfelty, D. E., & Schomburg, C. J. (1989). Volatilization of pesticides from soil. *Reactions and Movement of Organic Chemicals in Soils, 22*, 181–207.
Glotfelty, D. E., Schomburg, C. J., Sawhney, B. L., & Brown, K. (2015). Reactions and movement of organic chemicals in soils. Volatilization of Pesticides from Soil Wiley, 181–207.
Gong, P., Wang, X. P., Xue, Y. G., Sheng, J. J., Gao, S. P., & Yao, T. D. (2015). Influence of atmospheric circulation on the long-range transport of organochlorine pesticides to the western Tibetan Plateau. *Atmospheric Research, 166*, 157–164.
Gong, P., Xu, H., Wang, C., Chen, Y., Guo, L., & Wang, X. (2021). Persistent organic pollutant cycling in forests. *Nature Reviews Earth and Environment, 2*(3), 182–197.
Góngora-Echeverría, V. R., Martin-Laurent, F., Quintal-Franco, C., Lorenzo-Flores, A., Giácoman-Vallejos, G., & Ponce-Caballero, C. (2019). Dissipation and adsorption of 2, 4-D, atrazine, diazinon, and glyphosate in an agricultural soil from Yucatan State, Mexico. *Water, Air, and Soil Pollution, 230*, 1–15.
Green, J. (2000). Adjuvant outlook for pesticides. *Pesticide Outlook, 11*(5), 196–199.
Hassaan, M. A., & El Nemr, A. (2020). Pesticides pollution: Classifications, human health impact, extraction and treatment techniques. *The Egyptian Journal of Aquatic Research, 46*(3), 207–220.

He, D. C., Li, F. H., Wu, M., Luo, H. L., Qiu, L. Q., Ma, X. R., et al. (2022). Emission of volatile organic compounds (VOCs) from application of commercial pesticides in China. *Journal of Environmental Management, 314*, 115069.

Helfrich, L. A., Weigmann, D. L., Hipkins, P. A., & Stinson, E. R. (2009). Pesticides and aquatic animals: a guide to reducing impacts on aquatic systems.

Hofman, V., & Solseng, E. (2001). Reducing spray drift.

Holford, I. C. R., Haigh, B. M., & Ferris, I. G. (1989). Atrazine persistence and phytotoxicity on wheat as affected by nitrogen and rotation-induced changes in soil properties. *Australian Journal of Agricultural Research, 40*(6), 1143–1153. https://oec.world/en/profile/country/chn

Huang, H., Ding, Y., Chen, W., Zhang, Y., Chen, W., Chen, Y., et al. (2019). Two-way long-range atmospheric transport of organochlorine pesticides (OCPs) between the Yellow River source and the Sichuan Basin, Western China. *Science of the Total Environment, 651*, 3230–3240.

Hull, H. M. (1970). Leaf structure as related to absorption of pesticides and other compounds. *Residue Reviews: Residues of Pesticides and Other Foreign Chemicals in Foods and Feeds*, 1–150.

Jin, Z. P., Luo, K., Zhang, S., Zheng, Q., & Yang, H. (2012). Bioaccumulation and catabolism of prometryne in green algae. *Chemosphere, 87*(3), 278–284.

Jones, K. C., & De Voogt, P. (1999). Persistent organic pollutants (POPs): State of the science. *Environmental Pollution, 100*(1–3), 209–221.

Kabir, E. R., Rahman, M. S., & Rahman, I. (2015). A review on endocrine disruptors and their possible impacts on human health. *Environmental Toxicology and Pharmacology, 40*(1), 241–258.

Katagi, T. (2004). Photodegradation of pesticides on plant and soil surfaces. *Reviews of Environmental Contamination and Toxicology: Continuation of Residue Reviews*, 1–78.

Khan, M. S., & Rahman, M. S. (2017). *Pesticide residue in foods.* Springer.

Kjuus, H. (2001). Acrylamide in tunnel construction-new (or old) lessons to be learned? *Scandinavian Journal of Work, Environment and Health, 27*(4), 217–218.

Koch, D., Lu, C., Fisker-Andersen, J., Jolley, L., & Fenske, R. A. (2002). Temporal association of children's pesticide exposure and agricultural spraying: Report of a longitudinal biological monitoring study. *Environmental Health Perspectives, 110*(8), 829–833.

Kogan, M. (1998). Integrated pest management: Historical perspectives and contemporary developments. *Annual Review of Entomology, 43*(1), 243–270.

Kumar, A., Howard, C. J., Derrick, D., Malkina, I. L., Mitloehner, F. M., Kleeman, M. J., et al. (2011). Determination of volatile organic compound emissions and ozone formation from spraying solvent-based pesticides. *Journal of Environmental Quality, 40*(5), 1423–1431.

Larney, F. J., Cessna, A. J., & Bullock, M. S. (1999). *Herbicide transport on wind-eroded sediment* (Vol. 28, No. 5, pp. 1412–1421). American Society of Agronomy, Crop Science Society of America, and Soil Science Society of America.

Lei, Y. D., & Wania, F. (2004). Is rain or snow a more efficient scavenger of organic chemicals? *Atmospheric Environment, 38*(22), 3557–3571.

LeNoir, J. S., McConnell, L. L., Fellers, G. M., Cahill, T. M., & Seiber, J. N. (1999). Summertime transport of current-use pesticides from California's Central Valley to the Sierra Nevada Mountain Range, USA. *Environmental Toxicology and Chemistry: An International Journal, 18*(12), 2715–2722.

Li, J., Zhu, T., Wang, F., Qiu, X. H., & Lin, W. L. (2006). Observation of organochlorine pesticides in the air of the Mt. Everest region. *Ecotoxicology and Environmental Safety, 63*(1), 33–41.

Li, Y. F., Kallenborn, R., & Zhang, Z. (2024). Persistent organic pollutants and chemicals of emerging Arctic concern in the Arctic environment. *Environmental Science and Ecotechnology, 18.*

Liroff, R. A. (2000). Balancing risks of DDT and malaria in the global POPs treaty. *Pestic Safety News, 4*(3).

Lohmann, R., Breivik, K., Dachs, J., & Muir, D. (2007). Global fate of POPs: Current and future research directions. *Environmental Pollution, 150*(1), 150–165.

Majewski, M. S., & Capel, P. D. (1995). *Pesticides in the atmosphere*. Ann Arbor Press.
Majewski, M. S., & Capel, P. D. (1996). Pesticides in the atmosphere: Distribution, trends, and governing factors (Vol. 1, Pesticides in the Hydrologic System). Ann Arbor Press.
Malusá, E., Tartanus, M., Danelski, W., Miszczak, A., Szustakowska, E., Kicińska, J., & Furmanczyk, E. M. (2020). Monitoring of DDT in agricultural soils under organic farming in Poland and the risk of crop contamination. *Environmental Management, 66*(5), 916–929.
Mandal, M., Das, S., Roy, A., Rakwal, R., Jones, O. A., Popek, R., et al. (2023). Interactive relations between plants, the phyllosphere microbial community, and particulate matter pollution. *Science of The Total Environment, 890*, 164352. https://doi.org/10.1016/j.scitotenv.2023.164352
Mandal, M., Roy, A., Binha, S. K., Popek, R., Przybysz, A., Koczoń, P., et al. (2025). Waste dumps as microplastic hotspots: a comparative investigation at urban, suburban, and rural areas of Eastern India and associated risk assessment. *Environmental Toxicology and Chemistry, 44*(7), 1869–1882. https://doi.org/10.1093/etojnl/vgaf086
Mandal, M., Roy, A., Popek, R., & Sarkar, A. (2024). Micro-and nano-plastic degradation by bacterial enzymes: A solution to 'white pollution.' *The Microbe, 3*, Article 100072. https://doi.org/10.1016/j.microb.2024.100072
Mandal, M., Sarkar, M., Khan, A., Biswas, M., Masi, A., Rakwal, R., et al. (2022). Reactive Oxygen Species (ROS) and Reactive Nitrogen Species (RNS) in plants–maintenance of structural individuality and functional blend. *Advances in Redox Research, 5*, 100039. https://doi.org/10.1016/j.arres.2022.100039
Mao, K., Jin, R., Li, W., Ren, Z., Qin, X., He, S., et al. (2019). The influence of temperature on the toxicity of insecticides to Nilaparvata lugens (Stål). *Pesticide Biochemistry and Physiology, 156*, 80–86.
Masclet, P. (1997). Contamination of lake water by pesticides via atmospheric transport. In *Geosciences and water resources: environmental data modeling* (pp. 85–92). Springer.
Matsumura, F., & Boush, G. M. (1967). Dieldrin: Degradation by soil microorganisms. *Science, 156*(3777), 959–961.
Matsumura, F., & Boush, M. G. (1968). Degradation of insecticides by a soil fungus, Trichoderma viride. *Journal of Economic Entomology, 61*(3), 610–612.
Maybank, J., Yoshida, K., & Grover, R. (1978). Spray drift from agricultural pesticide applications. *Journal of the Air Pollution Control Association, 28*(10), 1009–1014.
Meeker, J. D., Barr, D. B., & Hauser, R. (2008). Human semen quality and sperm DNA damage in relation to urinary metabolites of pyrethroid insecticides. *Human Reproduction, 23*(8), 1932–1940.
Meire, R. O., Lee, S. C., Yao, Y., Targino, A. C., Torres, J. P. M., & Harner, T. (2012). Seasonal and altitudinal variations of legacy and current-use pesticides in the Brazilian tropical and subtropical mountains. *Atmospheric Environment, 59*, 108–116.
Mina, M., Rezaei, M., Sameni, A., Ostovari, Y., & Ritsema, C. (2022). Predicting wind erosion rate using portable wind tunnel combined with machine learning algorithms in calcareous soils, southern Iran. *Journal of Environmental Management, 304*, Article 114171.
Mishra, R., Lone, R., Manzoor, D., & Shuab, R. (2018). Imbalance due to pesticide contamination in different ecosystems. *International Journal of Theoretical and Applied Sciences, 10*, 239–246.
Mohanasrinivasan, V., Suganthi, V., Selvarajan, E., Subathra Devi, C., Ajith, E., Muhammed Farhan, N. P., & Gopalakrishnan, S. (2013). Bioremediation of endosulfan contaminated soil. *Research Journal of Chemistry and Environment, 17*(11), 93–101.
Niu, Z., Young, J., Johnson, W. G., Young, B., Wei, X., & Jin, J. (2023). Early setection of Dicamba and 2, 4-D herbicide drifting injuries on soybean with a new spatial-spectral algorithm based on LeafSpec, an accurate touch-based hyperspectral leaf scanner. *Remote Sensing, 15*(24), 5771.
Pariatamby, A., & Kee, Y. L. (2016). Persistent organic pollutants management and remediation. *Procedia Environmental Sciences, 31*, 842–848.
Park, J. Y., Choi, J. H., Kim, B. M., Park, J. H., Cho, S. K., Ghafar, M. W., et al. (2011). Determination of acetamiprid residues in zucchini grown under greenhouse conditions: application to behavioral dynamics. *Biomedical Chromatography, 25*(1–2), 136–146.

Pathak, V. M., Verma, V. K., Rawat, B. S., Kaur, B., Babu, N., Sharma, A., et al. (2022). Current status of pesticide effects on environment, human health and it's eco-friendly management as bioremediation: A comprehensive review. *Frontiers in Microbiology, 13*, 962619.

Peek, D. C., & Appleby, A. P. (1989). Phytotoxicity, adsorption, and mobility of metribuzin and its ethylthio analog as infuenced by soil properties. *Weed Science, 37*(3), 419–423.

Pezzoli, G., & Cereda, E. (2013). Exposure to pesticides or solvents and risk of Parkinson disease. *Neurology, 80*(22), 2035–2041.

Pokhrel, B., Gong, P., Wang, X., Khanal, S. N., Ren, J., Wang, C., et al. (2018). Atmospheric organochlorine pesticides and polychlorinated biphenyls in urban areas of Nepal: Spatial variation, sources, temporal trends, and long-range transport potential. *Atmospheric Chemistry and Physics, 18*(2), 1325–1336.

Powthong, P., Jantrapanukorn, B., & Suntornthiticharoen, P. (2016). Isolation, identification and analysis of DDT-degrading bacteria for agriculture area improvements. *Journal of Food, Agriculture and Environment, 14*(1), 131–136.

Pozo, K., Harner, T., Wania, F., Muir, D. C., Jones, K. C., & Barrie, L. A. (2006). Toward a global network for persistent organic pollutants in air: Results from the GAPS study. *Environmental Science and Technology, 40*(16), 4867–4873.

Qi, Z., Miller, G. W., & Voit, E. O. (2014). Rotenone and paraquat perturb dopamine metabolism: A computational analysis of pesticide toxicity. *Toxicology, 315*, 92–101.

Qu, J., Xu, Y., Ai, G. M., Liu, Y., & Liu, Z. P. (2015). Novel Chryseobacterium sp. PYR2 degrades various organochlorine pesticides (OCPs) and achieves enhancing removal and complete degradation of DDT in highly contaminated soil. *Journal of Environmental Management, 161*, 350–357.

Rajmohan, K. S., Chandrasekaran, R., & Varjani, S. (2020). A review on occurrence of pesticides in environment and current technologies for their remediation and management. *Indian Journal of Microbiology, 60*(2), 125–138.

Ramsey, R. J. L., Stephenson, G. R., & Hall, J. C. (2002). Effect of relative humidity on the uptake, translocation, and efficacy of glufosinate ammonium in wild oat (Avena fatua). *Pesticide Biochemistry and Physiology, 73*(1), 1–8.

Roy, A., Mandal, M., Das, S., Popek, R., Rakwal, R., Agrawal, G. K., et al. (2024). The cellular consequences of particulate matter pollutants in plants: Safeguarding the harmonious integration of structure and function. *Science of the Total Environment, 914*, 169763. https://doi.org/10.1016/j.scitotenv.2023.169763

Sakamoto, M., & Tsutsumi, T. (2004). Applicability of headspace solid-phase microextraction to the determination of multi-class pesticides in waters. *Journal of Chromatography A, 1028*(1), 63–74.

Salunkhe, V. P., Sawant, I. S., Banerjee, K., Wadkar, P. N., & Sawant, S. D. (2015). Enhanced dissipation of triazole and multiclass pesticide residues on grapes after foliar application of grapevine-associated bacillus species. *Journal of Agricultural and Food Chemistry, 63*(50), 10736–10746.

Scheringer, M. (2009). Long-range transport of organic chemicals in the environment. *Environmental Toxicology and Chemistry: An International Journal, 28*(4), 677–690.

Semeena, V. S., Feichter, J., & Lammel, G. (2006). Impact of the regional climate and substance properties on the fate and atmospheric long-range transport of persistent organic pollutants-examples of DDT and γ-HCH. *Atmospheric Chemistry and Physics, 6*(5), 1231–1248.

Sethunathan, N., Megharaj, M., Chen, Z. L., Williams, B. D., Lewis, G., & Naidu, R. (2004). Algal degradation of a known endocrine disrupting insecticide, α-endosulfan, and its metabolite, endosulfan sulfate, in liquid medium and soil. *Journal of Agricultural and Food Chemistry, 52*(10), 3030–3035.

Shine, R., Ward-Fear, G., & Brown, G. P. (2020). A famous failure: Why were cane toads an ineffective biocontrol in Australia? *Conservation Science and Practice, 2*(12), Article e296.

Shunthirasingham, C., Wania, F., MacLeod, M., Lei, Y. D., Quinn, C. L., Zhang, X., Scheringer, M., Wegmann, F., Hungerbühler, K., Ivemeyer, S., Heil, F., Klocke, P., Pacepavicius, G., &

Alaee, M. (2013). Mountain cold-trapping increases transfer of persistent organic pollutants from atmosphere to cow's milk. *Environmental Science & Technology, 47*(16), 9175–9181. https://doi.org/10.1021/es400851d
Singh, P., Saini, H. S., & Raj, M. (2016). Rhamnolipid mediated enhanced degradation of chlorpyrifos by bacterial consortium in soil-water system. *Ecotoxicology and Environmental Safety, 134*, 156–162.
Singh, T., & Singh, D. K. (2017). Phytoremediation of organochlorine pesticides: Concept, method, and recent developments. *International Journal of Phytoremediation, 19*(9), 834–843.
Stenersen, J. (2004). *Chemical pesticides mode of action and toxicology*. CRC Press.
Taylor, N., & Matthews, G. A. (1986). Effect of different adjuvants on the rainfastness of bendiocarb applied to Brussels sprout plants. *Crop Protection, 5*(4), 250–253.
The Observatory of Economic Complexity (OEC), Pesticides (HS 3808): Imports, 2019 [Interactive treemap], HS92 (2021).
Tudi, M., Daniel Ruan, H., Wang, L., Lyu, J., Sadler, R., Connell, D., et al. (2021). Agriculture development, pesticide application and its impact on the environment. *International Journal of Environmental Research and Public Health, 18*(3), 1112.
Ucar, T., & Hall, F. R. (2001). Windbreaks as a pesticide drift mitigation strategy: A review. *Pest Management Science: Formerly Pesticide Science, 57*(8), 663–675.
Vaksmaa, A., Guerrero-Cruz, S., Ghosh, P., Zeghal, E., Hernando-Morales, V., & Niemann, H. (2023). Role of fungi in bioremediation of emerging pollutants. *Frontiers in Marine Science, 10*, 1070905.
Varanasi, A., Prasad, P. V., & Jugulam, M. (2016). Impact of climate change factors on weeds and herbicide efficacy. *Advances in Agronomy, 135*, 107–146.
Varma, R., & Varma, D. R. (2005). The Bhopal disaster of 1984. *Bulletin of Science, Technology and Society, 25*(1), 37–45.
Verheyen, J., Delnat, V., & Theys, C. (2022). Daily temperature fluctuations can magnify the toxicity of pesticides. *Current Opinion in Insect Science, 51*, Article 100919.
Wang, L., Cao, G., Liu, L. Y., Zhang, Z. F., Jia, S. M., Fu, M. Q., & Ma, W. L. (2023). Cross-regional scale studies of organochlorine pesticides in air in China: Pollution characteristic, seasonal variation, and gas/particle partitioning. *Science of the Total Environment, 904*, Article 166709.
Wanwimolruk, S., Phopin, K., Boonpangrak, S., & Prachayasittikul, V. (2016). Food safety in Thailand 4: Comparison of pesticide residues found in three commonly consumed vegetables purchased from local markets and supermarkets in Thailand. *PeerJ, 4*, Article e2432.
Weber, J., Halsall, C. J., Muir, D., Teixeira, C., Small, J., Solomon, K., et al. (2010). Endosulfan, a global pesticide: a review of its fate in the environment and occurrence in the Arctic. *Science of the Total Environment, 408*(15), 2966–2984.
Werle, R., Oliveira, M. C., Jhala, A. J., Proctor, C. A., Rees, J., & Klein, R. (2018). Survey of Nebraska farmers' adoption of dicamba-resistant soybean technology and dicamba off-target movement. *Weed Technology, 32*(6), 754–761.
Woodrow, J. E., Gibson, K. A., & Seiber, J. N. (2019). Pesticides and related toxicants in the atmosphere. *Reviews of Environmental Contamination and Toxicology, 247*, 147–196.
World Food and Agriculture – Statistical Yearbook 2022 FAO.
Xiao, P., Mori, T., Kamei, I., & Kondo, R. (2011). Metabolism of organochlorine pesticide heptachlor and its metabolite heptachlor epoxide by white rot fungi, belonging to genus Phlebia. *FEMS Microbiology Letters, 314*(2), 140–146.
Yang, C., Lim, W., & Song, G. (2020). Mediation of oxidative stress toxicity induced by pyrethroid pesticides in fish. *Comparative Biochemistry and Physiology Part c: Toxicology and Pharmacology, 234*, Article 108758.
Yang, R., Yao, T., Xu, B., Jiang, G., & Xin, X. (2007). Accumulation features of organochlorine pesticides and heavy metals in fish from high mountain lakes and Lhasa River in the Tibetan Plateau. *Environment International, 33*(2), 151–156.

Yang, R., Zhang, S., Li, A., Jiang, G., & Jing, C. (2013). Altitudinal and spatial signature of persistent organic pollutants in soil, lichen, conifer needles, and bark of the southeast Tibetan Plateau: Implications for sources and environmental cycling. *Environmental Science and Technology, 47*(22), 12736–12743.

Yarwood, S. A. (2018). The role of wetland microorganisms in plant-litter decomposition and soil organic matter formation: a critical review. *FEMS Microbiology Ecology, 94*(11), fiy175.

Yuan, C. S., Cheng, W. H., & Huang, H. Y. (2022). Spatiotemporal distribution characteristics and potential sources of VOCs at an industrial harbor city in southern Taiwan: Three-year VOCs monitoring data analysis. *Journal of Environmental Management, 303*, Article 114259.

Zaffar, H., Ahmad, R., Pervez, A., & Naqvi, T. A. (2018). A newly isolated Pseudomonas sp. can degrade endosulfan via hydrolytic pathway. *Pesticide Biochemistry and Physiology, 152*, 69–75.

Zhang, J., Zhang, J., Liu, R., Gan, J., Liu, J., & Liu, W. (2016). Endocrine-disrupting effects of pesticides through interference with human glucocorticoid receptor. *Environmental Science and Technology, 50*(1), 435–443.

Zhao, Z., Zeng, H., Wu, J., & Zhang, L. (2013). Organochlorine pesticide (OCP) residues in mountain soils from Tajikistan. *Environmental Science: Processes and Impacts, 15*(3), 608–616.

Zhu, J., Zhao, Y., & Qiu, J. (2010). Isolation and application of a chlorpyrifos-degrading Bacillus licheniformis ZHU-1. *African Journal of Microbiology Research, 4*(24), 2716–2719.

Zongmao, C., & Haibin, W. (1988). Factors affecting residues of pesticides in tea. *Pesticide Science, 23*(2), 109–118.

Chapter 5
From Air to Earth and Water: The Biogeochemical Implications of Particulate Matter Deposition

Remya Krishnan and Kamlesh Kumari

Abstract The way that particulate matter deposition shapes the nutrient dynamics of aquatic and terrestrial ecosystems is important. An extensive evaluation of the mechanisms, properties, and consequences of particulate matter deposition on the health and functioning of ecosystems is given in this chapter. The source and particle matter transportations mechanisms are examined at the beginning of the chapter. Later, the effect of both natural and man-caused events on how particles spread across the atmosphere is also scrutinized. The importance of particulate matter in supporting ecosystems and its nutrient content are explored, emphasizing the role that particles play as carriers of vital nutrients such as micronutrients, phosphorus, and nitrogen. The chapter explores how particulate matter deposition affects terrestrial ecosystems, highlighting how it influences biodiversity, improves soil fertility, and encourages plant development. Furthermore, the properties of particulate matter on aquatic ecosystems are investigated, emphasizing its role in phytoplankton development, nutrient inputs, and aquatic productivity. To comprehend the effects of particulate matter deposition on biogeochemical cycling processes, such as the carbon, nitrogen, and phosphorus cycles, ecosystem functioning is examined. Particle matter deposition-related environmental factors, research techniques, and management approaches are also covered, offering insights into both the advantages and disadvantages of particle inputs to ecosystems. In the end, this chapter emphasizes how crucial it is to comprehend and control particulate matter deposition to maintain land and aquatic ecosystems' resilience/flexibility and well-being against the backdrop of anthropogenic climate alteration/change.

Keywords Particulate matter deposition · Nutrient dynamics · Ecosystem functioning · Biogeochemical cycling · Environmental resilience

---

R. Krishnan (✉) · K. Kumari
Department of Botany, Daulat Ram College, University of Delhi, Delhi, India
e-mail: remya@dr.du.ac.in

K. Kumari
e-mail: kamleshkumari@dr.du.ac.in

A. Sarkar et al. (eds.), *From Sky to Soil*, https://doi.org/10.1007/978-3-032-31991-3_5

5.1 Introduction

Particulate matter, which includes dust, pollen, soot, and other microscopic particles in the atmosphere, is a foremost source of nutrients for ecosystems which include both aquatic and terrestrial. The principal basis for classifying particulate matter (PM) is where the pollution is coming from. As a result, it is imperative to discuss the four primary/major sources that remain listed in the following order according to the cataloging system: main/major, indoor area, mobile/moveable, and natural sources (Mandal et al., 2023a).

There are hundreds of distinct compounds that make up PM; some particles are sufficiently big to be perceived with/through the unassisted/natural eye, whereas others/remaining ones are so tiny/small that an electron microscope (TEM or SEM) is required to detect them. PM consists of fine inhalable particles, which have a thickness of no more than 2.5 μm, and inhalable particles, which typically have a diameter of no more than 10 μm. Before settling down, PM or particles on land or in water, wind has the ability to transport them over extended distances.

A. *Major sources*: PM emissions from metallurgical and other industrial plants, petrochemical, power plant, fertilizer, chemical sectors, and public incineration/burning.
B. *Indoor/Enclosed area causes/sources* include dry cleaners/laundering, printing factories, gas stations, in addition to housekeeping services.
C. *Mobile/Transportable sources* comprise of vehicles of many kinds, including cars, trains, and airplanes.

Lastly, natural sources include physical calamities such as dust storms, volcano eruptions, forest fires, and agricultural burning.

One crucial mechanism for moving gases and particles from the atmosphere to the surfaces of the land and water is atmospheric deposition (Wangersky, 1977). This technique may or may not have positive environmental effects. We shall concentrate on the positive aspects of the PM in this chapter.

Among the advantages is the ability of particulate matter (PM) to be directly deposited in the soil to affect the rhizosphere bacteria and fungi, which in turn affects the nitrogen cycle. The soil's pH, which can be significantly changed by the deposition of different kinds of PM, determines the availability of aluminum and alkaline cations. Climate change is also associated with a regional impact of PM on both ecosystems. Elevated particulate matter has the potential to diminish sunlight absorption by plant canopy and decrease precipitation via diverse physical phenomena (Grantz et al., 2003). In addition, PM deposition facilitates the transfer of extra nutrients to plants and air purification. We have broken up the chapter into categories that highlight the several advantages of particulate matter (Fig. 5.1).

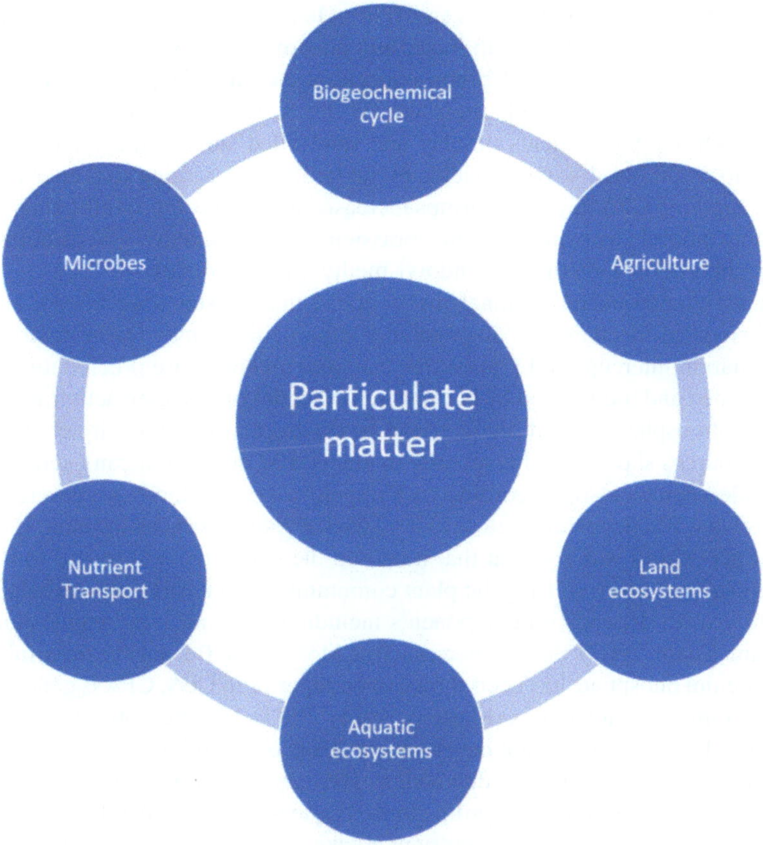

Fig. 5.1 Impact of particulate matter in nourishing the ecosystems

5.2 Nutrient Transport

Micronutrients and important nutrients like potassium, phosphorus, and nitrogen are carried in the form quantity of airborne particulate matter. In addition to human-caused activities including industrial processes, agriculture, and automobile emissions, natural sources of these particles include soil erosion, volcanic eruptions, and forest fires. Because PM deposition affects rhizosphere bacteria and fungus, it directly affects the soil nutrient cycle, particularly nitrogen cycling. The nitrogen (N_2) can be disseminated and absorbed in various plant sections/parts once it has been taken up by the leaves. During burning, PM can also transfer nitrogen from combustibles to plant leaves.

Hard coal fly ash also called CFA, is a powdery, fine-grained, gray, odorless material made mostly of glassy, spherical particles. They are helpful in nutrient transport to the soil. This is helpful in nutrient transport to the soil. SiO_2 (Silicon dioxide), Al_2O_3 (Aluminium oxide), and Fe_2O_3 (Iron (III) oxide) are the principal constituents of coal fly ash. According to Theis and Wirth (1977), the main constituents of CFA for PM were Al, Fe, and Si, with lesser amounts of Ca, K, Na, Ti, S, and a variety of traces of metals. Increased boron concentrations in CFA provide many benefits for the soil and help in plant growth (Wong & Su, 1997). According to Cline et al. (2000), CFA treatments on clay and sand-filled soils in the southern region of Ontario could boost soybean yields by up to 35%. According to Anderson et al. (1990), applying CFA to yellow turnips increased the levels of the well-being/health-promoting indolyl glucosinolates glucobrassicin (3-indolylmethyl glucosinolate) and neoglucobrassicin (1-methoxy-3-indolyl-methyl glucosinolate).

Climate change and the regional impact of PM on ecosystems are related. Through several physical mechanisms, augmented PM/particulates may lessen radioactivity contamination interception by plant/tree canopies and decrease precipitation. Here, PM is created and transported by burning and other dust-producing activities, as well as by the atmosphere's solid particles and liquid droplets (Castanheiro et al., 2021).

CFA sources show mixed responses in plants. PM deposition can occur through man-made sources also. According to Krippelova (1982), there are severe consequences from PM depositions close to a magnesite facility in Czeslowakia. Here, the pace of deposition was so great that it caused the pH of the soil to rise to 9.5 and surface crusts to form, shifting the plant communities from calcicoles to halophiles.

Plant growth depends on components including calcium, iron, magnesium, and potassium (Inam, 2007). Some others, such as Mo, Se, and Be, may be harmful. Since monocalcium phosphate has been shown to be superior to CFA, CFA is generally not the best source of phosphorus (Martens, 1971). But as legumes have demonstrated, Ca^{2+} and Mg^{2+} can boost plant development (Page et al., 1979).

Through both wet and dry deposition, PM can also introduce air metals like aluminum into ecosystems. According to Bojórquez-Quintal et al. (2017), aluminum stimulates root growth, increases nutrient uptake, increases enzyme activity, and has other benefits, particularly in acidic soil. The availability of Aluminum (Al) and the presence of alkaline cations depend on the pH of the soil, which can remain significantly altered/modified by the deposition of numerous categories of PM/particulates. This alkalinity can cause changes in the patterns of vegetation, such as the emergence of new vegetation cover on barren, acidic ground (Ulrichs et al., 2008).

Additionally, PM can lessen precipitation and radiation interception by plant canopies, which can affect ecosystems regionally and be linked to changes in the local weather (Luo et al., 2019). Additionally, plants can absorb atmospheric metals through their leaves or through their roots once they have been deposited on the ground. The particles' sizes and the blade's surface structure are the primary determinants of the blade's ability to absorb metal particles. The form and kind of metals are the primary factors influencing root uptake; atmospheric metals that have been deposited have a higher/large biotic activity/movement and are easily/more readily engrossed/absorbed by flora than metallic elements that are deposited in soil. Water bodies are primarily exposed to atmospheric alloys through deposition/collection. The presence of essential metallic elements (HMs) like cobalt/Co, copper/Cu, iron/Fe, manganese/Mn, molybdenum/Mo, nickel/Ni, and zinc/Zn promotes growth and development of vegetation (Morrissey & Guerinot, 2009). When these advantageous components are present in optimal amounts, they enhance the nutritional state of the flora and activate a number of processes necessary for healthy plant growth and increased output. When present in an ideal concentration as vital micronutrients, these metals are advantageous to plant growth, development, and production. Cobalt is present in vitamin B12, whereas iron aids in plant development by acting as a cofactor in cytochrome. Plants are able to take up essential heavy metals (HMs) like iron/Fe, zinc/Zn, copper/Cu, and manganese/Mn from the earth because of absorption/concentration gradients and careful selective absorption. Both cell metabolism and a variety of enzyme activities are significantly impacted by these ions. The creation of proteins, pigments, and nucleic acids—all of which are involved in photosynthetic processes—requires these metals as well. Additionally, they preserve the functional and structural integrity of cell membranes. Copper is one necessary heavy element that participates actively in photosynthesis. Many metabolic enzymes, such as malic dehydrogenase and oxalo-succinic decarboxylase, depend on manganese as a critical component. Even while arbitrary human activity and other natural processes significantly/majorly amend the amount of these heavy metallic ions in soils (Arif et al., 2016).

PM can travel great distances on wind and settle on land or in water, where its chemical structure may have an impact on the environment. These consequences include acidifying lakes and streams, altering soil nutrients, shifting agricultural crops and delicate forests, and reducing ecosystem diversity in broad river basins and coastal waters. Huge amounts of metallic element are released into the atmosphere by human activity, and these metallic elements then drift or settle with atmospheric activity and intrude the ecosystem (Mohapatra & Biswal, 2014) (Table 5.1).

5.3 Land Fertility

Particulate matter can directly enrich soil with nutrients as it settles on terrestrial ecosystems. In addition to supporting plant development and the general health of terrestrial ecosystems, this can improve soil fertility. Dust storms, for instance, could carry nutrients across continents and affect the fertility of receiving soils (Wu & Zhang, 2018). Additionally, the delivery of mineralogical nutrients by dust may boost the production of microbes in marine ecosystems.

An air supply through volcanoes that produce large amounts of particulate matter is sufficient for ecosystem establishment and maintenance, particularly in terms of the sulfur cycle, according to direct measurements of deposition in the older soils of Hawaiian basaltic shield volcanoes (Chadwick et al., 1999). Regular addition of organic residues (PM), especially those that have been composted, improved aggregate stability and decreased bulk density of the soil, which in turn increased physical fertility of the soil. In certain regions of the western United States, particulate matter deposition is the reason behind an upsurge in nitrogen/N_2 mineralization, which is the process of converting organic materials in the soil into nitrogen/N_2 that is readily

Table 5.1 Types and sizes of particulate matter (PM)

S. no.	Type		PM diameter [μm]
1	Particulate matter/ contaminants	Smog (fog + smoke)	0.01 to1
		Soot (amorphous carbon)	0.01 to 0.8
		Tobacco/nicotine smoke	0.01to 1
		Fly ash/pulverised fuel ash	1 to 100
		Cement Powder	8 to 100
2	Biological and microbial contaminants/Pollutants	Bacteria and bacterial spores	0.7 to10
		Viruses	0.01 to 1
		Fungi and molds	2 to 12
		Allergens (canines, felines, pollens, domestic dirt/dust)	0.1 to 100
3	Categories of dust	Atmospheric dust/dirt	0.01 to 1
		Weighty dust/particles	100 to 1000
		Subsiding dust/dirt	1 to100
		Gases - Different gaseous pollutants	0.0001 to 0.01

Adapted from Manisalidis et al. (2020)

obtainable/accessible (Grantz et al., 2003). Applying composted materials repeatedly raises the organic N_2/nitrogen content of the soil by up to 90%, storing it for mineralization in upcoming crop seasons, frequently without causing nitrate leaching to groundwater.

In less developed countries, burning agricultural crop residue (a type of particulate matter), such as stubble burning, is a very economical way to prepare land for the next harvest. It also provides the added benefits of reducing pest occurrence and getting rid of weeds (Ge et al., 2021; Maffia et al., 2020). It is widely used in agricultural systems for rice, wheat, and maize throughout South and East Asia. Crop residues left after burning are essential to the stability of agricultural ecosystems and are a valuable source of nutrients for plants. Crop residue burning is expected to increase by 150% globally between 1960 and 2015 (FAO, FAOSTAT web page). Fine dust in the soil can travel through air, causing the number of species to migrate to different places that live below the surface, such as mites and larvae, which help maintain soil fertility (Roy et al., 2024a).

Since it gives plants nourishment and soil organisms energy, particulate organic matter, or POM, is an essential part of soil fertility. POM is a heterogeneous amalgamation of organic matter made up of living things, particle debris, and structural components from plants and microbes (Liang et al., 2017). POM breaks down to release nutrients into the soil, which soil organisms then utilize as fuel. Plants could absorb nutrients that soil organisms are unable to employ. The biological and chemical properties of the POM determine how much nutrition is released.

5.4 Aquatic Productivity

Additionally, particulate matter can be very important to aquatic environments (Håkanson, 2006). Particulates can supply nutrients necessary for the growth of phytoplankton and other aquatic plants when they settle on water bodies. Particulate organic matter when flown from various sources is trapped by macrophytes for maximizing the retention of nutrients. As the foundation of aquatic food chains, phytoplankton sustains the entire aquatic ecosystem. Particulate matter deposition can therefore indirectly increase water biodiversity and production. Because particulate matter (SPM) increases turbidity, which hinders light penetration, it can influence aquatic environments by creating stratification inside water. This could increase biodiversity. Both photosynthetic activity and bioproduction may be impacted by this. Phytoplankton, silt, and clay are examples of suspended particles that are part of SPM. The concentrations of carbon-based/organic and mineral/inorganic dissolved and particulate substance (PM) determine the depth of the photic zone, that is the region where primary aquatic producers are confined (Sarà, 2007).

Water flowing via the hydrologic cycle directly connects aquatic and terrestrial ecosystems within the same terrain (Turner et al., 2002). Therefore, there is a functional relationship between terrestrial and aquatic ecosystems, and this relationship has significant ecological ramifications. The constant movement of energy and nutrients across the ecosystem is essential to its function. Because they provide nutrients for a variety of aquatic plants, PMs are crucial in this situation. An ecosystem/biosphere has a highly comprehensive budget report of the contributions in addition to outputs of nutrients; however, because detailed info about these inputs and productivities, as well as the subtle alterations that keep the ecosystem in balance, is lacking, it is challenging to evaluate the effects of human activity on landscapes or the ecosystems.

The pace at which nitrogen regenerates in surface waters largely determines the efficiency of the seas, except for upwelled water and spring blooms in temperate and boreal waters (Ittekkot & Zhang, 1989). The quantity of actively processing nitrogen-fixing bacteria present, which consequently seems to be a consequence of the water's particle composition, determines the rate of regeneration. Particle content like that of silicate also may be the primary factor regulating the productivity of open waters. For areas like, for example, the Sargasso Sea that are known for having low particle counts, it may be possible to artificially raise the water's particle concentration to increase productivity. Given the fact that silicate can only be replenished by the deep-dwelling process of diatom test re-solution (Krause et al., 2010; Turner & Millward, 2002).

Particulate matter, particularly allochthonous carbon, is essential for the total primary production and flow of organic matter in a wetland subjugated by mangroves

(Gillis et al., 2015). In a tidal pond in north-west (NW) Hong Kong, for instance, the expected production and detritus dynamics for the years 1986–1988 (9.1 ha) were determined. Out of 12.47 tonnes dry weight ha^{-1} per year, emergent macrophytes (the reed *Phragmites communis* and the mangrove *Kandelia candel*) accounted for over 90% of the total productivity of flora. Additionally, benthic macroalgae and phytoplankton were discovered, with *Enteromorpha crinata* predominating (Boydet al., 1998). As a result, the emergent macrophytes in the pond tend to continue to increase in number. Furthermore, the transition of landward wetlands from being net exporters of carbon to being net importers is probably a sign of the development of nutrient-conservative terrestrial systems (Lee, 1990; Likens & Bormann, 1975). This is a crucial step in the progression of a hydrosere, a form of plant succession in streams and ponds.

The Arabian Peninsula is an important contributor to atmospheric dust and a significant cause of nutrients for the nautical life in the Gulf of Arabia. The Gulf basin is covered in dust, primarily during the annual, recurring local dust storm occurrences. It seems that the Arabian Gulf ecosystem collects additional nutrients, especially Iron, phosphorus, and nitrogen from sources other than dust deposition, such as deepwater upsurge, recycled nutrients, sewage discharge, and salts from desalination facilities. The projected total annual dust flux for the Gulf area (240,000 km^2), assuming 10% uniform deposition, is 6.385 $gm/m^2/day$, or roughly 5.5 × 106 tons, over an average of 33 days $year^{-1}$. Nutrient emancipation from dust/dirt, for instance, depends on several processes in the marine environment, including microbial mineralization and recycling, atmospheric photochemical reactions, and acid rain dissolution (Hamza, 2021).

5.5 PM Application to Agricultural Soils and Root Absorption

If direct wet and dry deposition of particulate matter is applied to agricultural soils over an extended length of time, it may result in the accumulation of PM or its transformation products in the soil (Hou et al., 2020). PM that contains micronutrients or minerals like zinc (Zn), iron (Fe), and manganese (Mn) or essential plant macronutrients like nitrogen (N_2), phosphorous (P), sulfur (S), and potassium (K) may be beneficial for vegetation if it is deposited in moderation and the crops are growing in nutrient-poor conditions. On the other hand, if the particles are absorbed by the roots and concentrate in the edible portions like that of Solanaceous crops, the deposition of particles carrying hazardous substances could pose a direct threat to crop safety (Bi et al., 2018).

Particulate matter is the only source of soil nutrients to agricultural soil other than manure (Fig. 5.2). In the field of three farms in Germany, dried manure and the ensuing PM contained much fewer identified harmful and antibiotic-resistant

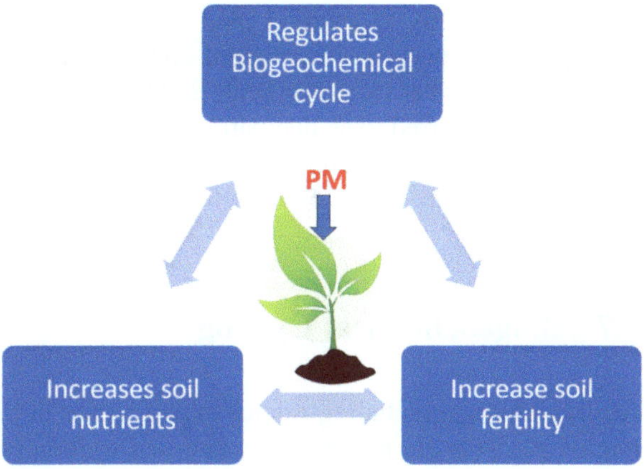

Fig. 5.2 Interdependence of particulate matter by plants

bacteria than fresh samples did. For crop productivity, applying trace elements to fertilizers is an efficient way to address micronutrient deficits (Kabelitz et al., 2020).

Agroecosystems contain trace metals at trace levels (mg kg^{-1} or less). Micronutrients/minerals for plants, such as Copper/Cu, Zinc/Zn, Manganese/Mn, Iron/Fe, Molybdenum/Mo, and Boron/B, are trace elements required for plant growth. Apart from boron (B), each of these elements is a heavy metal (HMs) that is toxic to crops in high doses (He et al., 2005). Trace elements are either inherited from parent materials in the soil or introduced by people into an agroecosystem (Mandal et al., 2023b).

5.6 Particles in the Atmosphere

Aerial "particulate matter," or PM, is a complicated concoction of solid particles and liquid condensations. The scientific characterization of "aerosols" denotes liquid or compact hard particles suspended in a gas mixture or carrier gas, though the term "aerosols" is also used (Kodas, 1990). The term "particles" is more appropriate when referring to the particle without reference to the carrying gas. It makes sense to group air particles according to how they form and their aerodynamic properties. Depending on how they form, airborne/aerial particles can be categorized as primary (1^0) or secondary (2^0) aerosols. The emission sources directly emit particulate primary aerosols into the atmosphere. However, secondary aerosols are particles that are formed in the atmosphere as a consequence of the evolution of a primary particle or as a result of the gas-particle conversion mode resulting from the concentration of natural and man-made vapors (Roy et al., 2024b).

Contingent on their dimensions, particles are referred to by diverse names. Total suspended particulates (TSP) in ambient air are all suspended particles in the air, regardless of size. PM1 particles, such as diesel particles, possess an aerodynamic thickness of less than 1 μm. Ultrafine particles are those that have an aerodynamic diameter/thickness of less than 0.1 μm, also called nanoparticles, or PM0.1. They are produced by poorly understood gas-to-particle conversion mechanisms. Because of their quick coagulation or haphazard impaction onto surfaces, their lifespan may be measured in minutes (Kulmala et al., 2004). Materials/Particles with an aerodynamic diameter are referred to as fine particles, or PM2.5, also of less than 2.5 μm. PM10, also known as coarse particles, are defined as those with an aerodynamic diameter of less than 10 μm. PM (2.5–10) refers to particles with a diameter of 2.5–10 μm.

5.7 Biogeochemical Cycling

All living things including microbes participate in the biogeochemical cycles, but microorganisms are particularly important to the operation and evolution of these cycles because of their great quantity, extraordinary metabolic potential, and capacity for adaptation. In soils, freshwater habitats the key biogeochemical sequences/cycles are carbon (C), nitrogen (N_2), sulfur (S_2), phosphorus (P), silicon, and metals. Many processes that are engaged in the turnover of biogeochemical cycles are described at the worldwide, regional/provincial, and homegrown/local scales, also in specific microenvironments such as particles/materials or microniches, and from gene to ecosystem (natural and man-made ecosystems). The metabolic pathways underpin the exchanges and biotransformations of organic and mineral/inorganic components within ecosystems/biosphere which are highlighted together with their biodiversity (Bertrand et al., 2015; Turner & Millward, 2002). Here mineral components play a crucial role in ecosystem resilience, resistance, and adaptation. However, for these living beings to worm PM can be a provider for optimum nutrients.

Once raised from the surface, particulate debris, like small dust, can travel great distances through the sky. This is a component of the biogeochemical cycle/series, which is the flow of substances/materials and elements flanked by the atmosphere, living things, and the crust of the Earth. Lithosphere (land and soil), the biosphere (organisms), the atmosphere (air), in addition to the ocean's hydrosphere are the four main systems that exchange materials during the cycle. From the words "bio" (biosphere), "geo" (geological components), and "chemical" (elements that flow through the cycle), the name "biogeochemical" is derived. In practically every habitat on the world, microorganisms are essential for controlling biogeochemical systems. For instance, bacterial metabolism of carbon, nitrogen, and sulfur may be impacted by PM2.5.

Particulate matter from the air deposition supports the nitrogen, phosphorus, and carbon cycles also Iron among other biogeochemical cycles, particularly those seen in Oceans or seas (Boyd & Ellwood, 2010; Yamanaka & Tajika, 1996). These cycles are essential mechanisms that control how energy and nutrients move through ecosystems. Particulate matter deposition affects ecosystem resilience and dynamics, as well as the rates at which nutrients are cycled.

Aerosols as particulate matter are an important part for the biogeochemical cycle. The composition and characteristics of atmospheric aerosols are complicated and varied. Although atmospheric aerosols' effects on climate and human health are of great scientific interest, it is not as clear how aerosols function in multiphase chemistry, which affects how they carry nutrients in ecosystems. N_2, P and Fe cycles, since these essential elements are recognized to restrict primary output based on their location in addition to availability. They are also vital nutrients for ecosystems (Kanakidou et al., 2018).

Phosphorus (P) is also required for the development of skeletons, the synthesis of energy currency of cell adenosine triphosphate (ATP), which powers cellular processes and stores energy in cells, the synthesis of DNA and RNA, and the manufacture of hemoglobin, which delivers oxygen through the blood (Smil, 2000). Photosynthetic organisms' fixation of inorganic carbon into cellular biomass is significantly inclined by the accessibility of P. P supply is a pivotal element influencing ecosystem productivity, particularly in the long run of ecosystem evolution (Benitez-Nelson, 2000). P (and Fe), in contrast to atmospheric reactive N, are mostly found in the aerosol phase of the atmosphere, apart from a small number of phosphines that remain relatively rare in the surrounding environment.

Phosphorus comes from a range of organic sources, such as burning biomass, dust, sea spray, volcanoes, and freshwater wetlands that release phosphine in the form of phosphate (Glindemann et al., 2005; Han et al., 2011; Walch et al., 2022). A multitude of human endeavors, like as deforestation/desertification, P mining from land, and farming practices including fertilizers/manures and insecticides, also contribute to its production. According to Mackey and Paytan (2009), emissions caused by humans make atmospheric P more soluble and lead to P buildup in soils and aquatic habitats. Based on extrapolations from limited measurements, the initial estimations of the entire total worldwide deposition/accumulation changes of P ranged from 3 to 4 Teragram P per year (Graham & Duce, 1979; Smil, 2000). Here, however, just a small portion of this total is soluble or falls within the range of nanoparticle sizes that ecosystems can easily access.

The first worldwide atmospheric P cycle models, according to Mahowald et al. (2008), focused on soil particles sources and assumed that P was fixedly soluble in 10% of dust particles, 100% of volcanic emanations, and fifty percentile of various sources. In the super rough mode, dust, combustion, and biological atomizers have been added as P sources in recent studies (Brahney et al., 2015; Tipping et al., 2014; Wang et al., 2015) to account for the overall amount of phosphorus in the surrounding atmosphere. Consequently, the bulk of P sources are aerosols and PM.

Nearly all living things require iron because it is essential to several metabolic activities, including oxygen transport, DNA synthesis, and electron transport (Abbaspour et al., 2014). Iron's relevance is accounted for by its involvement in redox (reduction/oxidation) reactions and its capacity to reversibly change its oxidation state by electron transfer, which changes depending on the ligand with which Fe is complexed (Guerinot, 1994). Since various enzymes involved in photosynthesis, respiration, and N fixation require iron, it is a crucial micronutrient for marine species (Twining et al., 2004). High-nutrient/low-chlorophyll (HNLC) zones are vast oceanic areas where primary productivity might be restricted by a lack in iron (Jickells et al., 2005; Martin et al., 1991). Global N fixing and biological C export can be significantly enhanced by Fe deposition through particulate matter (PM). Furthermore, increased Fe deposition has been proposed to lead to increased regional respiration and productivity as well as the expansion of minimal oxygen zones (Ito et al., 2016), all of which could have an impact on marine life. According to Twining and Baines (2013), the availability of Fe affects ecosystem biodiversity. Diatoms and cryptomonads are examples of species that seem to use a mechanism to adapt to a "bioavailability envelope," whereby they become less common in the presence of Fe.

Due to the fact that it is a part of amino acids, the structural units of proteins, nitrogen is necessary for life. Numerous other biological materials also include it, such as nucleotides and nucleic acids (ribonucleic acid, or RNA/deoxyribonucleic acid/ DNA) (Galloway et al., 2008). Moreover, molecules containing nitrogen include electrons and hydrogen; they are both essential for biological functions including photosynthesis and breathing. It is demonstrated that N is a part of chlorophyll when nitrogen atoms bond Mg^{2+} close to the chlorophyll central structure. According to estimates (Kanakidou et al. 2012, 2016), organic N is roughly 50% from human activities and makes up 20–40% of all soluble N deposition, with significant regional variability. This estimate is consistent with observations (Cornell, 2011). The HNO_3/NO_3–pH-driven partitioning to the aerosol phase and the formation of N-containing SOA (secondary organic aerosol) in the presence of sufficient accessible NOx have been considered in simulations and experiments. Previous global modeling studies suggested that around 10% of NOx emissions are deposited as organic nitrogen instead of inorganic nitrogen based on the latter method (Dentener et al., 2006; Lamarque et al., 2013). Additionally, they calculated that about forty percent of the total N deposition is caused by air aerosols.

Because of their capacity to suspend, transfer, deposit, in addition to re-suspension, subdivisions deliver a means of material transit flanked by numerous environmental mediums and sites. Like instance, soil-derived nitric oxide (NO) emissions represent a small part of the nitrogen cycle (Kulmala & Petäjä, 2011). Oxidation of NO to HNO_3 occurs in the environment. A similar mixture can produce ammonium nitrate salt: $NH_3 + HNO_3$. When this particle lands at a position downwind, the cycle amidst the surface and atmosphere of Earth is finished. PMs are therefore crucial in this situation as well.

The dynamic and distinctive coastal waters play a significant role as both sources and sinks of a number of atmospheric species that are vital to the climate and global biogeochemical cycles. Strong physical-biogeochemical interactions, which involve the atmosphere as a vehicle for the exchange of matter and energy, are characteristics of the transition zone between land and ocean. As a result, a system with inherent complexity emerges. The coastal/sea zone is the site of several important biogeochemical exchanges, including those involving water (H_2O), nutrients (nutrients like phosphorus/P, nitrogen/N_2, iron/Fe, and silica/Si), salts (chlorine/Cl, bromine/Br, and iodine/I_2), carbon/C carbon dioxide/CO_2, particulate organic carbon/POC, dissolved organic carbon/DOC, dissolved inorganic carbon/DIC, and reactive organic/carbon-based and inorganic/mineral trace gases (nitrous oxide/N_2O) (Kanakidou et al., 2018).

The reactivity, transport, and biological effects of compounds in aquatic settings are all greatly influenced by suspended particles, which also offer an essential channel for chemical contents between the water column, bed sediment, and food chain. To give information on the movement of heavy metals from the river to the sea, for instance, sediments and suspended debris from the mouth of the Adige River were examined. Grain-size spectra of suspended matter recorded vertically in front of the river mouth indicate that riverine particles are restricted to the top brackish layer. Mangrove vegetation increases biodiversity in these locations; hence this could influence those areas (Juračić et al., 1987).

5.8 Particulate Matter (PM) and Health

Tiny solid or liquid droplets, known as particulate matter, can cause lung damage if inhaled. Particles known as PM10, or particles smaller than 10 μm, can reach the circulation after being inhaled and enter the lungs. PM2.5, or fine particles, are more detrimental to human health. Dust, benzene, O_3, particulate matter (PMs), and other substances can cause significant injury to the respiratory system. Additionally, if you currently have a respiratory illness, such as asthma, your risk is higher. Long-term consequences are most likely to occur in people who have a predisposed sickness condition. Acute tracheal exposure to impurities might result in audible alterations in voice quality. Chronic obstructive pulmonary disease/COPD is a disorder that can be brought on by airborne pollution and increases rates of sickness and death. Industrial airborne pollution, transportation, and the long-term consequences of fuel burning are the main risk factors/issues for COPD/ chronic obstructive pulmonary disease. (Li et al., 2016).

Research focusing on either short-term/acute or long-term/chronic PM exposure has confirmed a link between particulate matter/PM and harmful health effects. Typically, the different pollutants react chemically to generate particulate matter (PM) in the atmosphere. Particle size has a direct impact on how well they penetrate The US Environmental Protection Agency identified particles as part of the category known as particulate matter i.e., PM. PM-10, or elements having a diameter of 10 μm (μm) or less, and extremely fine particles, which typically have a diameter of 2.5 μm (μm) or less, are included in the category of particulate matter (PM) pollution.

Tiny particulate matter contains liquid or solid droplets that can be consumed and have detrimental consequences on one's health. After inhalation, particles smaller than 10 μm in diameter (PM10) could infiltrate the lungs and potentially enter the passage. The well-being risk from very thin particles, or PM2.5, is advanced.

In addition, long-term chronic consequences include immunological system affection and respiratory illnesses. It is important to remember that those who have diabetes, asthma, pneumonia, or other respiratory or cardiovascular conditions are more vulnerable to the negative effects of PM. There is a high correlation between PM2.5 and a variety of respiratory system disorders, with PM10 coming in second

(Kappos et al., 2004), since they may permeate inner spaces due to their dimensions (Boschi, 1999). The properties of the particles, both chemical and physical, have detrimental effects. Both carbon-based organic (polycyclic aromatic hydrocarbons/HC, dioxins, benzene, and 1–3 butadiene) and inorganic (C/carbon, chlorides, nitrates/NO^{3-}, sulfates SO_4^{2-}, and metallic element) constituents can be found in PM10 and also PM2.5 (Cheung et al., 2011).

5.9 Environmental Impacts

Because it provides nutrients to ecosystems, particulate matter deposition can have together advantageous/beneficial and harmful/damaging effects. For instance, excessive pollution deposition from industrial processes, such as heavy metals or nitrogen compounds, can damage ecosystems and cause environmental deterioration (Bi et al., 2018). Particulate matter (PM) can harm the ecosystem in a variety of ways, such as PM can generate haze by scattering and absorbing light in the atmosphere, which can limit visibility. While certain PM constituents, like sulfate and nitrate, have a cooling impact, others, such as black carbon, contribute to global warming and climate change. PM has a role in producing acid rain, which erodes stone and other materials, destroying statues and monuments. Droughts can be brought on by PM and can alter weather patterns. PM has the potential to cause ocean acidification. Reduced vision, or haze, is mostly caused by fine particles (PM2.5) throughout a large portion of the US, including many of our beloved wilderness areas/regions and national parks. Additionally, PM can harm materials and ecosystems in the following ways: PM can cause the acidity of lakes and streams. River basins and coastal waters are two places where PM can alter the nutrient balance. PM can decrease the nutrients in the soil. PM can harm agricultural crops and delicate woodlands. PM has an impact on ecological diversity (Ge et al., 2021; Rai, 2015).

5.10 Environmental Damage

According to national studies, particulate matter increases the risk of illness and mortality. Global warming and climate change may make things worse (He et al., 2023). In addition, the elderly and those in vulnerable situations have been found to have greater rates of hospitalization (a measure of morbidity) for certain reasons. Because fine in addition to ultrafine/superfine particulate matter (PM) can more readily enter/pass the bloodstream and penetrate the deepest areas of the airways, it appears to be connected/linked to more serious disorders (Griffin, 2013).

Because of overcrowding, unconstrained suburbanization, and the advancement of industrial development, the issue is particularly severe in emerging nations. This leads to poor air quality, especially in countries where there are large socioeconomic differences and minimal awareness of ecologically friendly behaviors. People who use solid fuels like wood for domestic uses and have low incomes are subjected to polluted, low-quality air at home. China is one of the Asian nations most affected by air pollution because of its rapid industrialization and population growth. Fine particles are linked to the reported lung cancer mortality in China (Maji et al., 2017). Extended or continuous exposure has been associated with negative effects on the cardiovascular system. Extreme air pollution and extremely low air quality have been reported from India. New Delhi is amongst the most polluted cities in India. Air pollution causes frequent cancellations of aircrafts into and out of New Delhi International Airport. India's swift industrialization, urbanization, and growth in the use of motorcycles for transportation are all contributing factors to pollution in both urban and rural areas. However, one of the main causes of home air pollution in Nepal and India is the burning of biomass related to cooking and heating needs and activities. In India, there is a difference in the interior air quality between regions with varying climatic conditions, populations, and educational attainment. The northern parts of India have greater PM2.5 levels (557–601 $\mu g/m^3$) than the southern states (183–214 $\mu g/m^3$). This could be primarily due to the colder temperature in North India, where longer stays at home and more heating are required than in southern India's tropical environment. Major health repercussions are linked to household air pollution in India, particularly for mothers and young children who spend more time indoors. Lung melanoma and chronic or long-lasting obstructive respiratory disease (CORD) are frequent in females, but young children under five years old have a higher risk of acute lower respiratory illness (Tsurumi & Managi, 2020).

The number of deaths increased where in December 1952, 4 K was in London, and in 1963, 400 were in New York City. This is because of the air pollution accumulation/growth, particularly sulfur dioxide (SO_2) and fume/smoke, which reached 1,500 mg/m^3. Based on measurements of outdoor pollution in six major United States cities, a connection between contamination/pollution and death rate was established. Mortality appears to be more closely correlated in all cases with the amounts of smaller, sulfate-containing, inhalable particles compared to the aerosol acidity levels, nitrogen dioxide/NO_2, sulfur dioxide/SO_2, or total particulate pollution (Pope et al., 2018).

Here, for instance, we outline the source-level control strategies using automobile catalytic converters. These are devices/machines that use catalysis and reduction–oxidation processes to convert the hazardous gases and pollutants released by ignition locomotives into fewer hazardous contaminants. To lessen traffic during rush hour, private vehicle usage in Greece was limited by registering license plate numbers. Collectors and closed systems can reduce industrial emissions to the legally mandated minimum levels of air pollution (Moustris et al., 2016).

5.10.1 Effects of Acid Rain on Materials

Some particulate matter, like sulfates, cools the environment by reflecting sunlight back into space and encouraging the creation of clouds. Thus, while lowering the levels of sulfate pollution is good for human health, it can also increase the warming effect of greenhouse gasses (NASA report, 2020) Acidic deposition can be of wet or dry type (Singh & Agrawal, 2007). The nitric/NO_3 and sulfuric acids/H_2SO_4 cause acid shower and dry, acidic materials to fall on Earth, which may land on structures, sculptures, and other artifacts created by humans, damaging their surfaces. The metal is corroded and paint and stone damage is accelerated by the acidic elements. They also taint memorials and the outside surface of constructions and other edifices. The effects of this damage can be expensive, including lost detail on metal and stone statues, monuments, and tombstones, damaged materials that need to be replaced or repaired, and higher maintenance expenses.

5.11 Conclusions

Particulate matter is important in shaping the nutrient components of the aquatic and terrestrial ecosystems. PM supports nutrient transport from one place to another, thereby increasing soil fertility, supports plant growth, and augments root absorption. It also enhances aquatic productivity as deep oceans receive nutrients from PM. PM also supports various biogeochemical cycles through wind, aerosols, etc., thus maintaining the earth's supply of nutrients. Even though PM is considered negative for the atmosphere, however, there are many benefits of PM also which are left unnoticed.

References

Abbaspour, N., Hurrell, R., & Kelishadi, R. (2014). Review on iron and its importance for human health. *Journal of Research in Medical Sciences: THe Official Journal of Isfahan University of Medical Sciences, 19*(2), 164.

Anderson, J. L., Lisk, D. J., & Stoewsand, G. (1990). Glucosinolates in rutabaga grown in soil capped over coal fly ash. *Journal of Food Science, 55*, 556–557.

Arif, N., Yadav, V., Singh, S., Singh, S., Ahmad, P., Mishra, R. K., Sharma, S., Tripathi, D. K., Dubey, N. K., Chauhan, D. K. (2016). Influence of high and low levels of plant-beneficial heavy metal ions on plant growth and development. *Frontiers in Environmental Science, 4*(2016), 69.

Benitez-Nelson, C. R. (2000). The biogeochemical cycling of phosphorus in marine systems. *Earth-Science Reviews, 51*(1–4), 109–135.

Bertrand, J. C., et al. (2015). Biogeochemical cycles. In J. C. Bertrand, P. Caumette, P. Lebaron, R. Matheron, P. Normand, & T. Sime-Ngando (Eds.), *Environmental microbiology: Fundamentals and applications*. Springer.

Bi, C., Zhou, Y., Chen, Z., Jia, J., & Bao, X. (2018). Heavy metals and lead isotopes in soils, road dust and leafy vegetables and health risks via vegetable consumption in the industrial areas of Shanghai, China. *Science of the Total Environment, 619*, 1349–1357.

Bojórquez-Quintal, E., Escalante-Magaña, C., Echevarría-Machado, I., & Martínez-Estévez, M. (2017). Aluminum, a friend or foe of higher plants in acid soils. *Frontiers in Plant Science, 8*, 1767.

Boschi, N. (1999). Defining an educational framework for indoor air sciences education. *Education and training in indoor air sciences* (pp. 3–6). Springer, Netherlands.

Boyd, P. W., & Ellwood, M. J. (2010). The biogeochemical cycle of iron in the ocean. *Nature Geoscience, 3*(10), 675–682.

Boyd, P. W., Ibisanmi, E., Sander, S. G., Hunter, K. A., & Jackson, G. A. (2010). Remineralization of upper ocean particles: Implications for iron biogeochemistry. *Limnology and Oceanography, 55*(3), 1271–1288.

Boyd, C. E., Tucker, C. S., Boyd, C. E., & Tucker, C. S. (1998). Ecology of aquaculture ponds. *Pond Aquaculture Water Quality Management*, 8–86.

Brahney, J., Mahowald, N., Ward, D. S., Ballantyne, A. P., & Neff, J. C. (2015). Is atmospheric phosphorus pollution altering global alpine Lake stoichiometry? *Global Biogeochemical Cycles, 29*(9), 1369–1383.

Castanheiro, A., Wuyts, K., Hofman, J., Nuyts, G., De Wael, K., & Samson, R. (2021). Morphological and elemental characterization of leaf-deposited particulate matter from different source types: a microscopic investigation. *Environmental Science and Pollution Research, 28*(2021), 25716–25732.

Chadwick, O. A., Derry, L. A., Vitousek, P. M., Huebert, B. J., & Hedin, L. O. (1999). Changing sources of nutrients during four million years of ecosystem development. *Nature, 497*, 491–497.

Cheung, K., Daher, N., Kam, W., Shafer, M. M., Ning, Z., Schauer, J. J., & Sioutas, C. (2011). Spatial and temporal variation of chemical composition and mass closure of ambient coarse particulate matter (PM10–2.5) in the Los Angeles area. *Atmospheric Environment, 45*(16), 2651–2662.

Cline, J. A., Bijl, M., & Torrenueva, A. (2000). Coal fly ash as a soil conditioner for field crops in southern Ontario. *Journal of Environmental Quality, 29*, 1982–1989.

Cornell, S. E. (2011). Atmospheric nitrogen deposition: Revisiting the question of the importance of the organic component. *Environmental Pollution, 159*(10), 2214–2222.

Dentener, F., Drevet, J., Lamarque, J. F., Bey, I., Eickhout, B., Fiore, A. M., Hauglustaine, D., Horowitz, L. W., Krol, M., Kulshrestha, U. C., & Lawrence, M. (2006). Nitrogen and sulfur deposition on regional and global scales: A multimodel evaluation. *Global Biogeochemical Cycles, 20*(4).

Galloway, J. N., Townsend, A. R., Erisman, J. W., Bekunda, M., Cai, Z., Freney, J. R., Martinelli, L. A., Seitzinger, S. P., & Sutton, M. A. (2008). Transformation of the nitrogen cycle: Recent trends, questions, and potential solutions. *Science, 320*(5878), 889–892.

Ge, P., Chen, M., Cui, Y., & Nie, D. (2021). The research progress of the influence of agricultural activities on atmospheric environment in recent ten years: A review. *Atmosphere, 12*(5), 635–641.

Gillis, L. G., Bouma, T. J., Cathalot, C., Ziegler, A. D., & Herman, P. M. (2015). Particulate matter in mangrove forests and seagrass beds as a nitrogen source in tropical coastal ecosystems. *Biotropica, 47*(3), 286–291.

Glindemann, D., Edwards, M., Liu, J. A., & Kuschk, P. (2005). Phosphine in soils, sludges, biogases and atmospheric implications—a review. *Ecological Engineering, 24*(5), 457–463.

Graham, W. F., & Duce, R. A. (1979). Atmospheric pathways of the phosphorus cycle. *Geochimica Et Cosmochimica Acta, 43*(8), 1195–1208.

Grantz, D. A., Garner, J. H. B., & Johnson, D. W. (2003). Ecological effects of particulate matter. *Environment International, 29*(2–3), 213–239.

Griffin, R. J. (2013). The sources and impacts of tropospheric particulate matter. *Nature Education Knowledge, 4*(5), 1–5.

Guerinot, M. L. (1994). Microbial iron transport. *Annual Review of Microbiology, 48*(1), 743–772.

Håkanson, L. (2006). The relationship between salinity, suspended particulate matter and water clarity in aquatic systems. *Ecological Research, 21*, 75–90.

Hamza, W. (2021). Dust storms and its benefits to the marine life of the Arabian Gulf. *The Arabian Seas: Biodiversity, Environmental Challenges and Conservation Measures*, 141–160.

Han, H., Bosch, N., & Allan, J. D. (2011). Spatial and temporal variation in phosphorus budgets for 24 watersheds in the Lake Erie and Lake Michigan basins. *Biogeochemistry, 102*, 45–58.

He, Z. L., Yang, X. E., & Stoffella, P. J. (2005). Trace elements in agroecosystems and impacts on the environment. *Journal of Trace Elements in Medicine and Biology, 19*(2–3), 125–140.

He, L., Wang, S., Liu, M., Chen, Z., Xu, J., & Dong, Y. (2023). Transport and transformation of atmospheric metals in ecosystems: A review. *Journal of Hazardous Materials Advances, 9*, Article 100218.

Hou, D., O'Connor, D., Igalavithana, A. D., Alessi, D. S., Luo, J., Tsang, D. C., ... & Ok, Y. S. (2020). Metal contamination and bioremediation of agricultural soils for food safety and sustainability. *Nature Reviews Earth & Environment, 1*(7), 366–381.

Inam, A. (2007). Use of flyash on turnip (*Brassica rapa* L.) cultivation. *Pollution Research, 26*, 39–42.

Ittekkot, V., & Zhang, S. (1989). Pattern of particulate nitrogen transport in world rivers. *Global Biogeochemical Cycles, 3*(4), 383–391.

Ito, A., & Shi, Z. (2016). Delivery of anthropogenic bioavailable iron from mineral dust and combustion aerosols to the ocean. *Atmospheric Chemistry and Physics, 16*(1), 85–99. https://doi.org/10.5194/acp-16-85-2016

Jickells, T. D., An, Z. S., Andersen, K. K., Baker, A. R., Bergametti, G., Brooks, N., Cao, J. J., Boyd, P. W., Duce, R. A., Hunter, K. A., & Kawahata, H. (2005). Global iron connections between desert dust, ocean biogeochemistry, and climate. *Science, 308*(5718), 67–71.

Juračić, M., Vitturi, L. M., Rabitti, S., & Rampazzo, G. (1987). The role of suspended matter in the biogeochemical cycles in the Adige River estuary (Northern Adriatic Sea). *Estuarine, Coastal and Shelf Science, 24*(3), 349–362.

Kabelitz, T., Ammon, C., Funk, R., Münch, S., Biniasch, O., Nübel, U., Thiel, N., Rösler, U., Siller, P., Amon, B., & Aarnink, A. J. (2020). Functional relationship of particulate matter (PM) emissions, animal species, and moisture content during manure application. *Environment International, 143*, 105577–105579.

Kanakidou, M., Myriokefalitakis, S., Daskalakis, N., Fanourgakis, G., Nenes, A., Baker, A. R., Tsigaridis, K., & Mihalopoulos, N. (2016). Past, present, and future atmospheric nitrogen deposition. *Journal of the Atmospheric Sciences, 73*(5), 2039–2047.

Kanakidou, M., Myriokefalitakis, S., & Tsigaridis, K. (2018). Aerosols in atmospheric chemistry and biogeochemical cycles of nutrients. *Environmental Research Letters, 13*(6), Article 063004.

Kanakidou, M., Duce, R. A., Prospero, J. M., Baker, A. R., Benitez-Nelson, C., Dentener, F. J., Hunter, K. A., Liss, P. S., Mahowald, N., Okin, G. S., & Sarin, M. (2012). Atmospheric fluxes of organic N and P to the global ocean. *Global Biogeochemical Cycles, 26*(3).

Kappos, A. D., Bruckmann, P., Eikmann, T., Englert, N., Heinrich, U., Höppe, P., Koch, E., Krause, G. H., Kreyling, W. G., Rauchfuss, K., & Rombout, P. (2004). Health effects of particles in ambient air. *International Journal of Hygiene and Environmental Health, 207*(4), 399–407.

Kodas, T. (1990). Aerosol sampling, science and practice. In J. H. Vincent (Ed.) (p. xix, 390, bound, £ 80). Wiley. ISBN 0-471-92175-0.

Krause, J. W., Nelson, D. M., & Lomas, M. W. (2010). Production, dissolution, accumulation, and potential export of biogenic silica in a Sargasso Sea mode-water eddy. *Limnology and Oceanography, 55*(2), 569–579.

Krippelova, T. (1982). The influence of emissions from a magnesium factory on ruderal communities. In R. Bornkamm, J. A. Lee, & M. R. D. Seaward (Eds.), *Urban ecology* (pp. 334–335). Blackwell Scientific Publications.

Kulmala, M., & Petäjä, T. (2011). Soil nitrites influence atmospheric chemistry. *Science, 333*(6049), 1586–1587.

Kulmala, M., Suni, T., Lehtinen, K. E. J., Dal Maso, M., Boy, M., Reissell, A., Rannik, Ü., Aalto, P., Keronen, P., Hakola, H., & Bäck, J. (2004). A new feedback mechanism linking forests, aerosols, and climate. *Atmospheric Chemistry and Physics, 4*(2), 557–562.

Lamarque, J. F., Dentener, F., McConnell, J., Ro, C. U., Shaw, M., Vet, R., Bergmann, D., Cameron-Smith, P., Dalsoren, S., Doherty, R., & Faluvegi, G. (2013). Multi-model mean nitrogen and sulfur deposition from the Atmospheric Chemistry and Climate Model Intercomparison Project

(ACCMIP): Evaluation of historical and projected future changes. *Atmospheric Chemistry and Physics, 13*(16), 7997–8018.

Lee, S. Y. (1990). Primary productivity and particulate organic matter flow in an estuarine mangrove-wetland in Hong Kong. *Marine Biology, 106*, 453–463.

Li, M. H., Fan, L. C., Mao, B., Yang, J. W., Choi, A. M., Cao, W. J., & Xu, J. F. (2016). Short-term exposure to ambient fine particulate matter increases hospitalizations and mortality in COPD: A systematic review and meta-analysis. *Chest, 149*(2), 447–458.

Liang, C., Schimel, J. P., & Jastrow, J. D. (2017). The importance of anabolism in microbial control over soil carbon storage. *Nature Microbiology, 2*(8), 1–6.

Likens, G. E., & Bormann, F. H. (1975). Nutrient-Hydrologic Interactions (Eastern United States). In A. D. Hasler (Eds.), *Coupling of Land and water systems. Ecological studies* (Vol. 10). Springer.

Luo, X., Bing, H., Luo, Z., Wang, Y., & Jin, L. (2019). Impacts of atmospheric particulate matter pollution on environmental biogeochemistry of trace metals in soil-plant system: A review. *Environmental Pollution, 255*, 113138. https://doi.org/10.1016/j.envpol.2019.113138

Mackey, K. R. M., & Paytan, A. (2009). Phosphorus cycle. *Encyclopedia of Microbiology, 3*, 322–334.

Maffia, J., Dinuccio, E., Amon, B., & Balsari, P. (2020). PM emissions from open field crop management: Emission factors, assessment methods and mitigation measures–a review. *Atmospheric Environment, 226*, Article 117381.

Mahowald, N., Jickells, T. D., Baker, A. R., Artaxo, P., Benitez-Nelson, C. R., Bergametti, G., Bond, T. C., Chen, Y., Cohen, D. D., Herut, B., & Kubilay, N. (2008). Global distribution of atmospheric phosphorus sources, concentrations and deposition rates, and anthropogenic impacts. *Global Biogeochemical Cycles, 22*(4), 1–19.

Maji, K. J., Arora, M., & Dikshit, A. K. (2017). Burden of disease attributed to ambient PM 2.5 and PM 10 exposure in 190 cities in China. *Environmental Science and Pollution Research, 24*, 11559–11572.

Mandal, M., Popek, R., Przybysz, A., Roy, A., Das, S., & Sarkar, A. (2023b). Breathing fresh air in the city: Implementing avenue trees as a sustainable solution to reduce particulate pollution in urban agglomerations. *Plants, 12*(7), 1545. https://doi.org/10.3390/plants12071545

Mandal, M., Das, S., Roy, A., Rakwal, R., Jones, O. A., Popek, R., et al. (2023b). Interactive relations between plants, the phyllosphere microbial community, and particulate matter pollution. *Science of The Total Environment, 890*, 164352. https://doi.org/10.1016/j.scitotenv.2023.164352

Manisalidis, I., Stavropoulou, E., Stavropoulos, & Bezirtzoglou, E. (2020). Environmental and health impacts of air pollution: a review. *Frontiers in Public Health,* 505570–505583.

Martens, D. C. (1971). Availability of plant nutrients in flyash. *Compost Science, 12*, 15–19.

Martin, J. H., Gordon, M., & Fitzwater, S. E. (1991). The case for iron. *Limnology and Oceanography, 36*(8), 1793–1802.

Mohapatra, K., & Biswal, S. K. (2014). Effect of particulate matter (PM) on plants, climate, ecosystem and human health. *The International Journal of Advanced Technology in Engineering and Science, 2*(4), 2348–7550.

Morrissey, J., & Guerinot, M. L. (2009). Iron uptake and transport in plants: the good, the bad, and the ionome. *Chemical Reviews, 109*, 10(2009), 4553–4567.

Moustris, K. P., Proias, G. T., Larissi, I. K., Nastos, P. T., Koukouletsos, K. V., & Paliatsos, A. G. (2016). Health impacts due to particulate air pollution in Volos City, Greece. *Journal of Environmental Science and Health, Part A, 51*(1), 15–20.

NASA. (2020). https://science.nasa.gov/earth/earth-atmosphere/aerosols/getting-to-the-heart-of-the-particulate-matter/

Page, Al., Elseewi, A. A., & Straughan, I. R. (1979). Physical and chemical properties from flyash from coal-fired power-plants with special reference to environmental impacts. *Residue Reviews, 71*, 83–120.

Pope, C. A., Ezzati, M., Cannon, J. B., Allen, R. T., Jerrett, M., & Burnett, R. T. (2018). Mortality risk and PM 2.5 air pollution in the USA: An analysis of a national prospective cohort. *Air Quality, Atmosphere and Health, 11*, 245–252.

Rai, P. K. (2015). Multifaceted health impacts of particulate matter (PM) and its management: An overview. *Environmental Skeptics and Critics, 4*(1), 1–26.

Roy, A., Das, S., Singh, P., Mandal, M., Kumar, M., Rajlaxmi, A., et al. (2024a). Summer-time monitoring and source apportionment study of both coarse, fine, and ultra-fine particulate pollution in Eastern Himalayan Darjeeling: A hint to health risk during peak tourist season. *Mapan, 39*(4), 995–1009. https://doi.org/10.1007/s12647-024-00776-9

Roy, A., Mandal, M., Das, S., Popek, R., Rakwal, R., Agrawal, G. K., et al. (2024b). The cellular consequences of particulate matter pollutants in plants: Safeguarding the harmonious integration of structure and function. *Science of the Total Environment, 914*, Article 169763. https://doi.org/10.1016/j.scitotenv.2023.169763

Sarà, G. (2007). Ecological effects of aquaculture on living and non-living suspended fractions of the water column: A meta-analysis. *Water Research, 41*(15), 3187–3200.

Singh, A., & Agrawal, M. (2007). Acid rain and its ecological consequences. *Journal of Environmental Biology, 29*(1), 15–24.

Smil, V. (2000). Phosphorus in the environment: Natural flows and human interferences. *Annual Review of Energy and the Environment, 25*(1), 53–88.

Tipping, E., Benham, S., Boyle, J. F., Crow, P., Davies, J., Fischer, U., Guyatt, H., Helliwell, R., Jackson-Blake, L., Lawlor, A. J., & Monteith, D. T. (2014). Atmospheric deposition of phosphorus to land and freshwater. *Environmental Science: Processes and Impacts, 16*(7), 1608–1617.

Theis, T., & Wirth, J. L. (1977). Sorptive behaviour of trace metals on flyash in aqueous systems. *Environmental Science and Technology, 11*, 1096–1100.

Tsurumi, T., & Managi, S. (2020). Health-related and non-health-related effects of PM2. 5 on life satisfaction: Evidence from India, China and Japan. *Economic Analysis and Policy, 67*, 114–123.

Turner, A., & Millward, G. E. (2002). Suspended particles: Their role in estuarine biogeochemical cycles. *Estuarine, Coastal and Shelf Science, 55*(6), 857–883.

Twining, B. S., & Baines, S. B. (2013). The trace metal composition of marine phytoplankton. *Annual Review of Marine Science, 5*, 191–215.

Twining, B. S., Baines, S. B., & Fisher, N. S. (2004). Element stoichiometries of individual plankton cells collected during the Southern Ocean Iron Experiment (SOFeX). *Limnology and Oceanography, 49*(6), 2115–2128.

Ulrichs, C., Welke, B., Mucha-Pelzer, T., Goswami, A., & Mewis, I. (2008). Effect of solid particulate matter deposits on vegetation: A review. *Functional Plant Science and Biotechnology, 2*(1), 56–62.

Walch, H., von der Kammer, F., & Hofmann, T. (2022). Freshwater suspended particulate matter—Key components and processes in floc formation and dynamics. *Water Research, 220*, 118655–118684.

Wang, R., Balkanski, Y., Boucher, O., Ciais, P., Peñuelas, J., & Tao, S. (2015). Significant contribution of combustion-related emissions to the atmospheric phosphorus budget. *Nature Geoscience, 8*(1), 48–54. https://doi.org/10.1038/ngeo2324

Wangersky, P. J. (1977). The role of particulate matter in the productivity of surface waters. *Helgolander Wiss Meeresunters, 30*, 546–564.

Wong, J. W. C., & Su, D. C. (1997). Reutilization of coal fly-ash and sewage sludge as an artificial soil-mix: Effects of preincubation on soil physico-chemical pro- perties. *Bioresource Technology, 59*, 97–102.

Wu, W., & Zhang, Y. (2018). Effects of particulate matter (PM2. 5) and associated acidity on ecosystem functioning: response of leaf litter breakdown. *Environmental Science and Pollution Research, 25*(30), 30720–30727.

Yamanaka, Y., & Tajika, E. (1996). The role of the vertical fluxes of particulate organic matter and calcite in the oceanic carbon cycle: Studies using an ocean biogeochemical general circulation model. *Global Biogeochemical Cycles, 10*(2), 361–382.

Chapter 6
Linking Atmospheric Particulate Deposition to Alterations in Plant Community Structure and Biodiversity

Nivedita Chaudhary, Hemant Kumar Virt, Jaya Goyal, Kumari Rinku, and Vimla Singh

Abstract Air pollution caused by particulate matter (PM) has significantly increased in the past century. Increasing concentrations of PM affect plant's biodiversity and interactions of biotic and abiotic components in the environment, and raise threats to various ecosystems. PM pollution is a significant environmental concerning ecosystems and humans, particularly in countries like India, where industrialization and urbanization are escalating. PMs interact with various plant systems during potential dry or wet deposition, including urban, tree, and agricultural soil. Besides root uptake, leaf deposition and foliar absorption processes are essential in causing adversarial effects on plants. Many phenological characteristics and seasonal variations influence the accumulation and deposition of PMs on the surface of the plants. The uptake of PMs has altered plants' capacity to neutralize reactive oxygen species by causing lipid breakdown, biomolecule oxidation, reduction of enzyme activity, and genetic alterations. Therefore, the plant's physiological, biochemical, and morphological characteristics are affected. The present chapters show the transport of PMs and their effects on vegetation after the deposition, mechanism of action influences biomass production, ultimately impacting plant biodiversity.

Keywords Particulate matter · Deposition · Physiology · Productivity · Biodiversity

N. Chaudhary (✉) · H. K. Virt · J. Goyal · K. Rinku
Department of Environmental Science, School of Earth Sciences, Central University of Rajasthan, Ajmer, Rajasthan, India
e-mail: nivedita@curaj.ac.in

V. Singh
Miranda House, University of Delhi, New Delhi, India

A. Sarkar et al. (eds.), *From Sky to Soil*, https://doi.org/10.1007/978-3-032-31991-3_6

6.1 Introduction

Over the past century, the escalation of particulate matter (PM) air pollution has appeared as a substantial environmental concern with global implications for ecosystems. In terms of dust pollution, PM has a significant impact on the biodiversity, and both the abiotic and biotic constituents of the ecosystem are affected by PM pollution. Accelerated urbanization and industrialization are the major causes for PM pollution dispersion and effect (Shukla et al., 2020). Besides other factors, for instance, dust from quarries, vehicular emissions, and agricultural activities release PM into the environment in high concentrations and cause severe hazards to humans and the environment (Rai, 2016; Rai & Panda, 2014).

Significant attributes of PM comprise differences in their size, shape, and composition, which influence their interaction with other gaseous pollutants and meteorological parameters such as sunlight and water vapor. Further, these interactions impact the climate and the environment (Zeb et al., 2018). Atmospheric suspended PM significantly alters the Earth's heat budget (Tiwari et al., 2015). In terms of composition, PM is a heterogeneous mixture of solid and liquid particles made up of a broad range of minerals, soil particles, biological components, trace metals, organic and inorganic salts, ammonia, nitrate, and sulfate (Satsangi & Yadav, 2014). Yang et al. (2018) reported spatio-temporal variation in PM 2.5 and PM10 concentrations in various nations and areas. China and India had far higher PM levels than the United States. However, season-wise, winter has the highest values, while summer has the lowest. On the other hand, the United States PM pollutant distribution revealed comparatively high summertime concentrations and comparatively low wintertime levels. McDuffie et al. (2021) suggested that the source of PM comprising direct emissions from forest fires, agricultural burning and inefficient burning of fuels further leads to secondary emissions from the chemical reaction in the atmosphere, which depends on the local environmental conditions, leading to crossing political boundaries. Effective mitigation requires addressing transboundary impacts at local, national, and global levels. The background climate variables, such as temperature, wind direction, air stability, mixing height, and wind speed, greatly influence the PM concentrations. Plants can uptake atmospheric PM through the stomatal opening on the leaves, which differs depending on the leaf size, the tree's canopy, plant height, leaf density, trichomes, and the number of stomata, which leads to the significant surface area for the deposition, absorption, and accumulation of PM (Beckett et al., 2000; Tallis et al., 2011). Despite the effects of PM on vegetation, it helps to reduce the level of PM in the ambient air by altering humidity levels, volatile organic compounds, wind velocity, and temperature, and also acting as a barrier in hindering the entry of pollutants into a particular region by effectively trapping PM in foliage (Jeanjean et al., 2016). A variety of species of trees have different capabilities to capture the PM due to variation in their leaf characteristics, for instance, leaf surface, waxy cuticles, trichomes, adhesiveness, and roughness (Hofman et al., 2014; Wang et al., 2016). PM dynamic constituents comprising solid and liquid particles cause negative impacts on plant morphology, physiology, and biochemistry through their dry and wet deposition (Mandal et al., 2023a, 2023b) on plant surfaces, directly by either foliar absorption or through root uptake (Rai, 2016). Therefore, the impact of PM can be understood by assessing the possible sources, its deposition, and the ecological effects. The present chapter provides an insight into the sources, their deposition, as well as effects on the vegetation.

6.2 Sources and Composition of Particulate Matter

The composition of particulate matter (PM) can vary from diverse suspended particles (solid and liquid) in the ambient air, which may or may not be visible but can cause a significant effect on biodiversity (Mandal et al., 2023a, 2023b). These particles are dynamic and complex and also find various ways to interact with other species in the atmosphere, where they usually have a solid core surrounded by liquid particles and have differing sizes, shapes, and chemical constituents; therefore, PM is constituted of both organic and inorganic substances, such as metal oxides, dust, pollen, and soot, etc. PM extensively influences air pollution globally as it is present broadly in the environment. PM can be generally classified based on the size of the particle as coarse and fine particles. Coarse PM has an aerodynamic diameter of $>10\ \mu m$, which can settle faster. Due to their larger size and higher settling velocity compared to fine particles, coarse particles do not remain suspended in the air for a long time (Caffrey et al., 1998). Fine PM have aerodynamic diameters of $<2.5\ \mu m$. Fine particles have lower mass and slower settling velocities. As a result, they take longer time to settle and remain suspended in the air for longer durations (Patil et al., 2013; Roy et al., 2024a).

6.2.1 Sources of Particulate Matter on the Basis of Origin

Sources of PM are mainly categorized into two classes: natural and anthropogenic activities, including homes, energy production, transportation, garbage, industry, sandstorms, volcanic eruptions, and agriculture. Due to increasing human interference, emissions from anthropogenic activities exceed those from natural activities (Table 6.1). Sources as natural activities comprise resuspension of soil (Huertas et al., 2014), forest fire, volcanic eruption, and lightning are the common causes of wildfire, anthropogenic activities too contribute to some extent (Busico et al., 2019), soil dust erosion (De Sales et al., 2016), volcanism (Silva et al., 2020). Sea sprawls (Von Glasow, 2013). Certain biogenic activities from non-fossil carbon sources include fungal spores, plant detritus, pollen, and biogenic secondary organic aerosols (Glasius et al., 2011). Macroalgae release iodine-containing gases in the intertidal region (Keng et al., 2020). Anthropogenic activities comprise emissions from *i*ndustrial processes (Yatkin & Bayram, 2008) and PM composition differs in terms of water-soluble ions, mineral elements, and carbonaceous species in different industries (Jia et al., 2018; Mandal et al., 2023a, 2023b). Traffic emissions (Rissler et al., 2012). Non-exhaust emissions (Yatkin & Bayram, 2008). PM from non-exhaust sources produced by road surface abrasion, tyre and brake wear, corrosion or deterioration of other vehicle parts like the clutch, and the movement of dust off the road consists of dust particles (Pant & Harrison et al., 2013). Fossil combustion (Vohra et al., 2021). Wood burning (Sippula et al., 2009).

6.2.2 Composition

PM can contain a variety of elements and ions, including SO_4^{2-}, NO_3–, and NH_4^+, as well as elements like Cd, Bi, As, Sb, Cr, Mo, Pb, Ni, Zn, Co, and Cu (Roy et al., 2023). Elements, for instance, Al, Ca, Fe, and K are predominantly found in soil particles and coal, mainly in lignite. Coal burning also releases trace elements in significant quantities, for instance, Cr, Cu, Ni, Pb, and V. Industrial process emissions are mainly composed of heavy or tracer elements, Zn, Fe, Pb, Cd, Cr, Cu, Ni, etc. Burning of biomass releases substantial concentrations of K and Na in the atmosphere (Yatkin & Bayram, 2008). Oxides, in addition to sulfate and nitrate groups, also emit Cr, Al, Ni, Sr, Cd, Fe and Zn in free state as well as in the combined state (Chernyshev et al., 2019). Crustal elements are predominantly present as oxides and form crustal aerosol. Sea salt aerosol mainly comprises sodium and chloride, but sulfate, magnesium, calcium, and potassium are also found in significant amounts—elemental carbon from combustion forms primary anthropogenic aerosol (Perrino et al., 2009). The elevated levels of Al, Ca, Mg, Fe, Mn, Sr, Ti, and Zn are mostly generated by dust generated from soil and roads (Watson et al., 2001). Incomplete combustion is the primary source of emitted elemental carbon. Organic carbon originates mainly from combustion and secondary organic carbons, formed through chemical reactions of organic gases. These reactions occur alongside combustion processes. SOC significantly impacts the organic C and elemental C concentration ratios (Jia et al., 2018).

Alkali metal sulfates are the primary constituents of the tiny particles released after wood burning. Distinguished release factors for Pb, Zn, Cd, Cu, and As (Sippula et al., 2009). Particles originated by combustion are of numerous sizes and chemicals. The fuel's ash-forming components that evaporate during combustion make up the fine fly ash portion (Jokiniemi et al., 1994). The incomplete combustion conditions led to the formation of soot and organic substances. Organic vapors that escape from the heating system and subsequently condense to the particle phase comprise fine particles' organic material contents (Bockhorn, 2013). Depending on the fuel characteristics, such as lubrication oil, diesel and oil boilers, the fine C- particles with an elemental carbon core and an organic material layer on top are released in the environment (Sippula et al., 2009). The low-grade fuels considered as dense or residual fuel oils, are used in marine diesel engines, machinery in power plants, and heating systems. They include high levels of sulfur and various ash-forming elements, significantly increasing particle emissions (Kittelson, 1998).

Table 6.1 Natural and anthropogenic sources of particulate matter

Source category	Source	Description	References
Natural sources	Volcanic eruptions	Volcanic ash is released causing impacts on climate and quality of air. Volcanic ash can spread far and wide and affect areas far from the eruption site	Ke et al. (2024)
	Forest fires	The smoke and ash emitted contain a wide range of particulate matter, affecting visibility and air quality. Humans or natural causes can start wildfires and significantly impact many locations downwind from the fire	Warneke et al. (2023)
	Sea spray	Wind and waves hold salt particles and organic matter from the ocean surface into the atmosphere. These particles can affect weather patterns and cloud formation	Jacob et al. (2016)
	Soil and dust storms	Wind erosion affects particle matter levels, which raises dust and fine soil particles into the atmosphere. Dust storms can travel large distances, affecting areas far from the source	Darvishi Boloorani et al. (2023)
	Biological sources	Spores, pollen, and naturally occurring microorganisms (such as fungi and bacteria) enter the atmosphere. These particles can affect ecosystems and human health, especially during certain seasons	Mack et al. (2019)
	Erosion and resuspension	Natural weathering processes, human activities, and wind-induced particle resuspension	Wu et al. (2023)
Anthropogenic sources	Industrial emissions	Factories and power plants emit PM while burning fossil fuels and other industrial activities. These emissions may contain various metals, chemical compounds, and other contaminants	Manisalidis et al. (2020)
	Vehicle emissions	Fine particles are emitted by exhaust fumes from cars, trucks, and other vehicles, especially diesel-powered vehicles. These particles causes adverse both on human health and make urban air polluted	Pan et al. (2021)
	Construction activities	Construction, demolition, and other construction-related activities generate dust and particulate matter. These particles can affect both visibility and local air quality	Berlinger et al. (2024)

(continued)

Table 6.1 (continued)

Source category	Source	Description	References
	Agricultural activities	PM is released during tillage, crop residue burning, fertilizers, and pesticide use. Ammonia emissions from agricultural practices can also promote the formation of secondary PM	Turner (2021)
	Residential heating	PM is created when fuels such as coal, wood, or other materials are burned to heat homes. In winters PM increases significantly both in urban and rural areas	Price-Allison (2022)
	Waste burning	PM produced by burning of waste. Depending on the composition of the waste, emissions can also contain various harmful components	Patil et al. (2013)
	Mining and quarrying	PM and dust are released during the extraction and processing of minerals. These operations pose an increased risk to the occupational health of workers and have a significant impact on the quality of local air	Tiwary and Williams (2018)
	Shipping and marine activities	Emissions from ships and other watercraft, such as, cargo handling and fuel combustion particles. These sources are significant in port and coastal communities	Boviatsis (2023)
	Road dust	PM resulting from road surface abrasion, tyre wear, and brake wear. This source contributes significantly to urban particulate pollution, especially in areas with heavy traffic	Zhu et al. (2024)
	Chemical reactions	Chemical processes in the atmosphere produce secondary particles. Sulfur dioxide (SO_2) and nitrogen oxides (NOx) are two examples of gaseous pollutants that can combine to create delicate particulate matter that can impact regional air quality	Dizengremel (2021)

6.3 Factors Effecting Transportation and Deposition of Particulate Matter

Various factors influence the transportation and deposition of PM in the atmosphere. Meteorological elements such as wind velocity and atmospheric stability perform an essential role in influencing the dispersion and movement of PM (Huertas et al., 2021). Also, topographical features like mountains and valleys can alter airflow patterns, leading to variations in PM concentration and deposition (Watts et al., 2020).

6.3.1 Environmental and Meteorological Factors

The dispersion of PM influences its concentration in the troposphere, where a major role is played by air temperature, wind velocity, atmospheric humidity, rainfall, and global radiation, causing its dispersion and deposition. Meteorological conditions diverge seasonally and affect the occurrence of dust particles in the atmosphere. The transition of PM from one region to another is guided by wind speed and wind direction. Flat topography with the constant flowing wind will not allow air pollution to build up. Warm air is conducive to dispersion rather than cold air (Owoade et al., 2012). The air quality deteriorates in spring mainly due to the atmospheric pressure, which also affects the atmosphere in winter and relative humidity. In autumn, relative humidity and atmospheric pressure play a crucial role. The atmospheric pollution is low; where atmospheric pressure is low, they are correlated positively up to a specific limit. In the spring season, relative humidity substantially influences the environment. Higher relative humidity leads to higher pollution, but in summer, autumn, and winter, it adds up to the air quality; otherwise, it shows an inverse relation. Wind speed also impacts the presence of particulate matter; it disperses the pollution, but the probability of pollution from other regions with the flowing wind would increase. Near the earth's surface, the accumulation of PM is low because high temperatures lead to turbulence and an unstable atmosphere. The atmospheric temperature decreases with altitude. Wind direction also impacts the presence of pollution in the air (Tian et al., 2014).

The PM concentration of atmospheric PM decreases as temperature increases. The high temperature helps form convection current, thus enhancing the dispersion. In low-temperature regions, an inversion layer is formed near the Earth's surface, which reduces the effect of convection current and dispersion. Precipitation impacts PM's ability to adsorb, collide, and settle down as wet deposition. The concentration of PM decreases substantially after precipitation. When the relative humidity is high, the water droplets in the atmosphere provide substratum to PM or hygroscopicity growth in a stable atmosphere, causing pollution. The relation of relative humidity with PM dispersion is not apparent; other factors' influence is evident (Li et al., 2015). There are seasonal differences in the PM that is deposited on plant leaves. In winter, plants have a greater capacity to retain dust due to the retention of dew. Because of environmental conditions, they have a lower capacity to do so in summer and the rainy season. Meravi et al. (2021) reported that there were differences in the concentration of PM that could accumulate on different species of plants, some plants could accumulate more dust than others; for instance, the greatest dust buildup takes place on *Butea monosperma, Calotropis procera, Ficus benghalensis,* and *Ricinus communis.* However, there are species with less dust accumulation, such as Eucalyptus sp., *Acacia auriculiformis,* and *Dendrocalamus strictus.* For accumulation, petiole length, trichomes, leaf orientation, leaf lamina shape and size, and leaf texture participated in collecting and retaining dust on a leaf.

Xu et al. (2018) indicated that extreme weather conditions, for instance, heatwaves and high-intensity rainfall, significantly affect PM concentrations by enhancing dispersion or promoting deposition. Due to changing wind patterns and stable environments, urban heat islands in metropolitan areas aggravate PM pollution (Yang et al., 2021). Huertas et al. (2021) further suggested that the activities of PM also depend on solar radiation, where low solar radiation, during cloudy conditions, causes accumulation of PM closer to the surface of the Earth. In contrast, incoming high solar radiation causes vertical mixing and dispersion of pollutants. Singh et al. (2023) reported that the interaction of PM atmospheric boundary layers affects local climatic conditions. It was found that stable boundary layer conditions, which often occur at night or in winter, can potentially retain pollutants near the surface, leading to increased concentrations. Particulate matter can also initiate slight atmospheric cooling when in higher concentration and a more extended duration by reducing the solar radiation intensity reaching the Earth (Liu et al., 2018). Vegetation acts as a biological filter by trapping them on their surfaces. According to the plant species, leaf size, and environmental factors all influence vegetation's effectiveness in reducing PM concentrations locally (Sgrigna et al., 2020).

6.3.2 Spatio-Temporal Variations

In recent years, landscapes have been transformed, causing alterations in the heat budget and atmospheric composition due to urbanization and industrialization, which is the primary cause of PM loads in the atmosphere at local, regional, and global levels (Gautam & Brema, 2020). However, the concentration varies with space, and time, for instance, in mountain regions, the cool air drainage during nighttime increases pollutant accumulation due to a stable atmosphere and low wind speed (Othman et al., 2022). In urban regions, less vegetation coverage due to landscape fragmentation is responsible for the source of PM and its high concentration (Wang et al., 2016). Seasonal and regional disparities influence PM concentration in the atmosphere. Usually, its peak during winter is evaluated to other seasons and lower in rural than urban areas (Liu et al., 2013; Popek et al., 2013; Wang et al., 2019).

Liu et al. (2003) suggested the important phenomenon of pollution transport during the winter season in East Asia by geostrophic winds. Pollutants can also build up by the extensive meteorological characteristics of the atmosphere, such as weakened horizontal transport and relatively less intense northerly winds at the surface due to negative anomalies in sea level pressure over the high-pressure center. Winds stimulated by wide-ranging meteorological patterns weaken at various altitudes, hindering the typical pollution transport. A north–south dipole pattern characterizes the spatial distribution of PM, with higher concentrations observed in the northern regions. Specifically, during the formation of a typical mid-latitude cyclone, regions experiencing cold air advection exhibit anomalous sinking motion, producing stable boundary layer conditions in the northern areas (Wang et al., 2019). Low wind speeds and boundary heights indicate stagnant weather conditions favorable for air pollution (Liu et al., 2013); a weakened high-pressure center, coupled with large-scale atmospheric movements, results in downward airflow locally and prevents the planetary boundary layer development; these factors play a significant function in the spatio-temporal dynamics of PM in the atmosphere (Wang et al., 2015).

PM distribution under geographic heterogeneity and seasonal variations was also assessed by using aerosol optical depth (satellite-based) to estimate PM2.5 concentrations (Xu et al., 2021). Evaluate the regional patterns of PM spatial distribution is majorly influenced by anthropogenic activities, for instance, due to the increasing emission sources from highly industrialized and populated regions (Wang et al., 2015). Chung (2014) suggested that localized factors contributed to high PM concentrations due to the complex urban environment, such as infrastructures, which affect airflow and dispersion of pollutants (Liu et al., 2022). Over a period of time, due to proper implementation of air quality regulations—such as limiting emissions from industries and using clean energy sources—significantly reduced PM concentrations in certain countries (Henneman et al., 2017). However, spatio-temporal variations affect the concentrations of PM, such as rising temperatures, changing rainfall patterns and precipitation rates, and extreme weather events involved in the dispersion and deposition of PM; for example, higher temperatures can enable the production of secondary aerosols in air and precipitation can affect wet deposition rates, respectively (Li et al., 2023).

6.4 Absorption and Response of PM by Plant

Plants can absorb and adsorb PM through the leaves and roots. Absorption involves the PM uptake into the plant tissues and adsorption due to the adherence of PM onto external surfaces (Ge et al., 2022). This process of both dry and wet deposition of PM on vegetation can affect the plant diversity (Fig. 6.1).

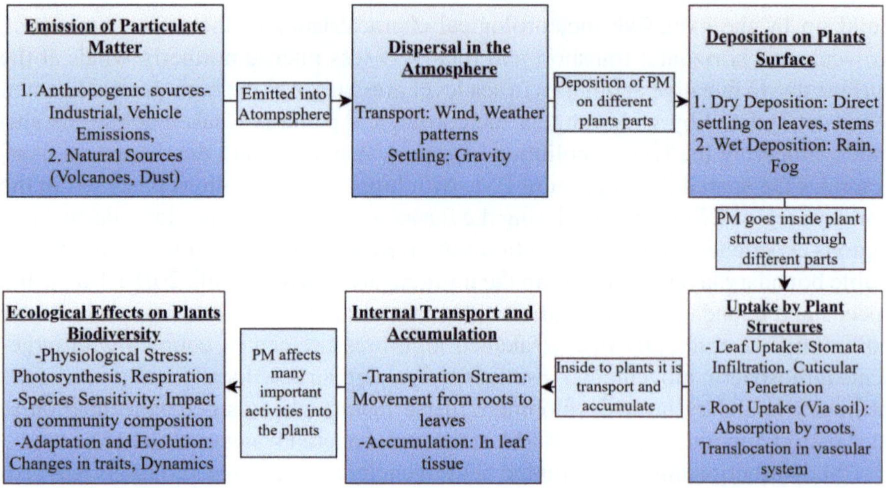

Fig. 6.1 Particulate matter (PM2.5, PM10) transport from source to plants

6.4.1 Absorption on Plants Leaves, Branches and Trunk Surfaces

Leaves surface, branches, and tree trunks provide a surface in removing PM from the atmosphere and improving air quality. Extensive contact surface area and rough surfaces contribute to both absorption and adsorption. Tree density and canopies increase humidity and help reduce wind speed, which contributes to settling on leaves, branches, and tree trunks (Beckett et al., 2000). Finely-haired leaves capture particles effectively, and the enhanced surface area promotes particle deposition. Significant deposition of leaves is promoted by the substantial surface area for collecting particles due to the folded epidermal cells and wax granules. Along with the particle size, chemical composition and electric charge of PM also substantially influence the deposition on vegetation (Heshmatol Vaezin et al., 2021).

Wang et al., (2016) suggested that the number and stomatal density increase the probability of PM deposition, though the cuticular arches and waxy rings surrounding the stomata guard them from particles but also participate in the retention of particles. A humid environment encourages PM to contact the leaves and the water droplets' contact angle, providing a more hydrophilic surface. Therefore, trees have better absorbing capabilities and hold airborne PM; however, an opposite phenomenon is observed in the waxy coating on the leaf surfaces with hydrophobic qualities, which minimizes the area where particles can adhere (Klamerus-Iwan et al., 2018). Plants include less waxy structures, abundant stomata, and uneven surfaces having better PM-capturing capacity (Wang et al., 2016). Their surface roughness significantly influences the capability of broadleaf leaves to capture more PM and microclimatic conditions. Both conifer and broadleaf species show sensitivity to structural appearances such as leaf stomata, epicuticular wax deposits, and cuticles (Zhao et al., 2013).

The density of ridges, grooves, and micro-configurations of epidermal cells usually determines the leaf surface roughness. Broadleaf species generally exhibit greater surface roughness compared to conifer species. However, despite their smoother surfaces, conifer leaves exhibit higher dust-retention ability than broadleaf leaves (Finer et al., 2008). In conifer species, the ability to capture particles is primarily influenced by epidermal wax deposits, stomata, and cuticles. On the other hand, in broadleaf species, the trapping ability is directly related to surface roughness and other morphological features. These features suggestively contribute to PM-capture capacity (Zhang et al., 2018).

6.4.2 Effects of Particulate Matter on Plants

Particulate matter pollution has caused various detrimental effects on plants, animals, and humans as listed in Table 6.2. There are several direct and indirect effects of PM, especially in terms of agricultural crop productivity, due to the change in the plant's physiological and biochemical characteristics responses (Fig. 6.2).

6.4.2.1 Morphological Attributes

Plants leaf area, number of stomata structure, flowering, growth, reproduction. Leaves are the leading atmospheric indicators for both gaseous and particle pollution. PM pollution has been shown to impact several plant morphological traits, including modifications to stomatal structure (Mandal et al., 2023a, 2023b). Before entering the leaf tissues, these pollutants interact on the leaf surface, changing the structure of the leaf tissues. Particle accumulation on leaf surfaces, containing both coarse and ultrafine in size, has been perceived to inhibit plant growth by disturbing leaf gaseous exchange, rate and time of flowering, reproduction, and the number and area of leaves. Reduced leaf growth and production rates and hastened senescence are the causes of the drop in leaf area and quantity (Rai, 2016). Furthermore, the accumulation of dust hampers the ability of stomata to open and close and the frequency of stomatal closure; stomatal closure mechanisms can protect plants from the damaging effects of particulate matter pollution (Roy et al., 2024a, 2024b).

6.4.2.2 Physiological Characteristics

PM can block the stomatal pores and leads to affecting the plant's photosynthetic system (Meravi et al., 2021). Plants growing in the region of high PM concentrations tend to be exposed to more PM on their leaves and show reduced photosynthetic efficiency, photosynthesis rates, and increased stomatal resistance. Plants exposed to PM also showed structural changes in the leaf shape, size, and surface area. PM deposition, accumulation, and impact are detected more in plant leaves with rough

Table 6.2 Effects of PM10, PM2.5 and PM1 on plants, animals and humans

Type of PM	Description	Source	Lifetime in atmosphere	Effect on plants	Effect on animals	Effect on humans	Reference
PM10 (Coarse Particles)	Particle size with diameter ≤10 μm	Soil dust suspension from operations such as farming, mining, industry generated dust, building, coal, oil combustion, and ocean spraying	The action ranges from a few minutes to a few hours	PM10 particles, closer to their origin, affect plants in the immediate vicinity, which is correlated with specific leaf areas and inversely related to chlorophyll content, carotenoids, and water content	Livestock houses significantly impact air quality due to increased PM10 prevalence, with poultry and pig houses having higher levels, while dairy husbandry emits less dust	An increase in PM10 of 10 g m^{-3} raises the chance of myocardial infarction hospitalization	Atkinson et al. (2010), Burnett et al. (1999), Morris (2001), Zanobetti (2001), Zanobetti and Schwartz (2005), von Klot et al. (2005), Zanobetti and Schwartz (2006), Lai et al. (2014), Dzierżanowski et al. (2021)
				Re-suspended road dust and crustal debris greatly impact the plant ecosystem	It is still the element of air pollution that is most consistently linked to illnesses in both humans and animals	Lung cancer incidences rise by 22% when exposed to 10 μg m^{-3} of PM10 particles	Raaschou-Nielsen et al. (2013), Prajapati and Tripathi (2007), Anderson et al. (2012)
					Its size and composition significantly impact its ability to harm animals' respiratory systems, induce oxidative stress, and cause inflammation	It can intensify allergic inflammation and respiratory conditions, including asthma and COPD	Barlow et al. (2008), Zanobetti et al. (2009)

(continued)

Table 6.2 (continued)

Type of PM	Description	Source	Lifetime in atmosphere	Effect on plants	Effect on animals	Effect on humans	Reference
					The inhalation of PM10 particles results in the development of pulmonary diseases like bronchitis and asthma	Because PM10 contains high levels of PAHs and heavy metals, it poses a serious risk to public health	Prajapati and Tripathi (2008a, 2008b), Spickett et al. (2023)
					Fibrosis, a disorder marked by lung tissue scarring, can also occur in animals due to its oxidative and inflammatory effects on the respiratory system	Ambient PM10 introduction can lead to various health issues, including immune system, lung irritation, asthma, respiratory and cardiovascular diseases, bronchitis, and cancer	Jones and Richeldi (2014)

(continued)

Table 6.2 (continued)

Type of PM	Description	Source	Lifetime in atmosphere	Effect on plants	Effect on animals	Effect on humans	Reference
PM2.5 (Fine particles)	Particle size with diameter <2.5 μm	Burning of coal, oil, and petroleum fuels such as gasoline; including biogenic organics such as terpene); smelters, and steel mills	From a few days to weeks	It inhibits plant respiration, and obstructs stomata in leaves, which ends up decreasing plant photosynthesis	In a high-risk population, and even minimal exposure to high levels of PM2.5 raises the health risk within a few hours	PM2.5 poses significant public health risks due to its easy penetration through alveoli, and global climate change is exacerbated by pollution issues like reduced visibility, forest and crop damage, crop productivity loss, and ecosystem loss	Cheung et al. (2011), Hirano et al. (1995), Finer et al. (2008)
				It affects leaf surface morphology, leading to stomatal blocking and decreased conductance, negatively impacting gas exchange, and water absorption, and altering plants' photosynthetic characteristics	Respiratory diseases, cardiovascular issues, oxidative stress, inflammation, and adverse effects on lung function in animals	It significantly impacts mortality, and increasing heart and lung diseases	Adrees et al. (2016), Pan et al. (2017), Baccini et al. (2011), Raaschou-Nielsen et al. (2012), Pascal et al. (2013), Beelen et al. (2014), Cachon et al. (2014)

(continued)

Table 6.2 (continued)

Type of PM	Description	Source	Lifetime in atmosphere	Effect on plants	Effect on animals	Effect on humans	Reference
				Plants can adapt to changes in their surroundings by shrinking their stomata in response to PM2.5 pollution	The finest fraction deposits in deeper parenchyma of animals, access alveolar ducts and alveolus, causing cell signaling and inflammatory mediators	Causes oxidative stress and pulmonary inflammation, endothelial dysfunction, and an imbalance in the nervous system	Zhang (2015), Shusterman (2011), Deng et al. (2013), Kelly and Fussell (2015)
				Changes in the Earth's radiation balance, nutrient transformation in watersheds, and reduction of soil nutrients It can also impact a plant's water regulation by covering a significant amount of stomata, which lowers the rate of transpiration and evaporative cooling	Prolonged exposure to PM2.5 can weaken mucociliary defenses, potentially leading to respiratory infections in experimental animals	PM2.5 particles cause 3.2 million premature deaths worldwide	Myhre (2009), Velasco et al. (2006), Grantz et al. (2003), Lim et al. (2010), Sharifi et al. (1997), Saldiva et al. (1992)

(continued)

Table 6.2 (continued)

Type of PM	Description	Source	Lifetime in atmosphere	Effect on plants	Effect on animals	Effect on humans	Reference
				PM2.5 pollution hurts plant and human health and the quality of air. Anthropogenic sources of PM include agricultural practices such as livestock production and manure management	Animals exposed to PM2.5 can experience lung damage, increased lung functions, alveolar collapse, lung tissue inflammation, and respiratory health issues	Smog, characterized by high aerosol or PM2.5 particle concentrations, significantly increases the morbidity of asthma and acute respiratory tract infections	Yadav et al. (2003), Zhang et al. (2013), Funk et al. (2008), Riva et al. (2011)
				Its accumulation on the plant surfaces causes stomatal blockage, decreased photosynthetic activities, leaf damage, among other detrimental impacts on plant physiology	The size, severity, concentration and composition of aortic plaques are all impacted in mice with atherosclerotic lesions by sub-chronic exposure to ambient particles	In humans, the frequency of atherosclerosis and carotid intima-media thickness	Künzli et al. (2005), Hoffmann et al. (2007), Garg et al. (2000), Shrivastava and Joshi (2002), Chen and Nadziejko (2005)

(continued)

Table 6.2 (continued)

Type of PM	Description	Source	Lifetime in atmosphere	Effect on plants	Effect on animals	Effect on humans	Reference
PM1 (Ultrafine particles)	Particles with diameter ≤ 0.1 μm	smoke from oil and diesel engines, metal fumes, and sea salt nuclei	From a week to months	Plants under physiological stress caused by ultrafine particle exposure may exhibit altered leaf morphology, decreased chlorophyll content, and altered water-use efficiency	In animal and cellular models, its toxicity is higher. In rat lungs, for instance, low-toxicity and -solubility PM1 cause more inflammation than PM2.5 of the same substance	Ultrafine Particles, cause health risks due to their small size, high toxicity, and easy inhalation, affecting respiratory, cardiovascular, and nervous systems	Srimuruganandam and Nagendra (2012), Abdillah and Wang (2023), Zhang et al. (2024), Schraufnagel, (2020), Cassee et al. (2002), Donaldson et al. (2002), Beckett et al. (2000)
				Plant stress due to decreased photosynthesis, and potential contamination of edible plant parts result in lowering of crop yield	Polymer Fume Fever, initiated by inhaling heated poly-tetra-fluoro-ethylene can cause pulmonary inflammation	Linked to pulmonary inflammation, cardiovascular effects, systemic inflammation, endothelial dysfunction, coagulation changes, and increased risks of diseases like asthma, COPD, cardiovascular diseases, pre-term birth, and cancer	Schraufnagel, (2020), Nordenhall et al. (2000), Rani et al. (2021)

(continued)

Table 6.2 (continued)

Type of PM	Description	Source	Lifetime in atmosphere	Effect on plants	Effect on animals	Effect on humans	Reference
				Modify plant metabolic processes, such as nutrient uptake and secondary metabolite production	Exposure to ultrafine PM 0.1 μm size can reduce the total circulating leukocytes in the blood of animals, specifically causing a decrease in neutrophil numbers	PM0.1 μm size has longer delayed effects and a bigger impact on mortality as compared to PM2.5 having immediate consequences	Ibald-Mulli et al. (2002), Cozzi et al. (2006)
					Ultrafine PM exposure can cause in larger myocardial infarctions after ischemia–reperfusion injury in animal models, indicating a negative impact on heart health	Metal fume fever (MFF) is a disease primarily caused by PM0.1, produced by welding	Chang et al. (2013), Cozzi et al. (2006)

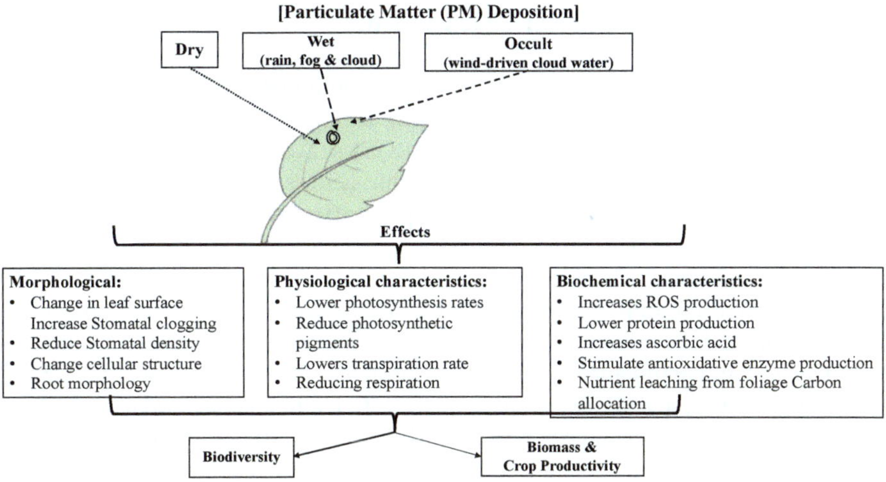

Fig. 6.2 Particulate matter effects on different plant characteristics

surfaces than smooth-surface leaves (Roy et al., 2024a, 2024b). Furthermore, density of trichomes, veins, leaf shape, and other microstructural elements of leaves also significantly influence the PM accumulation on the leaf's surface. The leaf shape influences trapping and accumulating PM with its particular microstructures (Roy et al., 2024a, 2024b). Upon entering PM leaves through their stomata, it affects the plant's physiological activities leading to effects the growth and development, for instance, reduction in assimilate translocation and leaf area reduction. Rai (2016) also reported hormonal disbalance and delayed flowering in plants exposed to PM, additionally composition of PM containing heavy metals and polycyclic aromatic hydrocarbons affect the cellular structure and function, further leads to reduced root development, reduced photosynthetic pigments, and photosynthetic activities in *Arabidopsis thaliana* and rice cultivars (Roy et al., 2024a, 2024b).

Epicuticular wax on leaf surfaces degrades due to the accumulation of cement and road-generated dust composition in PM. This deterioration changes the transpiration and accelerates wax structure destruction. Significant physiological repercussions from these modifications may include inhibiting gaseous exchange and photosynthesis and solute leakage from leaf cells. The structure and composition of the epicuticular wax influence the PM deposition more than the concentration (Liu et al., 2012); alternatively, the quantity of epicuticular wax also affects PM deposition. Furthermore, hairy leaves were beneficial in preventing PM deposition and accumulation (Mandal et al., 2023a, 2023b; Rai, 2016; Weerakkody et al., 2018).

6.4.2.3 Effects on Biochemical Characteristics

The long-term atmospheric suspension of fine dust pollution particles, such as PM2.5, can directly and indirectly affect plant health, estimated in terms of reduction in pigment content, reactive oxygen species (ROS) scavenging capacity, enzymatic activities, changes in protein structure, ascorbic acid, sugar, etc. PM causes ROS production, which can oxidize membrane lipids, affect the photosynthetic system, and limit protoplast regeneration, furthermore reducing in the activities of catalase activity and superoxide dismutase enzymatic activity. ROS include H_2O_2, O_2^-, OH radicals damage proteins, lipids, and DNA. These processes include photochemical reactions and the redox cycling of contaminants. Moreover, plants may allocate resources to the manufacture of antioxidant proteins, for instance, catalase activity, superoxide dismutase activity, and peroxidase activity, to combat oxidative damage. This might alter the percentage of proteins (Mandal et al., 2023a, 2023b). The antioxidant ascorbic acid in plants is important for pollution tolerance and determines the plant's adaption to air pollution. Stress generated by PM not only stimulates defensive mechanisms, but as an antioxidant, ascorbate that can help sustaining plants in stress to reduce air pollution. Ascorbic acid also controls cell division and proliferation and its lower amounts during stress correlate with the sensitivity of plants towards air pollution, though, a high pH value may improve the effectiveness of hexose sugar conversion to ascorbic acid.

Furthermore, exposure to PM-induced ROS can cause an increase in the synthesis of enzymes that scavenge free radicals, such as peroxidases, catalase, ascorbate peroxidases, and superoxide dismutase. PM exposure in plants can also impact on non-enzymatic components, such as different sugars, amino acids, carotenoids, flavonoids, tocopherols, and ascorbic acid. However, proline helps in scavenging ROS by acting as osmo-protectants in response to PM-induced ROS in plants (Mandal et al., 2023a, 2023b). When PM is absorbed through the roots or deposited on the surfaces of plants, it can trigger stress reactions that also change the gene expressions. Stress-prone genes may be overexpressed, including those encoding heat-shock proteins, chaperones and defence-related proteins. Conversely, if plant growth and development-associated genes are suppressed, then protein content will also be reduced. Some of the PM combined with heavy metals like Cd, Pb, and Hg can also directly interfere with the ribosome responsible for protein synthesis (Mandal et al., 2023a, 2023b).

6.4.2.4 Productivity of Crop

PM exposure to vegetation causes various physiological and biochemical changes, which leads to the reduction of productivity in terms of biomass and yield. Mostly the chronic exposure comprises long-term exposure with less concentrations affecting the plant's growth and development and finally productivity (Rai, 2016). PM deposition affects the leaf surface properties due to persistent humidity on the leaves and amplified chance of pathogen exposure, resulting in a reduction in plant productivity

(Cape, 2008). Among the trees, productivity depends on attributes such as leaf trait changes with season if deciduous or evergreen and phenological stages are related to the PM accumulation and its effects (Steinparzer et al., 2023).

6.5 Effect of PM on Plant's Biodiversity

The effects of PM on various characteristics of plants can lead to an overall adverse impact on biodiversity. Rai (2016) suggested that the fine PM deposition in forests alters the leaf surface features, prolonging wetness and changing the environment for the epiphytic organisms, which further increases the rest of the contamination by PM constituents causing toxicity. Furthermore, PM can directly impact photosynthesis by obstructing the gas exchange through stomata and causing abrasion, particularly in trees. Due to the large canopy surface area, stem, and branches in trees, along with the air turbulence created by their structure, stress more effectively at absorbing the PM than shorter plants (Lovett et al., 2009). The impacts of PM deposition on vegetation and ecosystems depend largely on the exposure of individual plants and ecosystems to different PM components types and concentrations. Effects on biodiversity can be observed at regional or ecosystem levels due to the cumulative impacts of heterogeneous PM deposition. Different vegetation types and topographies influence deposition modes in varied ways. PM deposition velocity, micrometeorological factors, and landscapes play a crucial role in its deposition on vegetation affecting biodiversity in different landforms (Unsworth & Wilshaw, 1989). Variation in species diversity can arise from altered competition due to the deposition of the highly variable nature of both PM and potential vegetation, receptors, as well as regional nature of PM, any significant progress protecting plant population, vegetation surface and ecosystem functions (Grantz et al., 2003). For a flourished biodiversity proper photosynthesis, nutrient uptake, and carbon allocation are required, which are important for nutrient cycling, food chain structure, and nutrient balance. Natural disturbances can be recovered (Odum, 1993), but stresses like PM deposition in larger concentrations alter the ecosystem function and structure (Grantz et al., 2003).

6.6 Future Perspectives

In terms of the fate of PM, once it is absorbed in plants and the mechanism of effects, there are notably limited studies available. Plants can capture the PM through their stomata, some by their cuticles, and act as the temporary sink for the ambient PM. However, the dynamic nature of PM can lead to unclear responses inside the plants. The question is about the internal cellular structural responses, where they are either broken down, change in chemical composition, or stored, which can lead to specific tissue potential to show the impact. Apart from the biological impact, there exist micrometeorological factors that the plant boundary layer gives ideas about the efficiency of plant's uptake of PM causing long-term effects, indirect effects via root absorption, from PM absorption leaf falling and its degradation, and nutrient cycling also important to understand the ecological role of vegetation in managing the particle air pollution.

6.7 Conclusion

Particulate matter (PM) in the air and its wet and dry dispersion and deposition affect the ecosystem. Due to the numerous sources, PM's composition, shape and size varied. Climatic and local micrometeorological conditions participated in PM dispersion and deposition in vegetation. The interaction of plant traits with PM's temporal and spatial distribution helps to understand the mechanism of removing PM from the atmosphere. Plants have significant exposure to PM through stomata and uptake from soil. Apart from the concentration and deposition rate, leaves' physical and chemical characteristics also participate in the accumulation of PM. Plant trait interaction with the PM in the ambient air provides an approach to understanding the occurrence of PM deposition, and its effects. Plants' morphological, physiological and biological characteristics are affected by PM, further affecting the productivity of the plants. PM deposition on the leaves' surfaces obstructs the gaseous exchange by clogging stomata, inhibiting the photosynthetic activity and cellular effects; however, plant defence mechanisms regulate the physiological and biochemical alterations for the plant's survival and alleviate the removal of PM from the atmosphere.

Acknowledgements The authors are thankful to Anusandhan National Research Foundation (ANRF), Govt. of India, Assistant Grant No. SCP/2022/000339 for resources and support.

References

Abdillah, S. F., & Wang, Y. F. (2023). Ambient ultrafine particle (PM0. 1): Sources, characteristics, measurements and exposure implications on human health. *Environmental Research, 218*, 115061.

Adrees, M., Ibrahim, M., Shah, A. M., Abbas, F., Saleem, F., Rizwan, M., et al. (2016). Gaseous pollutants from brick kiln industry decreased the growth, photosynthesis, and yield of wheat (Triticum aestivum L.). *Environmental Monitoring and Assessment, 188*, 1–11.

Atkinson, R. W., Fuller, G. W., Anderson, H. R., Harrison, R. M., & Armstrong, B. (2010). Urban ambient particle metrics and health: A time-series analysis. *Epidemiology, 21*(4), 501–511.

Baccini, M., Biggeri, A., Grillo, P., Consonni, D., & Bertazzi, P. A. (2011). Health impact assessment of fine particle pollution at the regional level. *American Journal of Epidemiology, 174*(12), 1396–1405.

Barlow, P. G., Brown, D. M., Donaldson, K., MacCallum, J., & Stone, V. (2008). Reduced alveolar macrophage migration induced by acute ambient particle (PM 10) exposure. *Cell Biology and Toxicology, 24*, 243–252.

Beckett, K. P., Freer-Smith, P. H., & Taylor, G. (2000). Particulate pollution capture by urban trees: Effect of species and windspeed. *Global Change Biology, 6*(8), 995–1003.
Beelen, R., Raaschou-Nielsen, O., Stafoggia, M., Andersen, Z. J., Weinmayr, G., Hoffmann, B., et al. (2014). Effects of long-term exposure to air pollution on natural-cause mortality: an analysis of 22 European cohorts within the multicentre ESCAPE project Grantz. *The Lancet, 383*(9919), 785–795.
Berlinger, B., Fehérvári, P., Kővágó, C., Lányi, K., Mátis, G., Mackei, M., & Könyves, L. (2024). There is still a need for a comprehensive investigation of the health consequences of exposure to urban air with special regard to Particulate Matter (PM) and cardiovascular effects. *Atmosphere, 15*(3), 296.
Bockhorn, H. (Ed.). (2013). *Soot formation in combustion: mechanisms and models* (Vol. 59). Springer Science & Business Media.
Boviatsis, M. (2023). A proactive and diligent approach to institutional developments for sustainable vessel operation.
Brook, R. D., Rajagopalan, S., Pope III, C. A., Brook, J. R., Bhatnagar, A., Diez-Roux, A. V., et al. (2010). Particulate matter air pollution and cardiovascular disease: an update to the scientific statement from the American Heart Association. *Circulation, 121*(21), 2331–2378.
Bukaveckas, P. A., MacDonald, A., Aufdenkampe, A., Chick, J. H., Havel, J. E., Schultz, R., et al. (2011). Phytoplankton abundance and contributions to suspended particulate matter in the Ohio, Upper Mississippi and Missouri Rivers. *Aquatic Sciences, 73*, 419–436.
Burnett, R. T., Smith-Doiron, M., Stieb, D., Cakmak, S., & Brook, J. R. (1999). Effects of particulate and gaseous air pollution on cardiorespiratory hospitalizations. *Archives of Environmental Health: An International Journal, 54*(2), 130–139.
Busico, G., Giuditta, E., Kazakis, N., & Colombani, N. (2019). A hybrid GIS and AHP approach for modelling actual and future forest fire risk under climate change accounting water resources attenuation role. *Sustainability, 11*(24), 7166.
Cachon, B. F., Firmin, S., Verdin, A., Ayi-Fanou, L., Billet, S., Cazier, F., et al. (2014). Proinflammatory effects and oxidative stress within human bronchial epithelial cells exposed to atmospheric particulate matter (PM2. 5 and PM > 2.5) collected from Cotonou, Benin. *Environmental Pollution, 185*, 340–351.
Caffrey, P. F., Ondov, J. M., Zufall, M. J., & Davidson, C. I. (1998). Determination of size-dependent dry particle deposition velocities with multiple intrinsic elemental tracers. *Environmental Science and Technology, 32*(11), 1615–1622.
Cape, J. N. (2008). Interactions of forests with secondary air pollutants: Some challenges for future research. *Environmental Pollution, 155*(3), 391–397.
Cascio, W. E. (2018). Wildland fire smoke and human health. *Science of the Total Environment, 624*, 586–595.
Cassee, F. R., Muijser, H., Duistermaat, E., Freijer, J. J., Geerse, K. B., Marijnissen, J. C., & Arts, J. H. (2002). Particle size-dependent total mass deposition in lungs determines inhalation toxicity of cadmium chloride aerosols in rats. Application of a multiple path dosimetry model. *Archives of Toxicology, 76*, 277–286.
Chang, C., Demokritou, P., Shafer, M., & Christiani, D. (2013). Physicochemical and toxicological characteristics of welding fume derived particles generated from real time welding processes. *Environmental Science: Processes and Impacts, 15*(1), 214–224.
Chen, L. C., & Nadziejko, C. (2005). Effects of subchronic exposures to concentrated ambient particles (CAPs) in mice: V. CAPs exacerbate aortic plaque development in hyperlipidemic mice. *Inhalation Toxicology, 17*(4–5), 217–224.
Chernyshev, V. V., Zakharenko, A. M., Ugay, S. M., Hien, T. T., Hai, L. H., Olesik, S. M., et al. (2019). Morphological and chemical composition of particulate matter in buses exhaust. *Toxicology Reports, 6*, 120–125.
Cheung, K., Daher, N., Kam, W., Shafer, M. M., Ning, Z., Schauer, J. J., & Sioutas, C. (2011). Spatial and temporal variation of chemical composition and mass closure of ambient coarse particulate matter (PM10–2.5) in the Los Angeles area. *Atmospheric Environment, 45*(16), 2651–2662.

Chow, J. C., Watson, J. G., Kuhns, H., Etyemezian, V., Lowenthal, D. H., Crow, D., et al. (2004). Source profiles for industrial, mobile, and area sources in the Big Bend Regional Aerosol Visibility and Observational study. *Chemosphere, 54*(2), 185–208.

Chung, A. (2014). Characterization of the Legionella pneumophila Translocated Substrate, Ceg15, and its Interaction with the Mitochondrial Protein, Pyrroline-5-carboxylate reductase 2 (PYCR2). University of Toronto (Canada).

Cozzi, E., Hazarika, S., Stallings III, H. W., Cascio, W. E., Devlin, R. B., Lust, R. M., et al. (2006). Ultrafine particulate matter exposure augments ischemia-reperfusion injury in mice. *American Journal of Physiology-Heart and Circulatory Physiology, 291*(2), H894–H903.

Darvishi Boloorani, A., Soleimani, M., Papi, R., Neysani Samany, N., Teymouri, P., & Soleimani, Z. (2023). Sources, drivers, and impacts of sand and dust storms: A global view. *Dust and Health: Challenges and solutions* (pp. 31–49). Springer International Publishing.

Das, C. R., Bhaumik, R., & Mondal, N. K. (2012). Effect of iron dusts on physiological responses of gram seedlings (Cicer arietinum L.) under laboratory conditions. *Journal of Stress Physiology and Biochemistry, 8*(3), 37–45.

De Sales, F., Xue, Y., & Okin, G. S. (2016). Impact of burned areas on the northern African seasonal climate from the perspective of regional modeling. *Climate Dynamics, 47*, 3393–3413.

Deng, X., Zhang, F., Rui, W., Long, F., Wang, L., Feng, Z., et al. (2013). PM2. 5-induced oxidative stress triggers autophagy in human lung epithelial A549 cells. *Toxicology in Vitro, 27*(6), 1762–1770.

Dizengremel, P. (2021). Chapter one history of air quality and plants: The fundamental role of ozone. *Tropospheric Ozone: A Hazard for Vegetation and Human Health, 20*, 1.

Donaldson, K., Brown, D., Clouter, A., Duffin, R., MacNee, W., Renwick, L., et al. (2002). The pulmonary toxicology of ultrafine particles. *Journal of Aerosol Medicine, 15*(2), 213–220.

Dos Santos, M., Dawidowski, L., Smichowski, P., Ulke, A. G., & Gómez, D. (2012). Factors controlling sea salt abundances in the urban atmosphere of a coastal South American megacity. *Atmospheric Environment, 59*, 483–491.

Dzierżanowski, K., Popek, R., Gawrońska, H., Sæbø, A., & Gawroński, S. W. (2011). Deposition of particulate matter of different size fractions on leaf surfaces and in waxes of urban forest species. *International Journal of Phytoremediation, 13*(10), 1037–1046.

Epa, U. (2016). *Particulate Matter (PM) basics*. United States Environmental Protection Agency.

Feng, Z., & Kobayashi, K. (2009). Assessing the impacts of current and future concentrations of surface ozone on crop yield with meta-analysis. *Atmospheric Environment, 43*(8), 1510–1519.

Finer, M., Jenkins, C. N., Pimm, S. L., Keane, B., & Ross, C. (2008). Oil and gas projects in the western Amazon: Threats to wilderness, biodiversity, and indigenous peoples. *PLoS ONE, 3*(8), Article e2932.

Funasaka, K., Miyazaki, T., Kawaraya, T., Tsuruho, K., & Mizuno, T. (1998). Characteristics of particulates and gaseous pollutants in a highway tunnel. *Environmental Pollution, 102*(2–3), 171–176.

Funk, R., Reuter, H. I., Hoffmann, C., Engel, W., & Öttl, D. (2008). Effect of moisture on fine dust emission from tillage operations on agricultural soils. *Earth Surface Processes and Landforms: THe Journal of the British Geomorphological Research Group, 33*(12), 1851–1863.

Garg, S. S., Kumar, N., & Das, G. (2000). Effects of the Bansal Ramraj mill dust on vegetation and health at Jaitwara, district Satna. *Indian Journal of Environmental Protection, 20*(5), 326–328.

Gautam, S., & Brema, J. (2020). Spatio-temporal variation in the concentration of atmospheric particulate matter: A study in fourth largest urban agglomeration in India. *Environmental Technology and Innovation, 17*, Article 100546.

Ge, Y., Vieno, M., Stevenson, D. S., Wind, P., & Heal, M. R. (2023). Global sensitivities of reactive N and S gas and particle concentrations and deposition to precursor emissions reductions. *Atmospheric Chemistry and Physics, 23*(11), 6083–6112.

Giri, S., Shrivastava, D., Deshmukh, K., & Dubey, P. (2013). Effect of air pollution on chlorophyll content of leaves. *Current Agriculture Research Journal, 1*(2), 93–98.

Giugliano, M., Lonati, G., Butelli, P., Romele, L., Tardivo, R., & Grosso, M. (2005). Fine particulate (PM2. 5–PM1) at urban sites with different traffic exposure. *Atmospheric Environment, 39*(13), 2421–2431.

Glasius, M., La Cour, A., & Lohse, C. (2011). Fossil and nonfossil carbon in fine particulate matter: A study of five European cities. *Journal of Geophysical Research: Atmospheres, 116*(D11).

Grantz, D. A., Garner, J. H. B., & Johnson, D. W. (2003). Ecological effects of particulate matter. *Environment International, 29*(2–3), 213–239.

Hallquist, M., Wenger, J. C., Baltensperger, U., Rudich, Y., Simpson, D., Claeys, M., et al. (2009). The formation, properties and impact of secondary organic aerosol: current and emerging issues. *Atmospheric Chemistry and Physics, 9*(14), 5155–5236.

Henneman, L. R., Liu, C., Mulholland, J. A., & Russell, A. G. (2017). Evaluating the effectiveness of air quality regulations: A review of accountability studies and frameworks. *Journal of the Air and Waste Management Association, 67*(2), 144–172.

Heshmatol Vaezin, S. M., Juybari, M. M., Daei, A., Avatefi Hemmat, M., Shirvany, A., Tallis, M. J., et al. (2021). The effectiveness of urban trees in reducing airborne particulate matter by dry deposition in Tehran, Iran. *Environmental Monitoring and Assessment, 193*, 1–14.

Hirano, T., Kiyota, M., & Aiga, I. (1995). Physical effects of dust on leaf physiology of cucumber and kidney bean plants. *Environmental Pollution, 89*(3), 255–261.

Hoffmann, B., Moebus, S., Mohlenkamp, S., Stang, A., Lehmann, N., Dragano, N., et al. (2007). Residential exposure to traffic is associated with coronary atherosclerosis. *Circulation, 116*(5), 489–496.

Hofman, J., Bartholomeus, H., Calders, K., Van Wittenberghe, S., Wuyts, K., & Samson, R. (2014). On the relation between tree crown morphology and particulate matter deposition on urban tree leaves: A ground-based LiDAR approach. *Atmospheric Environment, 99*, 130–139.

Hogrefe, C., Isukapalli, S. S., Tang, X., Georgopoulos, P. G., He, S., Zalewsky, E. E., et al. (2011). Impact of biogenic emission uncertainties on the simulated response of ozone and fine particulate matter to anthropogenic emission reductions. *Journal of the Air and Waste Management Association, 61*(1), 92–108.

Huertas, J. I., Huertas, M. E., Cervantes, G., & Díaz, J. (2014). Assessment of the natural sources of particulate matter on the opencast mines air quality. *Science of the Total Environment, 493*, 1047–1055.

Huertas, J. I., Martinez, D. S., & Prato, D. F. (2021). Numerical approximation to the effects of the atmospheric stability conditions on the dispersion of pollutants over flat areas. *Scientific Reports, 11*(1), 11566.

Ibald-Mulli, A., Wichmann, H. E., Kreyling, W., & Peters, A. (2002). Epidemiological evidence on health effects of ultrafine particles. *Journal of Aerosol Medicine, 15*(2), 189–201.

Jacob, D. J., Turner, A. J., Maasakkers, J. D., Sheng, J., Sun, K., Liu, X., et al. (2016). Satellite observations of atmospheric methane and their value for quantifying methane emissions. *Atmospheric Chemistry and Physics, 16*(22), 14371–14396.

Jaffe, D. A., Wigder, N., Downey, N., Pfister, G., Boynard, A., & Reid, S. B. (2013). Impact of wildfires on ozone exceptional events in the western US. *Environmental Science and Technology, 47*(19), 11065–11072.

Jeanjean, A. P., Monks, P. S., & Leigh, R. J. (2016). Modelling the effectiveness of urban trees and grass on PM2. 5 reduction via dispersion and deposition at a city scale. *Atmospheric Environment, 147*, 1–10.

Jia, J., Cheng, S., Yao, S., Xu, T., Zhang, T., Ma, Y., et al. (2018). Emission characteristics and chemical components of size-segregated particulate matter in iron and steel industry. *Atmospheric Environment, 182*, 115–127.

Jokiniemi, J. K., Lazaridis, M., Lehtinen, K. E., & Kauppinen, E. I. (1994). Numerical simulation of vapour-aerosol dynamics in combustion processes. *Journal of Aerosol Science, 25*(3), 429–446.

Jones, M. G., & Richeldi, L. (2014). Air pollution and acute exacerbations of idiopathic pulmonary fibrosis: Back to miasma? *European Respiratory Journal, 43*(4), 956–959.

Ke, Z., Tang, Q., Golaz, J. C., Liu, X., & Wang, H. (2024). Enhancing climate model performance through improving volcanic aerosol representation. *Egusphere, 2024*, 1–28.

Kelly, F. J., & Fussell, J. C. (2015). Linking ambient particulate matter pollution effects with oxidative biology and immune responses. *Annals of the New York Academy of Sciences, 1340*(1), 84–94.

Keng, F. S. L., Phang, S. M., Abd Rahman, N., Leedham Elvidge, E. C., Malin, G., & Sturges, W. T. (2020). The emission of volatile halocarbons by seaweeds and their response towards environmental changes. *Journal of Applied Phycology, 32*, 1377–1394.

Kittelson, D. B. (1998). Engines and nanoparticles: A review. *Journal of Aerosol Science, 29*(5–6), 575–588.

Klamerus-Iwan, A., Błońska, E., Lasota, J., Waligórski, P., & Kalandyk, A. (2018). Seasonal variability of leaf water capacity and wettability under the influence of pollution in different city zones. *Atmospheric Pollution Research, 9*(3), 455–463.

Kloog, I., Coull, B. A., Zanobetti, A., Koutrakis, P., & Schwartz, J. D. (2012). Acute and chronic effects of particles on hospital admissions in New-England. *PLoS ONE, 7*(4), Article e34664.

Kulshrestha, U. C., Saxena, A., Kumar, N., Kumari, K. M., & Srivastava, S. S. (1998). Chemical composition and association of size-differentiated aerosols at a suburban site in a semi-arid tract of India. *Journal of Atmospheric Chemistry, 29*, 109–118.

Künzli, N., Jerrett, M., Mack, W. J., Beckerman, B., LaBree, L., Gilliland, F., et al. (2005). Ambient air pollution and atherosclerosis in Los Angeles. *Environmental Health Perspectives, 113*(2), 201–206.

Kurosaki, Y., & Mikami, M. (2007). Threshold wind speed for dust emission in east Asia and its seasonal variations. *Journal of Geophysical Research: Atmospheres, 112*(D17).

Lai, H. T. L., Aarnink, A. J. A., Cambra-López, M., Huynh, T. T. T., Parmentier, H. K., & Koerkamp, P. G. (2014). Size distribution of airborne particles in animal houses. *Agricultural Engineering International: CIGR Journal, 16*(3), 28–42.

Lanki, T., Pekkanen, J., Aalto, P., Elosua, R., Berglind, N., D'Ippoliti, D., et al. (2006). Associations of traffic related air pollutants with hospitalisation for first acute myocardial infarction: the HEAPSS study. *Occupational and Environmental Medicine, 63*(12), 844–851.

Li, Y., Chen, Q., Zhao, H., Wang, L., & Tao, R. (2015). Variations in PM10, PM2. 5 and PM1. 0 in an urban area of the Sichuan Basin and their relation to meteorological factors. *Atmosphere, 6*(1), 150–163.

Li, W., Zong, X., He, Y. S., Meng, T., Tang, Y., Yang, Q., et al. (2023). Association between short-term exposure to ambient air pollution and outpatient visits for pulpitis in Hefei, China: a time series study. *Environmental Science and Pollution Research, 30*(32), 78607–78618.

Lim, S. S., Vos, T., Flaxman, A. D., Danaei, G., Shibuya, K., Adair-Rohani, H., et al. (2012). A comparative risk assessment of burden of disease and injury attributable to 67 risk factors and risk factor clusters in 21 regions, 1990–2010: A systematic analysis for the Global Burden of Disease Study 2010. *The Lancet, 380*(9859), 2224–2260.

Liu, L., Guan, D., Peart, M. R., Wang, G., Zhang, H., & Li, Z. (2013). The dust retention capacities of urban vegetation—a case study of Guangzhou, South China. *Environmental Science and Pollution Research, 20*, 6601–6610.

Liu, H., Jacob, D. J., Bey, I., Yantosca, R. M., Duncan, B. N., & Sachse, G. W. (2003). Transport pathways for Asian pollution outflow over the Pacific: Interannual and seasonal variations. *Journal of Geophysical Research: Atmospheres, 108*(D20).

Liu, Q., Jia, X., Quan, J., Li, J., Li, X., Wu, Y., et al. (2018). New positive feedback mechanism between boundary layer meteorology and secondary aerosol formation during severe haze events. *Scientific Reports, 8*(1), 6095.

Liu, S., Yang, X., Duan, F., & Zhao, W. (2022). Changes in air quality and drivers for the heavy PM2. 5 pollution on the North China plain pre-to post-COVID-19. *International Journal of Environmental Research and Public Health, 19*(19), 12904.

Lovett, G. M., Tear, T. H., Evers, D. C., Findlay, S. E., Cosby, B. J., Dunscomb, J. K., et al. (2009). Effects of air pollution on ecosystems and biological diversity in the Eastern United States. *Annals of the New York Academy of Sciences, 1162*(1), 99–135.
Lu, F., Xu, D., Cheng, Y., Dong, S., Guo, C., Jiang, X., & Zheng, X. (2015). Systematic review and meta-analysis of the adverse health effects of ambient PM2. 5 and PM10 pollution in the Chinese population. *Environmental Research, 136*, 196–204.
Mack, S. M., Madl, A. K., & Pinkerton, K. E. (2019). Respiratory health effects of exposure to ambient particulate matter and bioaerosols. *Comprehensive Physiology, 10*(1), 1.
Malek, M. A., Eom, H. J., Hwang, H., Do Hur, S., Hong, S., Hou, S., & Ro, C. U. (2019). Single particle mineralogy of microparticles from Himalayan ice-cores using SEM/EDX and ATR-FTIR imaging techniques for identification of volcanic ash signatures. *Chemical Geology, 504*, 205–215.
Mandal, M., Popek, R., Przybysz, A., Roy, A., Das, S., & Sarkar, A. (2023b). Breathing fresh air in the city: Implementing avenue trees as a sustainable solution to reduce particulate pollution in urban agglomerations. *Plants, 12*(7), 1545. https://doi.org/10.3390/plants12071545
Mandal, M., Das, S., Roy, A., Rakwal, R., Jones, O. A., Popek, R., et al. (2023). Interactive relations between plants, the phyllosphere microbial community, and particulate matter pollution. *Science of the Total Environment, 890*, 164352.
Manisalidis, I., Stavropoulou, E., Stavropoulos, A., & Bezirtzoglou, E. (2020). Environmental and health impacts of air pollution: A review. *Frontiers in Public Health, 8*, 14.
McDuffie, E. E., Martin, R. V., Spadaro, J. V., Burnett, R., Smith, S. J., O'Rourke, P., et al. (2021). Source sector and fuel contributions to ambient PM2. 5 and attributable mortality across multiple spatial scales. *Nature Communications, 12*(1), 1–12.
McEachran, A. D., Blackwell, B. R., Hanson, J. D., Wooten, K. J., Mayer, G. D., Cox, S. B., & Smith, P. N. (2015). Antibiotics, bacteria, and antibiotic resistance genes: Aerial transport from cattle feed yards via particulate matter. *Environmental Health Perspectives, 123*(4), 337–343.
Meravi, N., Singh, P. K., & Prajapati, S. K. (2021). Seasonal variation of dust deposition on plant leaves and its impact on various photochemical yields of plants. *Environmental Challenges, 4*, Article 100166.
Morris, R. D. (2001). Airborne particulates and hospital admissions for cardiovascular disease: A quantitative review of the evidence. *Environmental Health Perspectives, 109*(suppl 4), 495–500.
Myhre, G. (2009). Consistency between satellite-derived and modeled estimates of the direct aerosol effect. *Science, 325*(5937), 187–190.
Naumova, E. N., Yepes, H., Griffiths, J. K., Sempértegui, F., Khurana, G., Jagai, J. S., et al. (2007). Emergency room visits for respiratory conditions in children increased after Guagua Pichincha volcanic eruptions in April 2000 in Quito, Ecuador observational study: time series analysis. *Environmental Health, 6*, 1–11.
Nordenhall, C., Pourazar, J., Blomberg, A., Levin, J. O., Sandstrom, T., & Adelroth, E. (2000). Airway inflammation following exposure to diesel exhaust: A study of time kinetics using induced sputum. *European Respiratory Journal, 15*(6), 1046–1051.
Nowak, D. J., Hirabayashi, S., Bodine, A., & Greenfield, E. (2014). Tree and forest effects on air quality and human health in the United States. *Environmental Pollution, 193*, 119–129.
Oberdörster, G., Celein, R. M., Ferin, J., & Weiss, B. (1995). Association of particulate air pollution and acute mortality: Involvement of ultrafine particles? *Inhalation Toxicology, 7*(1), 111–124.
Odum, E. P. (1993). *Ecology and our endangered life-support systems* (2nd ed.). Sinauer Associates.
Othman, M., Latif, M. T., Hamid, H. H. A., Uning, R., Khumsaeng, T., Phairuang, W., et al. (2022). Spatial–temporal variability and health impact of particulate matter during a 2019–2020 biomass burning event in Southeast Asia. *Scientific Reports, 12*(1), 7630.
Owoade, O. K., Olise, F. S., Ogundele, L. T., Fawole, O. G., & Olaniyi, H. B. (2012). Correlation between particulate matter concentrations and meteorological parameters at a site in Ile-Ife, Nigeria. *Ife Journal of Science, 14*(1), 83–93.

Pan, S., Fulton, L. M., Roy, A., Jung, J., Choi, Y., & Gao, H. O. (2021). Shared use of electric autonomous vehicles: Air quality and health impacts of future mobility in the United States. *Renewable and Sustainable Energy Reviews, 149*, Article 111380.

Pant, P., & Harrison, R. M. (2013). Estimation of the contribution of road traffic emissions to particulate matter concentrations from field measurements: A review. *Atmospheric Environment, 77*, 78–97.

Pascal, M., Corso, M., Chanel, O., Declercq, C., Badaloni, C., Cesaroni, G., et al. (2013). Assessing the public health impacts of urban air pollution in 25 European cities: Results of the Aphekom project. *Science of the Total Environment, 449*, 390–400.

Patil, R. S., Kumar, R., Menon, R., Shah, M. K., & Sethi, V. (2013). Development of particulate matter speciation profiles for major sources in six cities in India. *Atmospheric Research, 132*, 1–11.

Perrino, C., Canepari, S., Catrambone, M., Dalla Torre, S., Rantica, E., & Sargolini, T. (2009). Influence of natural events on the concentration and composition of atmospheric particulate matter. *Atmospheric Environment, 43*(31), 4766–4779.

Pope, C. A., III., & Dockery, D. W. (2006). Health effects of fine particulate air pollution: Lines that connect. *Journal of the Air and Waste Management Association, 56*(6), 709–742.

Popek, R., Gawrońska, H., Wrochna, M., Gawroński, S. W., & Sæbø, A. (2013). Particulate matter on foliage of 13 woody species: Deposition on surfaces and phytostabilisation in waxes–a 3-year study. *International Journal of Phytoremediation, 15*(3), 245–256.

Prajapati, S. K., & Tripathi, B. D. (2007). Biomonitoring trace-element levels in PM10 released from vehicles using leaves of Saraca indica and Lantana camara. *AMBIO: A Journal of the Human Environment, 36*(8), 704–705.

Prajapati, S. K., & Tripathi, B. D. (2008a). Biomonitoring seasonal variation of urban air polycyclic aromatic hydrocarbons (PAHs) using Ficus benghalensis leaves. *Environmental Pollution, 151*(3), 543–548.

Prajapati, S. K., & Tripathi, B. D. (2008b). Seasonal variation of leaf dust accumulation and pigment content in plant species exposed to urban particulates pollution. *Journal of Environmental Quality, 37*(3), 865–870.

Price-Allison, A. M. (2022). *The use of solid biomass stoves for heating: An investigation into the effects of fuel properties and operational practices on pollutant emissions* (Doctoral dissertation, University of Leeds).

Przybysz, A., Popek, R., Gawrońska, H., Grab, K., Łoskot, K., Wrochna, M., & Gawroński, S. W. (2014). Efficiency of photosynthetic apparatus of plants grown in sites differing in level of particulate matter. *Acta Scientiarum Polonorum Hortorum Cultus, 13*(1), 17–30.

Raaschou-Nielsen, O., Andersen, Z. J., Jensen, S. S., Ketzel, M., Sørensen, M., Hansen, J., et al. (2012). Traffic air pollution and mortality from cardiovascular disease and all causes: a Danish cohort study. *Environmental Health, 11*, 1–12.

Raaschou-Nielsen, O., Andersen, Z. J., Beelen, R., Samoli, E., Stafoggia, M., Weinmayr, G., et al. (2013). Air pollution and lung cancer incidence in 17 European cohorts: prospective analyses from the European Study of Cohorts for Air Pollution Effects (ESCAPE). *The Lancet Oncology, 14*(9), 813–822.

Rai, P. K. (2016). Impacts of particulate matter pollution on plants: Implications for environmental biomonitoring. *Ecotoxicology and Environmental Safety, 129*, 120–136.

Rai, P. K., & Panda, L. L. (2014). Leaf dust deposition and its impact on biochemical aspect of some roadside plants of Aizawl, Mizoram, North East India. *International Research Journal of Environment Sciences, 3*(11), 14–19.

Rani, L., Thapa, K., Kanojia, N., Sharma, N., Singh, S., Grewal, A. S., et al. (2021). An extensive review on the consequences of chemical pesticides on human health and environment. *Journal of Cleaner Production, 283*, 124657.

Reid, C. E., Brauer, M., Johnston, F. H., Jerrett, M., Balmes, J. R., & Elliott, C. T. (2016). Critical review of health impacts of wildfire smoke exposure. *Environmental Health Perspectives, 124*(9), 1334–1343.

Rissler, J., Swietlicki, E., Bengtsson, A., Boman, C., Pagels, J., Sandström, T., et al. (2012). Experimental determination of deposition of diesel exhaust particles in the human respiratory tract. *Journal of Aerosol Science, 48*, 18–33.

Riva, D., Magalhães, C. B., Lopes, A., Lanças, T., Mauad, T., Malm, O., et al. (2011). Low dose of fine particulate matter (PM2. 5) can induce acute oxidative stress, inflammation and pulmonary impairment in healthy mice. *Inhalation Toxicology, 23*(5), 257–267.

Roy, A., Das, S., Singh, P., Mandal, M., Kumar, M., Rajlaxmi, A., et al. (2024a). Summer-time monitoring and source apportionment study of both coarse, fine, and ultra-fine particulate pollution in Eastern Himalayan Darjeeling: A hint to health risk during peak tourist season. *MAPAN, 39*(4), 995–1009. https://doi.org/10.1007/s12647-024-00776-9

Roy, A., Mandal, M., Das, S., Popek, R., Rakwal, R., Agrawal, G. K., et al. (2024). The cellular consequences of particulate matter pollutants in plants: Safeguarding the harmonious integration of structure and function. *Science of the Total Environment,* 169763.

Sæbø, A., Popek, R., Nawrot, B., Hanslin, H. M., Gawronska, H., & Gawronski, S. W. (2012). Plant species differences in particulate matter accumulation on leaf surfaces. *Science of the Total Environment, 427*, 347–354.

Saldiva, P. H., Zin, W. A., Santos, R. L., Eidelman, D. H., & Milic-Emili, J. (1992). Alveolar pressure measurement in open-chest rats. *Journal of Applied Physiology, 72*(1), 302–306.

Satsangi, P. G., & Yadav, S. (2014). Characterization of PM 2.5 by X-ray diffraction and scanning electron microscopy–energy dispersive spectrometer: Its relation with different pollution sources. *International Journal of Environmental Science and Technology, 11*, 217–232.

Schraufnagel, D. E. (2020). The health effects of ultrafine particles. *Experimental and Molecular Medicine, 52*(3), 311–317.

Seinfeld, J. H., & Pandis, S. N. (2016). *Atmospheric chemistry and physics: From air pollution to climate change*. Wiley.

Setsungnern, A., Treesubsuntorn, C., & Thiravetyan, P. (2017). The influence of different light quality and benzene on gene expression and benzene degradation of Chlorophytum comosum. *Plant Physiology and Biochemistry, 120*, 95–102.

Sgrigna, G., Baldacchini, C., Dreveck, S., Cheng, Z., & Calfapietra, C. (2020). Relationships between air particulate matter capture efficiency and leaf traits in twelve tree species from an Italian urban-industrial environment. *Science of the Total Environment, 718*, Article 137310.

Sharifi, M. R., Gibson, A. C., & Rundel, P. W. (1997). Surface dust impacts on gas exchange in Mojave Desert shrubs. *Journal of Applied Ecology*, 837–846.

Shrivastava, N., & Joshi, S. (2002). Effect of automobile air pollution on the growth of some plants at Kota. *Geobios, 29*(4), 281–282.

Shukla, K., Kumar, P., Mann, G. S., & Khare, M. (2020). Mapping spatial distribution of particulate matter using kriging and inverse distance weighting at supersites of megacity Delhi. *Sustainable Cities and Society, 54*, Article 101997.

Shusterman, D. (2011). The effects of air pollutants and irritants on the upper airway. *Proceedings of the American Thoracic Society, 8*(1), 101–105.

Silva, L. F., Pinto, D., Neckel, A., Oliveira, M. L., & Sampaio, C. H. (2020). Atmospheric nanocompounds on Lanzarote Island: Vehicular exhaust and igneous geologic formation interactions. *Chemosphere, 254*, Article 126822.

Singh, R., Sinha, B., Hakkim, H., & Sinha, V. (2023). Source apportionment of volatile organic compounds during paddy-residue burning season in north-west India reveals large pool of photochemically formed air toxics. *Environmental Pollution, 338*, Article 122656.

Sippula, O., Hokkinen, J., Puustinen, H., Yli-Pirilä, P., & Jokiniemi, J. (2009). Comparison of particle emissions from small heavy fuel oil and wood-fired boilers. *Atmospheric Environment, 43*(32), 4855–4864.

Spickett, J., Katscherian, D., & Harris, P. (2013). The role of health impact assessment in the setting of air quality standards: An Australian perspective. *Environmental Impact Assessment Review, 43*, 97–103.

Srimuruganandam, B., & Nagendra, S. S. (2012). Application of positive matrix factorization in characterization of PM10 and PM2. 5 emission sources at urban roadside. *Chemosphere, 88*(1), 120–130.

Steinparzer, M., Schaubmayr, J., Godbold, D. L., & Rewald, B. (2023). Particulate matter accumulation by tree foliage is driven by leaf habit types, urbanization-and pollution levels. *Environmental Pollution,* 122289.

Takai, H., Pedersen, S., Johnsen, J. O., Metz, J. H. M., Koerkamp, P. G., Uenk, G. H., et al. (1998). Concentrations and emissions of airborne dust in livestock buildings in Northern Europe. *Journal of Agricultural Engineering Research, 70*(1), 59–77.

Tallis, M., Taylor, G., Sinnett, D., & Freer-Smith, P. (2011). Estimating the removal of atmospheric particulate pollution by the urban tree canopy of London, under current and future environments. *Landscape and Urban Planning, 103*(2), 129–138.

Tian, G., Qiao, Z., & Xu, X. (2014). Characteristics of particulate matter (PM10) and its relationship with meteorological factors during 2001–2012 in Beijing. *Environmental Pollution, 192,* 266–274.

Tiwari, S., Pandithurai, G., Attri, S. D., Srivastava, A. K., Soni, V. K., Bisht, D. S., et al. (2015). Aerosol optical properties and their relationship with meteorological parameters during wintertime in Delhi, India. *Atmospheric Research, 153,* 465–479.

Tiwary, A., & Williams, I. (2018). *Air pollution: Measurement, modelling and mitigation.* CRC Press.

Turner, T., Walker, N., & Mitchell, K. (2018). Agricultural activities and particulate matter: Contributions and control measures. *Agricultural Systems, 165,* 34–45.

Turner, J. S. (2021). *Water Clarity and Suspended Particle Dynamics in the Chesapeake Bay: Local Effects of Oyster Aquaculture, Regional Effects of Reduced Shoreline Erosion, and Long-Term Trends in Remotely Sensed Reflectance* (Doctoral dissertation, The College of William and Mary).

Unsworth, M. H., & Wilshaw, J. C. (1989). Wet, occult and dry deposition of pollutants on forests. *Agricultural and Forest Meteorology, 47*(2–4), 221–238.

Vallabani, N. S., Gruzieva, O., Elihn, K., Juárez-Facio, A. T., Steimer, S. S., Kuhn, J., et al. (2023). Toxicity and health effects of ultrafine particles: Towards an understanding of the relative impacts of different transport modes. *Environmental Research, 231,* 116186.

Velasco, J., Lloret, J., Millán, A., Marin, A., Barahona, J., Abellán, P., & Sánchez-Fernández, D. (2006). Nutrient and particulate inputs into the Mar Menor lagoon (SE Spain) from an intensive agricultural watershed. *Water, Air, and Soil Pollution, 176,* 37–56.

Vohra, K., Vodonos, A., Schwartz, J., Marais, E. A., Sulprizio, M. P., & Mickley, L. J. (2021). Global mortality from outdoor fine particle pollution generated by fossil fuel combustion: Results from GEOS-Chem. *Environmental Research, 195,* Article 110754.

Von Glasow, R., Jickells, T. D., Baklanov, A., Carmichael, G. R., Church, T. M., Gallardo, L., et al. (2013). Megacities and large urban agglomerations in the coastal zone: interactions between atmosphere, land, and marine ecosystems. *Ambio, 42,* 13–28.

von Klot, S., Peters, A., Aalto, P., Bellander, T., Berglind, N., D'Ippoliti, D., et al. (2005). Ambient air pollution is associated with increased risk of hospital cardiac readmissions of myocardial infarction survivors in five European cities. *Circulation, 112*(20), 3073–3079.

Wang, L. F., & Dai, Z. D. (2016). Effects of the natural microstructures on the wettability of leaf surfaces. *Biosurface and Biotribology, 2*(2), 70–74.

Wang, D., Pakbin, P., Shafer, M. M., Antkiewicz, D., Schauer, J. J., & Sioutas, C. (2013). Macrophage reactive oxygen species activity of water-soluble and water-insoluble fractions of ambient coarse, PM2. 5 and ultrafine particulate matter (PM) in Los Angeles. *Atmospheric Environment, 77,* 301–310.

Wang, L., Wei, Z., Wei, W., Fu, J. S., Meng, C., & Ma, S. (2015). Source apportionment of PM2. 5 in top polluted cities in Hebei, China using the CMAQ model. *Atmospheric Environment, 122,* 723–736.

Wang, Y. R., Zhang, X. X., Zhao, J. Y., & Jiang, Q. O. (2016). Temporal and spatial distribution of PM 2.5 and its relationship with vegetation coverage in Beijing during the period of 2013–2014. *Ecology and Environmental Sciences, 25*(1), 103–111.

Wang, Z., Chen, G., Gu, Y., Zhao, B., Ma, Q., Wang, S., & Liou, K. N. (2019). Large-scale meteorological control on the spatial pattern of wintertime PM2. 5 pollution over China. *Atmospheric Science Letters, 20*(10), e938.

Wang, B., Kuang, S., Pfister, G. G., Biazar, A. P., Newchurch, M. J., Buchholz, R. R., & Langford, A. O. N. (2022). Vertical Accumulation of Ozone and Aerosol during the 2016 Southeastern US Wildfires. *Authorea Preprints.*

Warneke, C., Schwarz, J. P., Dibb, J., Kalashnikova, O., Frost, G., Al-Saad, J., et al. (2023). Fire influence on regional to global Environments and air quality (FIREX-AQ). *Journal of Geophysical Research: Atmospheres, 128*(2), e2022JD037758.

Watson, J. G., Chow, J. C., & Houck, J. E. (2001). PM2. 5 chemical source profiles for vehicle exhaust, vegetative burning, geological material, and coal burning in Northwestern Colorado during 1995. *Chemosphere, 43*(8), 1141–1151.

Watts, A. C., Samburova, V., & Moosmüller, H. (2020). Criteria-based identification of important fuels for Wildland fire emission research. *Atmosphere, 11*(6), 640.

Weerakkody, U., Dover, J. W., Mitchell, P., & Reiling, K. (2018). Evaluating the impact of individual leaf traits on atmospheric particulate matter accumulation using natural and synthetic leaves. *Urban Forestry and Urban Greening, 30*, 98–107.

World Health Organization. (2006). *Air quality guidelines: global update 2005: particulate matter, ozone, nitrogen dioxide, and sulfur dioxide.* World Health Organization.

Wu, N., Zhong, G., Niyazi, Y., Nugraha, H. D., & Steventon, M. J. (2023). Transformation of dense shelf water cascade to supercritical turbidity currents: Impact on seabed geomorphology and 2 implication for climate change 3.

Xu, Y., Xue, W., Lei, Y., Zhao, Y., Cheng, S., Ren, Z., & Huang, Q. (2018). Impact of meteorological conditions on PM2. 5 pollution in China during winter. *Atmosphere, 9*(11), 429.

Xu, Q., Chen, X., Yang, S., Tang, L., & Dong, J. (2021). Spatiotemporal relationship between Himawari-8 hourly columnar aerosol optical depth (AOD) and ground-level PM2. 5 mass concentration in mainland China. *Science of the Total Environment, 765*, 144241.

Yadav, A. K., Kumar, K., Hj Awg Kasim, M., Singh, M. P., Parida, S. K., & Sharan, M. (2003). Visibility and incidence of respiratory diseases during the 1998 haze episode in Brunei Darussalam. *Air Quality*, 265–277.

Yang, X., Jiang, L., Zhao, W., Xiong, Q., Zhao, W., & Yan, X. (2018). Comparison of ground-based PM2. 5 and PM10 concentrations in China, India, and the US. *International Journal of Environmental Research and Public Health, 15*(7), 1382.

Yang, G., Ren, G., Zhang, P., Xue, X., Tysa, S. K., Jia, W., et al. (2021). PM2. 5 influence on urban heat island (UHI) effect in Beijing and the possible mechanisms. *Journal of Geophysical Research: Atmospheres, 126*(17), e2021JD035227.

Yatkin, S., & Bayram, A. (2008). Determination of major natural and anthropogenic source profiles for particulate matter and trace elements in Izmir Turkey. *Chemosphere, 71*(4), 685–696.

Zanobetti, A., & Schwartz, J. (2005). The effect of particulate air pollution on emergency admissions for myocardial infarction: A multicity case-crossover analysis. *Environmental Health Perspectives, 113*(8), 978–982.

Zanobetti, A., & Schwartz, J. (2006). Air pollution and emergency admissions in Boston, MA. *Journal of Epidemiology and Community Health, 60*(10), 890–895.

Zanobetti, A., Franklin, M., Koutrakis, P., & Schwartz, J. (2009). Fine particulate air pollution and its components in association with cause-specific emergency admissions. *Environmental Health, 8*, 1–12.

Zeb, B., Alam, K., Sorooshian, A., Blaschke, T., Ahmad, I., & Shahid, I. (2018). On the morphology and composition of particulate matter in an urban environment. *Aerosol and Air Quality Research, 18*(6), 1431.

Zhang, W., Zhang, Z., Meng, H., & Zhang, T. (2018). How does leaf surface micromorphology of different trees impact their ability to capture particulate matter? *Forests, 9*(11), 681.

Zhang, X., Sun, J., Wang, Y., Li, W., Zhang, Q., Wang, W., et al. (2013). Factors contributing to haze and fog in China. *Chinese Science Bulletin, 58*(1178), 972013–1972050.

Zhang, H., Huang, L., Dai, Y., Liu, S., Hong, Y., Tian, L., et al. (2015). *Arabidopsis* AtERF15 positively regulates immunity against Pseudomonas syringae pv. tomato DC3000 and Botrytis cinerea. *Frontiers in Plant Science, 6*, 686.

Zhang, W. K., Wang, B., & Niu, X. (2015). Adsorption capacity of the air particulate matter in urban landscape plants in different polluted regions of Beijing. *Huan jing ke xue = Huanjing kexue, 36*(7), 2381–2388.

Zhang, J., Chen, Z., Shan, D., Wu, Y., Zhao, Y., Li, C., et al. (2024). Adverse effects of exposure to fine particles and ultrafine particles in the environment on different organs of organisms. *Journal of Environmental Sciences, 135*, 449–473.

Zhao, C. X., Wang, Y. J., Wang, Y. Q., & Zhang, H. L. (2013). Interactions between fine particulate matter (PM2. 5) and vegetation: A review. *Chinese Journal of Ecology, 32*(8), 2203.

Zhou, Q., Zheng, N., Liu, J., Wang, Y., Sun, C., Liu, Q., et al. (2015). Residents health risk of Pb, Cd and Cu exposure to street dust based on different particle sizes around zinc smelting plant, Northeast of China. *Environmental Geochemistry and Health, 37*, 207–220.

Zhu, H., Martin, R., van Donkelaar, A., Hammer, M., Li, C., Meng, J., et al. (2024). Global spatial variation in the PM 2.5 to AOD relationship strongly influenced by aerosol composition. *EGUsphere, 2024*, 1–32.

Chapter 7
Airborne Particulate Matter and Plant Reproductive Success: A Comprehensive Review

Priyanka Khanduri and Sudip Kumar Roy

Abstract Over the years, particulate matter (PM) has emerged as a global concern due to its harmful effects on human health. Like humans, PM also poses a major threat to plants, which are vulnerable to air pollutants. Therefore, understanding the effects of PM on the crucial plant functions such as reproduction is becoming increasingly important. Such research efforts are essential for making informed decisions about future agricultural and urban planning. While effects of other air pollutants on plant reproduction are quite well-known, the impacts of PM are relatively less studied. Nonetheless, data indicates that plant reproduction can be significantly affected by the negative impacts of particulate matter (PM). Some known effects of exposure to PM during reproductive phase of plant are reduced flowering potential, delayed flowering, reduced pollen function, disruption of plant-pollinator interaction, interference in pollen-stigma interaction and reduced seed growth and development. These effects ultimately hamper the reproductive output or yield of a plant. In this chapter, we aim to explore the potential abnormalities that can arise in the reproductive phase of plants when exposed to particulate matter (PM). Present chapter mainly focuses on cataloguing of available literature and providing future directions for the research in the area.

Keywords Flowering phenology · Fruit set · Particulate matter · Plant-pollinator interaction · Pollen abnormalities · Pollen development · Pollen-pistil interaction · Pollen viability · Plant reproduction · Reproductive success

Abbreviations

APM Air particulate matter

P. Khanduri (✉)
Department of Botany, Vidyasagar Metropolitan College, University of Calcutta, Kolkata, India
e-mail: priyanka.khanduri@gmail.com

S. K. Roy
Department of Botany, Charuchandra College, University of Calcutta, Kolkata, India

A. Sarkar et al. (eds.), *From Sky to Soil*, https://doi.org/10.1007/978-3-032-31991-3_7

CO	Carbon monoxide
HCs	Hydrocarbons
NO_2	Nitrogen dioxide
O_3	Ozone
PCD	Programmed cell death
PER	Proboscis extension response
PM	Particulate matter
RSPM	Respirable suspended particulate matter
SO_2	Sulphur dioxide
VOCs	Volatile organic compounds

7.1 Introduction

Particulate matter (PM) is a major class of air pollutants, that constitute a diverse mixture of particles from various origins and chemical compositions, which have varied effects on living organisms (Grantz et al., 2003). In general, PM is a mixture of fine solids such as soil, soil dust, dirt, pollen, ashes, molds and soot present in the atmosphere (Mukherjee & Agrawal, 2017). In the past few decades concerns have been raised about negative effects of PM, especially in urban areas. As a result, there has been a significant increase in research on the effects of PM pollution, particularly on human health. While research on atmospheric pollution and its impact on plants has been ongoing for a long time, the majority of studies have primarily focused on Nitrogen dioxide (NO_2), Ozone (O_3) and Sulphur dioxide (SO_2). Studies on the effects of PM on plants are few and mainly pertain to vegetative growth. PM can have indirect influence on plants by alteration of soil chemistry, effects on nutrient cycling and hampering plant nutrient uptake (*reviewed in* Ulrichs et al., 2008). Direct impact ranges from alteration in acidity, salinity, nutrients and surfactant properties of leaf to stomatal closure causing changes to cell/tissue to reduction in the photosynthetic efficiency of the plant (Grantz et al., 2003; Naidoo & Chirkoot, 2004; Nano & Ilias, 2007).

The impact of PM on plant yield through effects on reproductive processes is not well understood and has rarely been examined in isolation. The reproductive phase is the most crucial stage in a plant's life cycle. Disruption at any stage of reproductive development, including plant-pollinator interactions, fertilisation and seed development, can significantly impact the reproductive success of many agriculturally and ecologically important plant species. PM can either directly influence the reproductive organs and pollination or indirectly affect the vegetative organs which in turn changes assimilation distribution and the timing of flowering. In this chapter, we will provide an overview of this topic, emphasizing key areas for future research.

7.2 Effects of Airborne Particulate Matter on Various Stages of Plant Sexual Reproduction

7.2.1 Flowering Phenology

The study of the timing periodic life cycle events of a species is known as phenology. The timing of this recurring life events respond to the seasonal changes and climate. The timely occurrence of various life cycle phases has consequences on the fitness and survival of a species. Phenological studies help monitor these events and provide a valuable database. Changes in patterns observed over several years can indicate phenological shifts and serve as indicators of alterations in the abiotic environment of the species.

Research indicates that persistent or recurrent disruptions in abiotic factors lead to phenological changes in species. Consequently, these changes are often utilized to assess the impact of climate on a plant's fitness within a specific region. Flowering time is one of the most vital phenological event for a plant as it marks the beginning of reproductive phase. Alterations in flowering timing can impact the likelihood of successful pollination, subsequently influencing fruit set and overall yield (Fitter & Fitter, 2002; Rafferty & Ives, 2012). Consequently, first flowering dates are frequently used as indicators of plants' sensitivity to climate changes (Miller-Rushing et al., 2008).

Changes in flowering dates are mostly used to indicate the effects of global warming of a particular region. However, changes in weather closely interact with air pollution and collectively they pose a significant challenge for plants. Given the significant impact of pollution on plant physiology, such as the duration needed to emerge from winter dormancy (Grigorjev & Pakharkova, 2001), it is anticipated that phenological changes will occur in polluted regions (Kozlov et al., 2007). The first study to analyse the impact of pollution on plant seasonal development was done by Antipov (1957). He reported an average delay of 1–4 days, with a maximum of 17 days in budburst and flowering dates in 20 tree and shrub species growing near three unnamed industrial units (Antipov, 1957). Subsequent studies examined phenology in plants growing along roadside (Bhatti & Iqbai, 1988; Shafiq & Iqbal, 2004) and other polluted areas, including those near nickel-copper smelters and iron pellet (Kozlov et al., 2007). A delay in flowering time has been observed in certain annual plant species exposed to daily doses of urban dust (Rai et al., 2010). However, these studies focused on leaf growth patterns and did not include measurements of the first flowering dates. One of the few studies to examine flowering phenology was conducted by Honour et al. (2009). Delayed flowering was reported in the plants of *Solanum oleraceus* exposed to vehicle emissions (Honour et al., 2009). According to the study, control plants flowered after 5 weeks, whereas only 33% plants flowered in the polluted domes. Authors also proposed that pollutants delayed the advancement of flowering in the plants exposed to pollutants as 7 weeks later they were still at bud stage. In comparison control plants had undergone anthesis and a smaller proportion even had set seed. Recently, Kolek et al. (2021) documented earlier flowering in *Betula pendula* growing in Augsburg, Germany due to the influence of NO_2 and urban conditions. The study reported earlier flowering in the areas with higher levels of urbanization and it reached its peak and concluded sooner in locations with elevated NO_2 concentrations.

Most of the studies on flowering phenology and air pollution have taken into consideration a mixture of urban air pollutants. These studies have not focused the relative importance of individual pollutants in causing phenological changes in a species. Effect of PM on flowering phenology has been barely touched upon by investigators. Nonetheless, one such study was conducted by Jochner et al. (2015). In their extensive study, Jochner et al. (2015) carried out a vast field survey in Munich, Germany to analyse the effects of PM with an aerodynamic diameter <2.5 mm ($PM_{2.5}$) and <10 mm (PM_{10}) and other pollutants on flowering onset dates of silver birch (*Betula pendula*), common hazel (*Corylus avellana*) and horse chestnut (*Aesculus hippocastanum*). This study found that the full flowering of common hazel was delayed as particulate matter (PM) levels increased in urban areas. Stepwise regression analyses yielded superior models when temperature was excluded and only the estimates of PM_{10} or PM_{10} along with O_3 were considered indicating that the delay in flowering was a consequence of not only temperature but also pollution (Jochner et al., 2015). According to the authors, as common hazel is a deciduous tree and leaves develop after flowering, uptake of pollutants form leaves is not plausible. Thus, particulate matter (PM) may have affected flowering through its absorption in the soil.

Most of the work done in the last two decades has associated changes in phenological cycles with climate warming. However, the effect of air pollution on phenology cannot be negated. Air pollution is expected to become increasingly intense and severe in the coming years, particularly in urban areas. Therefore, understanding the effects of individual pollutants on phenology needs attention.

7.2.2 Pollen Grains

7.2.2.1 Pollen Development

The Stamen, the male reproductive organ in flowering plants, consists of two parts: the anther and the filament. The anther plays a crucial role in the development of the male gametophyte, i.e. pollen grain. Together anther and filament are responsible for releasing the pollen grains for pollination subsequently. In an anther, development of the pollen grains takes place in the microsporangium through two phases viz., microsporogenesis and microgametogenesis which are followed by pollen maturation. Pollen wall development or pollen maturation involves participation of different cell layers of anther and the associated molecules. The most important layer of the anther wall is tapetum. Tapetal cells supply the necessary nutrients for developing pollen grains and provide the precursors for pollen wall material by undergoing timely Programmed Cell Death (PCD) (Mangla et al., 2023). A poorly developed tapetum or its premature degeneration can result in pollen sterility. Thus, proper anther development is essential for successful reproduction in plants (Das et al., 2025).

The whole process of pollen development is highly synchronized and sensitive to environmental conditions. Adverse abiotic conditions can hamper the development of pollen inducing abortion and causing male sterility. Studies have shown that exposure to air pollution has a direct impact on the tapetal layers of the anther wall and the development of pollen. In all of these studies, anthers were harvested from plants growing in the areas with varying levels of pollution (mainly SO_2, NO_2, carbon monoxide (CO), hydrocarbons (HCs) and PM). Consequently, effect of individual air pollutant on pollen development has been obscure and the results show the effects of mixture of air pollutants. Hence, this section will discuss the role of air pollution as a whole on pollen development instead of only PM.

The detrimental effects of air pollution can be seen at various stages of pollen development, ultimately impacting the structure of the pollen grain. Rezanejad (2007) reported changes in the pollen structure and size as a result of abnormality in the tapetum in plants exposed to environmental pollution. In the anthers of *Spartium junceum* (Spanish broom) subjected to pollution stress, the tapetum layer exhibits early growth at the microspore mother cell stage and undergoes premature digestion at the tetrad stage (Rezanejad, 2009). As a result, pollution exposed anthers were abnormal and shrunken and pollen grains were much smaller, shrunken, clumped together and deformed in comparison to controls. Rezanejad (2013) investigated microsporogenesis, tapetum function and pollen development in *Petunia hybrida* exposed to air pollution. It was shown that under air pollution, microsporogenesis becomes irregular after tetrad stage leading to irregularly shaped, shrunken and fragile pollen grains. Pollen of *Platanus orientalis* were observed to have become swollen after several hours of fumigation with pollutants (Sedghy et al., 2017). The pollen morphology in response to air pollution was studied in 24 plant species in the city of Hyderabad, India (Ganga Kailas et al., 2015). It was shown that under conditions of high levels of PM particles, SO_2 and NO_x in the atmosphere, the exine of pollen grains shrink in species like *Cyperus rotundus*, *Cocos nucifera*, *Syzygium cumini*, *Acacia nilotica Hyptis suaveolens*, *Azadirachta indica*, *Eucalyptus tereticornis* and *Zea mays*. Similar changes on the shape of pollen grains upon exposure to air pollution have been reported from six legumes species (Malayeri et al., 2012), 'Golden Delicious' and 'Golab' apple cultivars (Ahmadi et al., 2019) and *Datura innoxia* (Ugale & Tidke, 2022).

Exposure to ambient air pollution has also been shown to increase the fragility of the exine. Pollutants have been reported to cause numerous cracks and collapse of exine surface. Accumulation of airborne PM on the surface of pollen grains has been reported to induce changes in the shape and tectum layer of exine in the pollen grains of *Zinnia* (Chehregani et al., 2004). Comparison of pollen of *Chenopodium album* from rural and urban zone of northern Portugal demonstrated that the opercula were well defined in former while there was a fine film covering exine with deformed opercula (Guedes et al., 2009). The pollen grains of *Spartium junceum* collected from polluted areas of Tehran displayed thinning of the exine (Rezanejad, 2009). Abnormalities in the exine, including irregularity, shrinkage, thinning and breakage, have also been reported in pollution-exposed anthers of *C. album* (Amjad & Shafighi, 2012).

7.2.2.2 Chemical Composition of Pollen Grains

Modifications in the chemical composition of pollen are known have been induced by air pollution. For instance, Dixon and Paiva (1995) demonstrated an increase in the quercetin content of pollen samples from a polluted site. Increased concentrations of total flavonoids were observed in the pollen grains of *Spartium junceum*, *Thuja orientalis* and *Lagerstroemia indica* when exposed to air pollution (Rezanejad, 2012). It was suggested that air pollutants, especially PMs, induce defence mechanisms like flavonoids biosynthesis in pollen grains (Rezanejad, 2012).

Protein is the most important component of pollen grains ranging from 2.5 to 61% of the pollen content. Most of the pollen proteins constitute the enzymes required for pollen tube growth and others play nutritional role for the floral visitors (Roulston et al., 2000). Pollutants have also been reported to cause variation of protein amount in pollen grains. For example, Shahali et al. (2009) observed a reduction in the total protein content of *Cupressus arizonica* pollen grains in areas affected by pollution. A comparison of protein profiles of rural versus urban pollen of *C. album* demonstrated decrease in protein content (Guedes et al., 2009). However, not all studies have reported similar results. For instance, Helander et al. (1997) found no significant differences in the protein composition of pollen grains collected from polluted and control areas. Similarly, Rezanejad (2009, 2013) reported that the SDS–PAGE patterns of pollen proteins in *Thuja orientalis* and *Petunia hybrida* showed no notable differences between polluted and control groups.

7.2.3 Pollination

Pollination is a crucial step in sexual plant reproduction, as it facilitates the union of male and female gametes. Plants utilize biotic and abiotic means for transfer of pollen grains to potential conspecific stigma. PM can influence pollination by directly contaminating the pollen during abiotic transfer or disrupting the plant-pollinator interactions in biotic pollination.

7.2.3.1 Particulate Matter Deposition on Pollen Grains in the Air

After anther dehiscence and release from the confinements of anther, pollen make their atmospheric journey to reach a potential stigma. During this time, the pollen grains can come into contact with airborne PM. Adhered atmospheric particles may lead to contamination of the exine of pollen grains and pollute them. This in turn may have effects on the reproductive potential of a plant. Studies have shown that pollen grains in polluted regions have higher levels of particulate matter (PM) deposition than those in areas with lower pollution (Amjad & Shafighi, 2012; Azzazy, 2016; Ribeiro & Abreu, 2014; Ribeiro et al., 2015; Shahali et al., 2009; Wang et al., 2012; Visez et al., 2020; Choël et al., 2022). However, most observations regarding the adhesion of PM on pollen grain surfaces have been qualitative. The extent of surface pollution by atmospheric particles remained largely unknown until recently. The first study to estimate the pollution of pollen grains by PM quantitively was conducted by Choël et al. (2022). Sampling of birch pollen grains in less-polluted urban air (average PM10 = 26.3 $\mu g\ m^{-3}$) reveals that an average of 0.86 particles deposit per pollen grain. Comparing the timing of PM deposition on pollen grains still on the parent plant with those during atmospheric transport indicates that the deposition of APM primarily occurs during windborne transport (Choël et al., 2022). Numerically, the number of particles per pollen grain was found to be 8 times higher on airborne pollen compared to pollen grains sampled from inflorescence (Choël et al., 2022).

7.2.3.2 Plant-Pollinator Interactions

Around 85% of plant species are dependent on biotic pollinators for pollination, with nearly 12,000 genera being insect pollinated (Ollerton, 2017; Renner & Ricklefs, 1995). Insect pollinators are the mainstay of our food production and approximately 35% of the world crop production for human consumption dependent on pollinators (Klein et al., 2007; Abrol, 2012). The estimated economic value of pollination worldwide was €153 billion in 2005, accounting for 9.5% of the value of global agricultural production used for human food (Gallai et al., 2009). Therefore, decline in the pollinator abundance and any disruption in the plant-pollinator interaction has major consequences not only for crop production but also for the survival of plant species. However, over the past century there has been a sustained decline in the abundance and diversity of insect pollinator (Potts et al., 2016; Powney et al., 2019). Research in past decade has shown common air pollutants contribute to pollinator decline and disruption in plant-pollinator interactions. Air pollutants like PM have effects either (i) directly on the plants (ii) on insect pollinator health (iii) pollinator foraging efficiency.

A. **Effects on Plants**

Plant-pollinator interactions involve transmission and reception of signals that are typically olfactory or visual. Olfactory signals are dependent on volatile organic compounds (VOCs), whereas visual cues are usually dependent on pigments responsible for the colours of flowers and appendages. Studies have shown that air pollutants including PM exert direct effects on olfactory signals produced by plants (Fuentes et al., 2016; Ryalls et al., 2022). Girling et al. (2013) demonstrated that diesel exhaust emissions, primarily composed of PM and NOX, contribute to a decline in floral volatiles that honeybees need for flower recognition. The study found that exposure to diesel exhaust significantly alters the blend of common flower volatiles and this alteration led to the reduced capability of honeybees to recognize floral volatiles (Girling et al., 2013). In a similar kind of study conducted by Lusebrink et al. (2015), synthetic blend of eight floral chemicals identified from oilseed rape was exposed to diesel exhaust pollution. Authors demonstrated that within one minute of exposure there was an alteration in the composition of odour mix which led to significant reduction in ability of the honeybees to recognize the altered odour. Such changes in the ability of the pollinators to recognize floral odour may have an impact on their foraging efficiency and thus pollination services provided.

Landing of pollen on the stigma is the culmination of pollination. Quantity and quality of pollen load on the stigma has a direct bearing on the plant reproductive function. Dust arising from natural sources and from human activities like agriculture, livestock husbandry and vehicular movements on unpaved roads are a source of PM. Due to the similarity in size of dust particles and pollen grains, dust particles can easily land on stigma. Deposition of dust onto the stigma can influence the pollination and reproduction of flowering plants. The impact of dust arising from unpaved roads on pollination and reproductive success of four species of montane wild flowers viz. *Delphinium nuttallianum, Ipomopsis aggregata, Linum lewisii* and *Castilleja sulphurea* was studied by Waser et al. (2017). Study demonstrated that flowers growing 1–2 m from road accumulated significantly more dust and received less pollen compared to those situated 40–50 m away. Authors suggested that increased dust load led to lower pollen load on stigma, as plants exposed to more dust received less pollen. However, despite these differences, seed set did not vary consistently for all the studied species.

B. **Effect on Pollinators**

In contrast to plants, numerous studies have examined the impact of PM on pollinators, particularly, *Apis mellifera* which is often used as a bio-monitor of PM (Negri et al., 2015; Pellecchia & Negri, 2018; Papa et al., 2021; Capitani et al., 2021). Foraging on pollen contaminated with PM may compromise the overall health of bees. Studies on the bees that feed on pollen exposed to PM indicate accumulation of PM in malpighian tubules, cytological and histological modifications of the gut epithelium and changes in the gut microbial community (Negri et al., 2015; Papa et al., 2021). Airborne fine PM has also been reported to have lethal effects on the developmental stages of larval bees when they are nursed by worker bees that have been contaminated (Siviter et al., 2023). Prolonged exposure to PM may hamper the development of the bees and decrease their life span resulting in colony-collapse (Ushitani et al., 2016). Tapparo et al (2012) demonstrated that PM released by the drilling machine during the sowing of corn seeds coated with neonicotinoid insecticides can directly contaminate the bees. The contamination can lead to colony breakdown which has been widely reported by beekeepers (Tapparo et al., 2012). As discussed in the previous section, anthropogenic PM may cause olfactory impairment in the pollinators disrupting their foraging efficiency and colony building abilities (Goulson et al., 2015; Klein et al., 2017). Recently Thimmegowda et al. (2020)

studied the sublethal effects of air pollutants on pollinators. They confirmed significant correlations between increased respirable suspended particulate matter (RSPM) deposition and various changes in bees including their survival rates, flower visitation frequency, heart rate, haemocyte levels and expression of genes associated with lipid metabolism, stress and immunity.

7.2.4 Pollen-Pistil Interaction

Once a pollen lands on stigma of a conspecific pistil, a series of coordinated events occurs: pollen germinates, the pollen tube grows through the length of the style, pollen tube enters into the embryo sac (female gametophyte) and finally fertilisation takes place. Successful fruit set is collective function of all these events and interactions that take place between pollen, pollen tube and pistil (Herrero, 2003). Air pollutants may interfere with pollen-stigma interactions by either affecting the biological vigour of pollen and the pollen tube growth or receptivity of stigma.

Gaseous pollutants are well known for interfering with pollen germination, often leading to decrease in pollen germination on stigma (Sénéchal et al., 2015). Deposition and accumulation of particulate matter on the stigmatic surface may affect the pollen directly by reducing viability and germination. Joshi and Sikka (2002) reported decreased pollen viability in *Cassia sp.* due to accumulation of Pb-enriched dust on the pollen grains. A study of pollen samples of *Pinus nigra* and *Corylus avellana* collected from three areas with varying levels of air pollution showed that pollen viability for both species was lower in high-traffic regions (Gottardini et al., 2004). In the same study, pollen viability of *P. nigra* was found to be negatively correlated with PM_{10} concentrations (Gottardini et al., 2004). Low pollen viability and germination in Catharanthus roseus, Martinia annua and *Tridax procumbens* was reported from heavy traffic sites with high levels of PM_{10} from Amravati city of Maharashtra, India (Ugale et al., 2023). Recently, Das et al (2025) studied the pollen viability of 40 plant species from an Indo-Gangetic urban sprawl exposed to high O_3, NO_2, SO_2 and PM_1. Authors reported decreased pollen viability in all the species except *Mangifera indica* as compared to the reference sites. Interference of dust in pollen-pistil interactions of California pistachio (*Pistacia vera*) was studied by Zhang et al (2019). The study demonstrated that the inflorescences which were treated with a high dust: pollen ratio exhibited lower pollen viability. However, no relationship between PM concentrations and the germination of pollen grains could be established in *Cichorium intybus* (Jaconis et al., 2017). Most conclusions regarding the impact of PM on pollen germination have been based on field studies. There are relatively few laboratory-based investigations assessing the effects of PM on in vitro pollen germination. One such study on cultivars of apple demonstrated that the rates of pollen germination and pollen tube growth are low flowers collected from polluted region than control site (Ahmadi et al., 2019).

The surface components of the stigma communicate with the surface components released by the pollen which significantly impacts pollen germination and pollen tube growth. High level of air pollutants might reduce these metabolites and result in inadequate pollen germination (Gillespie et al., 2015). There are few reports in the literature on the effects of PM on stigma structure and function. It is speculated that the chemical composition of PM can change acidity which in turn could affect the electrical charge of stigmatic surface, impacting its receptivity to pollen (Cox, 1984; Grantz et al., 2003). Wilting of stigma on application of dust particles and disappearance of stigma papilla on application of herbicide-contaminated particles has been reported in *P. vera* (Zhang et al., 2019). The dust particles loaded on the stigma can mask the stigmatic surface and block the interaction between the pollen and the papillae cells (Zhang et al., 2019). Decreased pollen germination due to dust deposition on the stigmatic surface and inhibition of contact between the pollen grains and stigmatic surface has been reported from *Phoenix dactylifera* (Torahi et al., 2017). However, no such changes were reported from *C. intybus* (Jaconis et al., 2017).

7.2.5 Fruit Set

The ultimate result of all the events participating in the sexual reproduction in plants is fruit set. Disruption at any stage is reflected as poor fruit set in terms of number of fruits, quality of fruit, number of seeds and quality of seeds (Fig. 7.1). Studies though few have shown that the most sensitive stage to be affected by PM is pollen germination, it being the most exposed stage during reproduction. Reduced pollen germination has been shown to affect fruit set in apple cultivars in polluted sites (7.50 and 7.66%) versus non-polluted sites (17.33%) (Ahmadi et al., 2019). Exposure to dust during pollen germination in pistachios has been documented to lead to parthenocarpy, fruit set without fertilisation (Zhang et al., 2019). Authors manually pollinated pistils with mixtures of dust and pollen and higher rate of fruits without seeds were seen with 1:1 pollen: dust mixture treatments. Prolonged exposure to high concentrations of PM has been shown to negatively impact the fruit set of mustard plants, resulting in decreased seed numbers and weight (Kumari, 2018). Soybean plants exposed to diesel exhaust in chambers produced significantly fewer fruits per plant and showed reduced seed mass, although they maintained the same number of seeds as the control plants (Jaconis et al., 2021). Whereas in the same study, chicory plants did not appear to have been negatively affected by diesel exhaust in the chambers. Authors suggested that, chicory being a roadside plant may already be adapted to stressful environments (Jaconis et al., 2021). Species growing in urban regions of Malda district of West Bengal, India, exhibited reduced seed numbers, seed weight, seed viability and *in-vivo* seed germination rates compared to those in reference sites (Das et al., 2025).

PM can (1) hinder the quality of pollen grains produced (2) disrupt plant-pollinator interaction through temporal asynchrony, altered attraction and perception (3) interfere with pollen-pistil interaction through reduced pollen germination and stigma

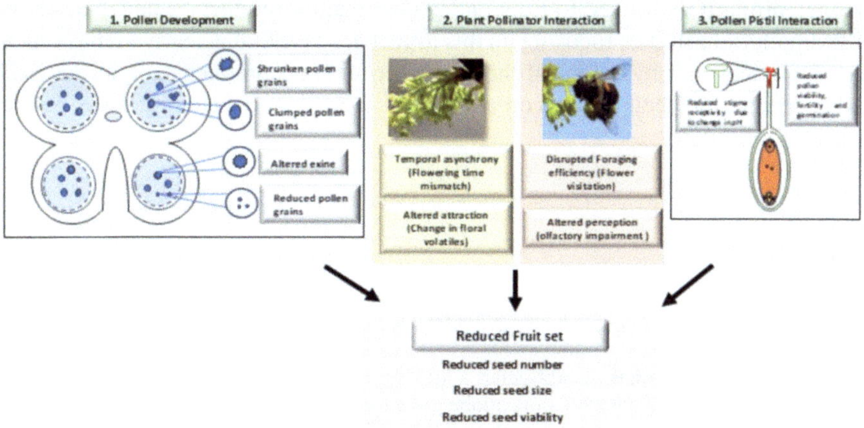

Fig. 7.1 Framework showing the effects of particulate matter on plant reproduction

receptivity. These effects have consequences for fruit set and ultimately for plant reproduction.

7.3 Conclusions and Future Perspectives

Overall, the discussion presented in this chapter suggests that PM affects all the major events or phases of plant reproduction. However, the effects may vary in degree depending on the differences in plant species' responses. Effects of PM are manifested in the form of delayed flowering, impedance in pollen development, disruption of plant-pollinator interactions and interference in pollen-pistil interaction. All these effects ultimately lead to reduction in the fruit set. It is evident that there are many knowledge gaps in our understanding of the impacts of PM on plant reproduction. Most studies on the impacts of air pollution on plant reproduction have focused on gaseous pollutants, while PM has received much less attention (Mandal et al., 2023; Roy et al., 2024). One of the major challenges in studying the impacts of PM on plant reproduction is accurately mimicking the realistic exposure of plants to PM, given their heterogeneous composition. There are several areas which warrant further research. Adhesion of PM on pollen is one the most reported aspect of plant reproduction under PM pollution. However, a very limited number of studies have discussed the extent of pollution of pollen grains or the number of particles adhering to pollen grains. Most of studies to understand the impacts of PM on pollen germination and viability are field based studies. Quantitative assessment of particles adhering to pollen grains is important for conducting which are almost lacking. Investigations in to the effects of PM on the other half for reproduction, i.e. the pistil are almost negligible. Accessing the female reproductive organ compared to the anther/pollen is much difficult. However, to fully understand plant reproduction under conditions of PM pollution, studies focusing on this aspect are essential. Clearly identifying the morphological and physiological changes in the female reproductive organ due to PM and understanding their impacts on yield loss is crucial for developing stress-resilient crops for the future.

References

Abrol, D. P. (2012). *Pollination biology: biodiversity conservation and agricultural production.* New York: Springer.

Ahmadi, F., ValizadehKaji, B., & Abbasifar, A. (2019). The effect of air pollution on developmental stages and pollen germination, pollen tube growth and fruit set of the apple cultivars 'golden delicious' and 'golab.' *Gesunde Pflanzen, 71*(4), 237–247.

Amjad, L., & Shafighi, M. (2012). The effect of air pollution on *Chenopodium album* L. pollen structure. *Journal of Agricultural Science and Technology, 2*(1A), 143–148.

Antipov, V. G. (1957). Impact of smoke and gases emitted by industrial enterprises on seasonal development of trees and shrubs. *Botanicheskii Zhurnal [Bot J], 42*, 92–95.

Azzazy, M. (2016). Environmental impacts of industrial pollution on pollen morphology of *Eucalyptus globulus Labill.* (Myrtaceae). *Journal of Applied Biology and Biotechnology, 05*, 57–62.

Bhatti, G., & Iqbai, M. Z. (1988). Investigations into the effect of automobile exhausts on the phenology, periodicity and productivity of some roadside trees. *Acta Societatis Botanicorum Poloniae, 57*(3), 395–399.

Capitani, G., Papa, G., Pellecchia, M., & Negri, I. (2021). Disentangling multiple PM emission sources in the Po Valley (Italy) using honey bees. *Heliyon, 1*(2), 7.

Chehregani, A., Majde, A., Moin, M., Gholami, M., Shariatzadeh, M. A., & Nassiri, H. (2004). Increasing allergy potency of *Zinnia* pollen grains in polluted areas. *Ecotoxicology and Environmental Safety, 58*(2), 267–272.

Choël, M., Ivanovsky, A., Roosc, A., Hamzé, M., Blanchenet, A. M., & Visez, N. (2022). Quantitative assessment of coagulation of atmospheric particles onto airborne birch pollen grains. *Journal of Aerosol Science, 161*, 105944.

Cox, R. M. (1984). Sensitivity of forest plant reproduction to long range transported air pollutants: In vitro and in vivo sensitivity of *Oenothera parviflora* L. pollen to simulated acid rain. *New Phytologist, 97*, 63–70.

Das, S., Barman, C., Roy, A., Mandal, M., Popek, R., Adit, A., & Sarkar, A. (2025). Air pollution may alter reproductive dynamics/efficacy in plants: Correlative evidences from an urban sprawl in Eastern Indo-Gangetic plain. *Aerobiologia, 41*, 35–53. https://doi.org/10.1007/s10453-024-09811-3

Dixon, R. A., & Paiva, N. L. (1995). Stress-induced phenylpropanoid metabolism. *Plant Cell, 7*, 1085–1097.

Fitter, A., & Fitter, R. (2002). Rapid changes in flowering time in British plants. *Science, 296*(5573), 1689–1691.

Fuentes, J. D., Chamecki, M., Roulston, T. A., Chen, B., & Pratt, K. R. (2016). Air pollutants degrade floral scents and increase insect foraging times. *Atmospheric Environment, 141*, 361–374.

Gallai, N., Salles, J. M., Settele, J., & Vaissière, B. E. (2009). Economic valuation of the vulnerability of world agriculture confronted with pollinator decline. *Ecological Economics, 68*(3), 810–821.

Ganga Kailas, J., Ramakrishna, H., & Seva, B. (2015). Pollution effect on pollen morphology in industrial areas of Hyderabad, Telangana state, India. *Journal of Industrial Pollution Control, 31*, 311–314.

Gillespie, C., Stabler, D., Tallentire, E., Goumenaki, E., & Barnes, J. (2015). Exposure to environmentally-relevant levels of ozone negatively influence pollen and fruit development. *Environmental Pollution, 206*, 494–501.

Girling, R. D., Lusebrink, I., Farthing, E., Newman, T. A., & Poppy, G. M. (2013). Diesel exhaust rapidly degrades floral odours used by honeybees. *Science and Reports, 3*(1), 2779.

Gottardini, E., Cristofolini, F., Paoletti, E., Lazzeri, P., & Pepponi, G. (2004). Pollen viability for air pollution bio-monitoring. *Journal of Atmospheric Chemistry, 49*, 149–159.

Goulson, D., Nicholls, E., Botías, C., & Rotheray, E. L. (2015). Bee declines driven by combined stress from parasites, pesticides, and lack of flowers. *Science, 347*(6229), 1255957.

Grantz, D. A., Garner, J. H., & Johnson, D. W. (2003). Ecological effects of particulate matter. *Environment International, 29*(2–3), 213–239.

Grigoryev, Y., & Pakharkova, N. V. (2001). Effect of industrial environmental pollution on winter dormancy in Scots pine. *Russian Journal of Ecology, 32*, 437–439.

Guedes, A., Ribeiro, N., Ribeiro, H., Oliveira, M., Noronha, F., & Abreu, I. (2009). Comparison between urban and rural pollen of *Chenopodium alba* and characterization of adhered pollutant aerosol particles. *Journal of Aerosol Science, 40*(1), 81–86.

Helander, M. L., Savolainen, J., & Ahlholm, J. (1997). Effects of air pollution and other environmental factors on birch pollen allergens. *Allergy, 52*(12), 1207–1214.

Herrero, M. (2003). Male and female synchrony and the regulation of mating in flowering plants. *Philosophical Transactions of the Royal Society B, 358*(1434):1019–1024.

Honour, S. L., Bell, J. N., Ashenden, T. W., Cape, J. N., & Power, S. A. (2009). Responses of herbaceous plants to urban air pollution: Effects on growth, phenology and leaf surface characteristics. *Environmental Pollution, 157*(4), 1279–1286.

Jaconis, S. Y., Culley, T. M., Keener, T., Odom, A. M., Elam, R., & Liang, M. S. (2021). Effects of short-term exposure to diesel exhaust on the ecophysiology, growth, and fecundity of soybean (*Glycine max* (L.) Merr.) and Chicory (*Cichorium intybus* L.). *The Ohio Journal of Science, 121*(2), 21–32.

Jaconis, S. Y., Culley, T. M., & Meier, A. M. (2017). Does particulate matter along roadsides interfere with plant reproduction? A comparison of effects of different road types on *Cichorium intybus* pollen deposition and germination. *Environmental Pollution, 222*, 261–266.

Jochner, S., Markevych, I., Beck, I., Traidl-Hoffmann, C., Heinrich, J., & Menzel, A. (2015). The effects of short-and long-term air pollutants on plant phenology and leaf characteristics. *Environmental Pollution, 206*, 382–389.

Joshi, O. P., & Sikka, J. (2002). Floral response of some tree species to air pollution. *Pollution Research, 21*(4), 417–419.

Klein, A. M., Vaissière, B. E., Cane, J. H., Steffan-Dewenter, I., Cunningham, S. A., Kremen, C., & Tscharntke, T. (2007). Importance of pollinators in changing landscapes for world crops. *Proceedings of the Biological Sciences, 274*(1608), 303–313.

Klein, S., Cabirol, A., Devaud, J. M., Barron, A. B., & Lihoreau, M. (2017). Why bees are so vulnerable to environmental stressors. *Trends in Ecology & Evolution, 32*(4), 268–278.

Kolek, F., Plaza, M. D., Leier-Wirtz, V., Friedmann, A., Traidl-Hoffmann, C., & Damialis, A. (2021). Earlier flowering of Betula pendula Roth in Augsburg, Germany, due to higher temperature, NO_2 and urbanity, and relationship with Betula spp. pollen season. *International Journal of Environmental Research and Public Health, 18*(19), 10325.

Kozlov, M. V., Eränen, J. K., & Zverev, V. E. (2007). Budburst phenology of white birch in industrially polluted areas. *Environmental Pollution, 148*(1), 125–131.

Kumari, N. (2018). Effects of pollutants on seed germinability, reproductive capacity and yield of mustard plant. *International Journal of Chemical Studies, 4*, 96.

Lusebrink, I., Girling, R. D., Farthing, E., Newman, T. A., Jackson, C. W., & Poppy, G. M. (2015). The effects of diesel exhaust pollution on floral volatiles and the consequences for honey bee olfaction. *Journal of Chemical Ecology, 41*, 904–912.

Malayeri, B. E., Noori, M., & Jafari, M. (2012). Using the pollen viability and morphology for fluoride pollution biomonitoring. *Biological Trace Element Research, 147*, 315–319.

Mandal, M., Popek, R., Przybysz, A., Roy, A., Das, S., & Sarkar, A. (2023). Breathing fresh air in the city: Implementing avenue trees as a sustainable solution to reduce particulate pollution in urban agglomerations. *Plants, 12*(7), 1545.
Mangla, Y., Khanduri, P., & Gupta, C. K. (2023). *Reproductive biology of angiosperms: Concepts and laboratory methods*. Cambridge University Press.
Miller-Rushing, A. J., Inouye, D. W., & Primack, R. B. (2008). How well do first flowering dates measure plant responses to climate change? The effects of population size and sampling frequency. *Journal of Ecology, 96*(6), 1289–1296.
Mukherjee, A., & Agrawal, M. (2017). World air particulate matter: Sources, distribution and health effects. *Environmental Chemistry Letters, 15*, 283–309.
Naidoo, G., & Chirkoot, D. (2004). The effects of coal dust on photosynthetic performance of the mangrove, *Avicennia marina* in Richards Bay, South Africa. *Environmental Pollution, 127*(3), 359–366.
Nanos, G. D., & Ilias, I. F. (2007). Effects of inert dust on olive (*Olea europaea* L.) leaf physiological parameters. *Environmental Science and Pollution Research, 14*, 212–214.
Negri, I., Mavris, C., Di Prisco, G., Caprio, E., & Pellecchia, M. (2015). Honey bees (*Apis mellifera*, L.) as active samplers of airborne particulate matter. *PLoS One, 10*(7), e0132491.
Ollerton, J. (2017). Pollinator diversity: Distribution, ecological function, and conservation. *Annual Review of Ecology Evolution and Systematics, 48*, 353–376.
Papa, G., Capitani, G., Capri, E., Pellecchia, M., & Negri, I. (2021). Vehicle-derived ultrafine particulate contaminating bees and bee products. *Science of the Total Environment, 750*, 141700.
Pellecchia, M., & Negri, I. (2018). Particulate matter collection by honey bees (*Apis mellifera*, L.) near to a cement factory in Italy. *PeerJ, 6*, e5322.
Potts, S. G., Imperatriz-Fonseca, V., Ngo, H. T., Aizen, M. A., Biesmeijer, J. C., Breeze, T. D., Dicks, L. V., Garibaldi, L. A., Hill, R., Settele, J., & Vanbergen, A. J. (2016). Safeguarding pollinators and their values to human well-being. *Nature, 540*(7632), 220–229.
Powney, G. D., Carvell, C., Edwards, M., Morris, R. K., Roy, H. E., Woodcock, B. A., & Isaac, N. J. (2019). Widespread losses of pollinating insects in Britain. *Nature Communications, 10*(1), 1–6.
Rafferty, N. E., & Ives, A. R. (2012). Pollinator effectiveness varies with experimental shifts in flowering time. *Ecology, 93*(4), 803–814.
Rai, A., Kulshreshtha, K., Srivastava, P. K., & Mohanty, C. S. (2010). Leaf surface structure alterations due to particulate pollution in some common plants. *The Environmentalist, 30*, 18–23.
Renner, S. S., & Ricklefs, R. E. (1995). Dioecy and its correlates in the flowering plants. *American Journal of Botany, 82*(5), 596–606.
Rezanejad, F. (2007). The effect of air pollution on microsporogenesis, pollen development and soluble pollen proteins in *Spartium junceum* L. (Fabaceae). *Turkish Journal of Botany, 31*, 183–191.
Rezanejad, F. (2009). Air pollution effects on structure, proteins and flavonoids in pollen grains of *Thuja orientalis* L. (Cupressaceae). *Grana, 48*(3), 205–213.
Rezanejad, F. (2012). Air pollution effects on flavonoids in pollen grains of some ornamental plants. *Turkish Journal of Botany, 36*, 49–54.
Rezanejad, F. (2013). The response of anther and pollen development, pollen cellular material release and pollen proteins to air pollution in Petunia hybrida Juss.(Solanaceae). *Iranian Journal of Science, 37*(1), 63–8.
Ribeiro, H., & Abreu, I. (2014). A 10-year survey of allergenic airborne pollen in the city of Porto (Portugal). *Aerobiologia, 30*, 333–344.
Ribeiro, H., Guimarães, F., Duque, L., Noronha, F., & Abreu, I. (2015). Characterisation of particulate matter on airborne pollen grains. *Environmental Pollution, 206*, 7–16.
Roulston, T. A., Cane, J. H., & Buchmann, S. L. (2000). What governs protein content of pollen: Pollinator preferences, pollen–pistil interactions, or phylogeny? *Ecological Monographs, 4*, 617–643.

Roy, A., Das, S., Singh, P., Mandal, M., Kumar, M., Rajlaxmi, A., et al. (2024). Summer-time monitoring and source apportionment study of both coarse, fine, and ultra-fine particulate pollution in Eastern Himalayan Darjeeling: A hint to health risk during peak tourist season. *MAPAN, 39*(4), 995–1009. https://doi.org/10.1007/s12647-024-00776-9

Ryalls, J. M., Langford, B., Mullinger, N. J., Bromfield, L. M., Nemitz, E., Pfrang, C., & Girling, R. D. (2022). Anthropogenic air pollutants reduce insect-mediated pollination services. *Environmental Pollution, 297*, 118847.

Sedghy, F., Sankian, M., Moghadam, M., Ghasemi, Z., Mahmoudi, M., & Varasteh, A. R. (2017). Impact of traffic-related air pollution on the expression of *Platanus orientalis* pollen allergens. *International Journal of Biometeorology, 61*, 1–9.

Sénéchal, H., Visez, N., Charpin, D., Shahali, Y., Peltre, G., Biolley, J. P., Lhuissier, F., Couderc, R., Yamada, O., Malrat-Domenge, A., Pham-Thi, N. (2015). A review of the effects of major atmospheric pollutants on pollen grains, pollen content, and allergenicity. *The Scientific World Journal, 2015*(1), 940243.

Shafiq, M. U., & Iqbal, M. Z. (2004). Effects of automobile pollution on the phenology and periodicity of some roadside plants. *Pakistan Journal of Botany, 35*(5), 931–938.

Shahali, Y., Pourpak, Z., Moin, M., Zare, A., & Majd, A. (2009). Impacts of air pollution exposure on the allergenic properties of Arizona cypress pollens. *Journal of Physics: Conference Series, 15*(1), 012027.

Siviter, H., Pardee, G. L., Baert, N., McArt, S., Jha, S., & Muth, F. (2023). Wild bees are exposed to low levels of pesticides in urban grasslands and community gardens. *Science of the Total Environment, 858*, 159839.

Tapparo, A., Marton, D., Giorio, C., Zanella, A., Soldà, L., Marzaro, M., Vivan, L., & Girolami, V. (2012). Assessment of the environmental exposure of honeybees to particulate matter containing neonicotinoid insecticides coming from corn coated seeds. *Environmental Science and Technology, 46*(5), 2592–2599.

Thimmegowda, G. G., Mullen, S., Sottilare, K., Sharma, A., Mohanta, R., Brockmann, A., Dhandapany, P. S., & Olsson, S. B. (2020). A field-based quantitative analysis of sublethal effects of air pollution on pollinators. *Proceedings of the National Academy of Sciences of the United States of America, 117*(34), 20653–20661.

Torahi, A., Arzani, K., & Moallemi, N. (2017). Studying the effects of dust on date palm (*Phoenix dactylifera* L.) pollination and fruit set. *Acta Horticulture, 1315*, 341–346.

Ugale, C., & Tidke, J. A. (2022). Air pollution effects on *Datura innoxia* Mill. pollen structure, protein and germination. *Aerobiologia, 38*(3):379–390

Ugale, C., Tidke, J. A., & Korekar, G. (2023). Pollen germination significantly affected by SO_2, NOx, PM10 and AQI in the Amravati City of Maharashtra, India. *Aerobiologia, 11*, 1–1.

Ulrichs, C., Welke, B., Mucha-Pelzer, T., Goswami, A., & Mewis, I. (2008). Effect of solid particulate matter deposits on vegetation: A review. *Functional Plant Science and Biotechnology, 2*(1), 56–62.

Ushitani, T., Perry, C. J., Cheng, K., & Barron, A. B. (2016). Accelerated behavioural development changes fine-scale search behaviour and spatial memory in honey bees (*Apis mellifera* L.). *The Journal of Experimental Biology, 219*(3), 412–418.

Visez, N., Ivanovsky, A., Roose, A., Gosselin, S., Sénéchal, H., Poncet, P., & Choël, M. (2020). Atmospheric particulate matter adhesion onto pollen: A review. *Aerobiologia, 36*, 49–62.

Wang, Q., Nakamura, S., Lu, S., Xiu, G., Nakajima, D., Suzuki, M., Sakamoto, K., & Miwa, M. (2012). Release behavior of small sized daughter allergens from *Cryptomeria japonica* pollen grains during urban rainfall event. *Aerobiologia, 28*, 71–81.

Waser, N. M., Price, M. V., Casco, G., Diaz, M., Morales, A. L., & Solverson, J. (2017). Effects of road dust on the pollination and reproduction of wildflowers. *International Journal of Plant Sciences, 178*(2), 85–93.

Zhang, L., Beede, R. H., Banuelos, G., Wallis, C. M., & Ferguson, L. (2019). Dust interferes with pollen–stigma interaction and fruit set in pistachio pistacia vera cv. Kerman. *Hortscience, 54*(11), 1967–1971.

Chapter 8
Airborne Particulates Threaten Pollinator Diversity and Crop Productivity: Implications for Food Security

Suranjana Sarkar

Abstract Pollinator-mediated crops are essential to providing key micronutrients in the human diet. Particulate matter (PM) has a considerable influence on pollinator diversity, which is necessary for food security. Air pollution especially PM, shows a significant influence on the pollination system, causing pollinator populations to drop and potentially jeopardizing the ability to maintain varied pollination systems. Bees are the most common crop visitors, whereas non-bee pollinators also play a significant role in visiting locally important crops. Insect pollinators are currently under threat from a variety of human-caused pressures, including climate change, habitat loss, and increased usage of pesticides. Anthropogenic activities have also caused significant increases in air pollution. Air pollution, particularly black carbon, can disrupt the gut microbiota of bees, which is critical for their survival. Additionally, PM deposition on pollinator habitats, particularly flowers, can introduce synthetic steroid growth boosters and antibiotics, endangering pollinators. Pollinator loss threatens ecological services, biodiversity, and food security. This article focuses on the negative effects of particulate matter on pollinators, how the pollinators are decreasing in number due to PM and subsequently its impact on human food chain and ecosystem. Pollinator ecosystem restoration should be undertaken as an organized remedy to the current situation.

Keywords Particulate matter (PM) · Air pollution · Pollinator diversity · Food security

S. Sarkar (✉)
Department of Botany, Surendranath College, Kolkata, West Bengal, India
e-mail: drsuranjanasarkar@gmail.com

A. Sarkar et al. (eds.), *From Sky to Soil*, https://doi.org/10.1007/978-3-032-31991-3_8

8.1 Introduction

Diversity of pollinators has great influence on safety of food supplies whereas pollinators are impacted by particulate matter. Air pollution especially originated by PMs causes loss of pollinators (Guenat & Dallimer, 2023), thereby posing a significant risk to the ecosystem together with food security and biodiversity (Mandal et al., 2023; Riaz et al., 2020). Thimmegowda et al., (2020) studied the effect of air pollutants on the pollinators. Bee is one of the important pollinators. Air pollution caused by black carbon affects the gut microbiome of bees thereby impacting the health and survival of bees (Penn et al., 2020). Also, pollinators are at risk due to PM deposition on pollinator habitats, such as flowers. PM deposition accelerates synthetic steroid growth boosters, growth of antibiotics and pesticides, thus making pollinators vulnerable (Green et al., 2023). Research carried out in Bangalore, India highlights how deposition of particulate matter (PM) is linked to change in length of bee survival, growth in flowers, their physiological parameters, and gene expression.

Pollinators viz., moths, flies, beetles, butterflies, and bees are insects that are necessary for the reproduction of plants and the preservation of a healthy bionetwork. This insect pollinator diversity is extremely essential for the maintenance of ecological resilience, for the increment of plant genetic diversity, and the assurance of food security worldwide. In the context of dispersion of pollen, each and every variety of pollinator has its individual, distinctive approach. For example, how butterflies make use of their long tongues in order to widen their grasp deep into flowers, while bees make use of their fuzzy bodies in order to collect pollen. Insect pollinators have a very important role in agriculture as they ensure food safety. Consumption of fruits, vegetables, and seeds would be far more challenging if they were not critical for the reproduction of several plants (Jones & Rader, 2022). Though, the usage of pesticides, habitat destruction and the consequences of climate change are some of the reasons putting them in jeopardy. As a direct result, a significant number of insect pollinators are seeing a decrease in population. Therefore, it is high time to take positive steps to protect them. This can be achieved if we promote sustainable agricultural practices such as reduction of pesticide usage, selecting gardens and agricultural fields favorable to pollinators (Kumar et al., 2022), and enhancement of public awareness about the significance of pollinators. The wide diversity of insect pollinators ensures plant productivity and variation, hence enhancing agricultural production, quality and improvement of global food security (Siviter et al., 2023).

8.2 Significance of Insect-Pollinator Diversity

Insect pollinators play a crucial role in the transfer of pollen from one flower to another, which is essential for the fertilization and subsequent production of seeds and fruits. The diverse array of pollinators ensures the efficient and successful pollination of different plant kinds. Different pollinators possess distinct adaptations and behaviors that impact their capacity to pollinate certain blooms. These parameters include characteristics like as body sizes, eating habits, and proboscis lengths. Variety of characteristics enables the efficient reproduction of a broad spectrum of plants, enhancing genetic diversity and adaptation within plant populations. Various pollinators have distinct preferences for flowers; hence, a variety of pollinators increases the chances of successful pollination for plants. This phenomenon arises due to the diverse pollinators that may carry distinct kinds of pollen. Consequently, when these pollinators fertilize a flower, it leads to the development of a new plant with distinctive genetic traits (Erickson et al., 2022).

Insect pollinators are encountering several challenges that jeopardize their existence. Various environmental hazards, such as pesticide use, climate change, and habitat depletion, pose significant threats to the ecosystem. The decline in accessible food and nesting grounds caused by habitat degradation presents a substantial danger to pollinators. Climate change endangers pollinators by disrupting flowering seasons and affecting the requirements of pollinators (Fisher et al., 2023). Ecosystem resilience is the capability of an ecosystem to endure and improve from disruptions while maintaining its essential functions and structures. Pollinators are crucial for plant pollination and seed production, especially when their populations are abundant. Recognizing the importance of plant variety in preserving ecosystem resilience is essential, as it aids in the creation of a more stable environment. The presence of a wide variety of plants decreases the likelihood of complete elimination of any one plant species (Hausmann et al., 2022). Insect pollinators, with their wide range of species, are critical for protecting the strength of ecosystems. Their presence helps to maintain and restore ecosystems after events like fires, droughts, or changes in land use. When certain pollinator species disappear, it can really mess up the whole plant-pollinator dynamic, causing a ripple effect on other organisms in the ecosystem (Bascompte & Scheffer, 2023).

8.3 Air Pollution is a Serious Threat to Pollinators

From the last decade, a lot of researchers have been done to study the effect of air pollution on pollinators (Doane, 1917). There are a lot of studies which investigated the impact of small-size particles on health starts from the small respiratory disorder to mortality (Agarwal et al., 2014; Awasthi et al., 2017). Biologists have observed that pollinators sometimes struggle with memory and learning, as well as experiencing negative impacts on their heart rates, immunity, and stress tolerance. Pollutant-contaminated air has been found to influence plants in various ways. It can decrease the diversity of plant communities, harm the leaves of herbaceous vegetation, hinder the growth of trees, and make plants more susceptible to other forms of stress (Roy et al., 2024a). Despite the understanding that air pollution can negatively affect pollen via modifications in its chemical composition, physical shape, and quantity, there has been limited investigation into the impact on plant reproduction (Choël & Visez, 2019). In the same way, some insects might face challenges in differentiating between floral volatiles and anthropogenic volatiles caused by air pollution (Jürgens & Bischoff, 2017).

8.3.1 The Particulate Matters (PM)

A combination of solid, liquid, or solid and liquid particles floating in the air is known as particulate aerosol. The size, makeup, and source of these suspended particles are all different. Aerodynamic properties are a convenient way to categorize particles because they control three things: (a) how particles are transported and removed from the air; (b) how they deposit in the respiratory system; and (c) how they are related to the sources and chemical makeup of particles. The aerodynamic diameter, or the size of a unit-density sphere with the same aerodynamic qualities, easily summarizes these parameters. Aerodynamic diameter, often known as particle size, is the basis for sampling and characterizing particles (Roy et al., 2024b).

PM is a specific form of pollutant that exists in the atmosphere and is made up of a wide variety of liquid and solid airborne particles (World Health Organization, 2013). Studies suggest that particle pollution may originate from a combination of natural and anthropogenic sources, including oceanic salts, volcanic residue, windborne soil particles, forest fire ash, pollen, fungal spores, and organic compounds generated by living organisms (Kelly & Fussell, 2012; Kim et al., 2015). It is via a variety of causes that human activities contribute to the pollution of air. The combustion of fossil fuels in vehicles and power plants, the wear and tear of pavement caused by traffic, the release of pollutants from brakes and tires, emissions from industrial processes such as the production of cement, metal, ceramics, and bricks, as well as activities such as construction, smelting, mining, quarrying, and certain agricultural practices are all aspects that contribute to the emission of greenhouse gases (Kumar et al., 2024; Perera, 2018). Additionally, smoking cigarettes and burning wood in stoves also contribute to air pollution (Kelly & Fussell, 2012; Kim et al., 2015).

Research shows that various sources, both human-made and natural, can release particles into the atmosphere (Juda-Rezler et al., 2011; Kelly & Fussell, 2012). Biological processes in the atmosphere give rise to secondary PM through photochemical gas-to-particle reactions. These reactions involve compounds with low volatility that transform into solid or liquid phases. These reactions involve non-methane volatile organic compounds and several gases like nitrogen oxides, sulfur dioxide, and ammonia. They have been extensively studied by various researchers (Harrison et al., 2000; Juda-Rezler et al., 2011).

PM is divided into three categories based on its aerodynamic diameter (da): ultrafine PM0.1 (da < 100 nm), fine $PM_{2.5}$ (da < 2.5 μm), and coarse PM_{10} (da ≥ 2.5 μm and da < 10 μm) (Brook et al., 2010). Particles with a diameter of 0.1–1 μm can remain in the atmosphere for extended periods, allowing them to be carried across long distances through the air (Das et al., 2023; Juda-Rezler et al., 2011). Average amounts of PMs worldwide are presented in Figs. 8.1, 8.2, and 8.3.

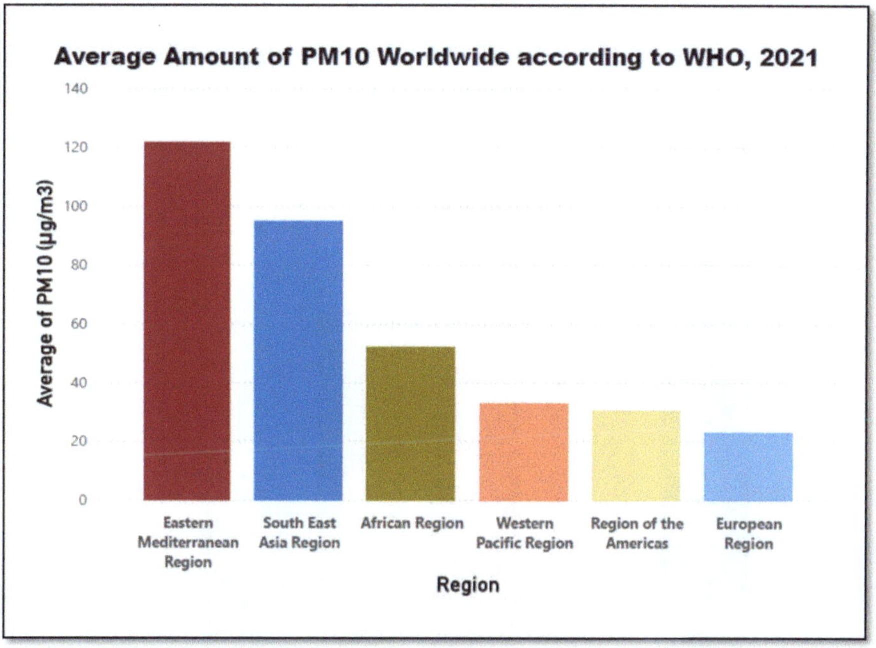

Fig. 8.1 World region wise average amount of PM10 (μg/m^3)

Metals, sulfates, nitrates, ammonium, minerals and various inorganic ions, such as calcium, magnesium, potassium, sodium, and chloride, along with elemental and organic carbon, and Polycyclic Aromatic Hydrocarbons (PAH), are some of the chemical components found in PM (Harrison, 2020; Kelly & Fussell, 2012; Kim et al., 2015). In addition, bacterial spores mold, viruses, fungus, bacteria, and pollen are the important biological constituents (Manisalidis et al., 2020). PM plays a significant role in impacting the health of ecosystems and is a leading contributor to illness and death worldwide (World Health Organization, 2013). Despite the extensive research conducted on atmospheric PM over the past few decades, the specific impact of each chemical component on its harmful effects remains uncertain.

8.3.2 *Threats of PM on Pollinators*

There are several ways that pollinators are exposed to pollution, and it is difficult or impossible for them to escape them altogether. Particulate matter can adhere to pollinators' bodies and wings during flight and accumulate over time. Additionally, as animals visit flowers or make their nests in the ground, particulate matter can collect on the surfaces of plants, flowers, and soil and transmit to the animals. In

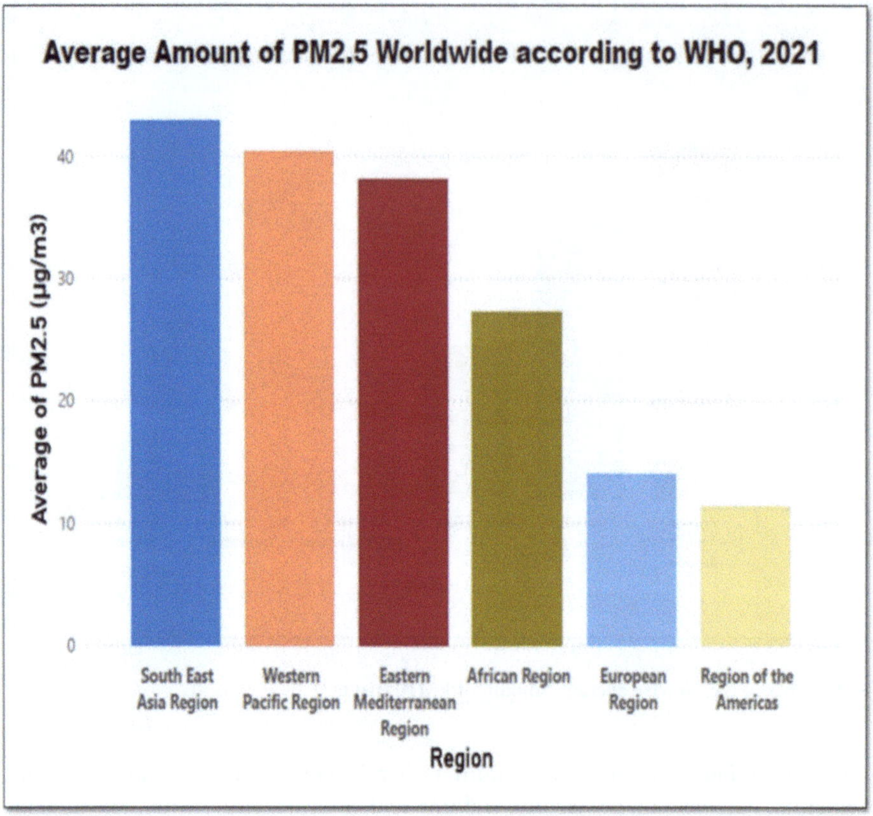

Fig. 8.2 World region wise average amount of PM2.5 ($\mu g/m^3$)

addition, pollinators and other insects inhale harmful gases in contaminated air, just like humans do.

The research's general findings are evident, even if people still need to understand more about how so many diverse pollinator species are affected. In the worst situations, air pollution can completely destroy pollinators by making it more difficult for them to locate flowers and effectively pollinate them, even at low exposure levels.

8.3.3 Particulate Matters Mess with Pollinators' Senses

PM interferes with bugs' capacity to smell, which has an impact on many aspects of their life. For many insects, but especially for pollinators, smell is an extremely essential sense. Scent-based pheromones are widely used by insects to find mates or warn one another of danger. Not only does air pollution mask these scents, but

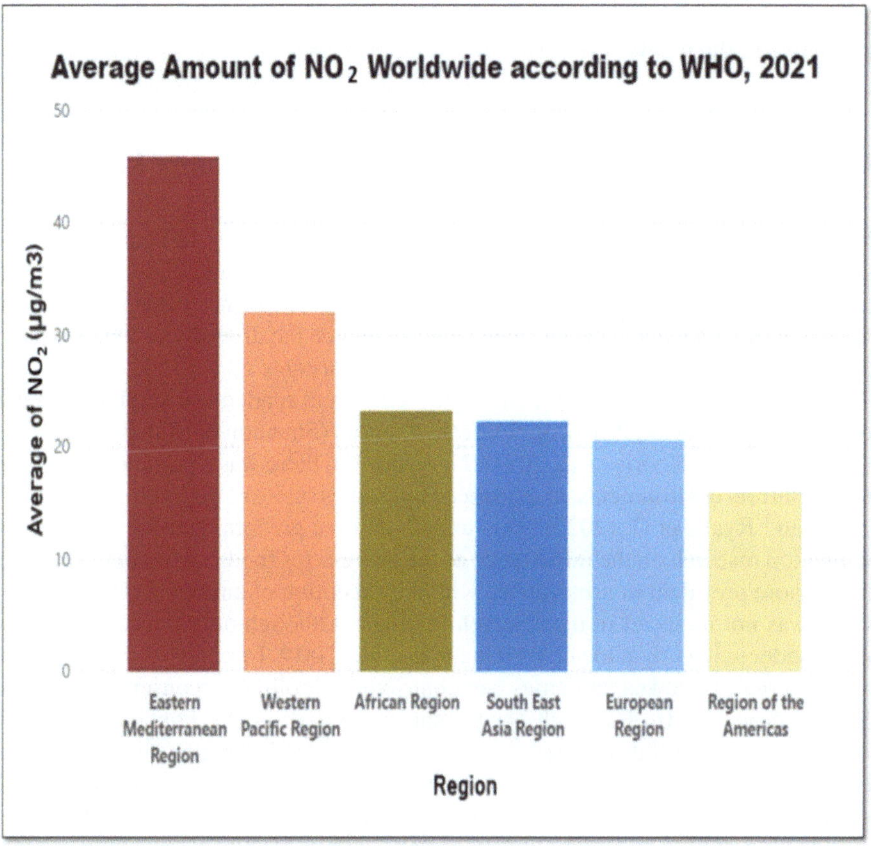

Fig. 8.3 World region wise average amount of NO_2 ($\mu g/m^3$)

it may also degrade the pheromone molecules, making this kind of communication much more difficult. The same is true of smells, such as those from flowers, which pollinators use to locate food sources or locations to reside. In one research, hawk-moth visits to flowers were reduced by as much as 50–70% when they were exposed to nitrogen oxide, a harmful pollutant.

According to a recent study, pollinators may find it more difficult to discover blooming plants due to the substances in PM that obscure their fragrance, thereby hindering plant reproduction. Due to PM content, it is difficult to detect the pollination as PM involves restricting plant reproduction by masking blossoming plants' scents. The study, which was published in the journal Science, demonstrated how ozone pollution and nitrate radicals (NO_2, NO_3), which can arise from combustion in gas-fueled cars and coal and gas power plants, can deteriorate the natural chemical fragrances of flowers. Because of all these issues, pollinators are forced to travel greater and greater distances in search of flowers, which results in fewer visits by pollinators to the plants.

8.3.4 Particulate Matters Harm the Plants Pollinators Rely on

Pollinators use plants as a home and a food source, but plants in contaminated air don't grow as big. This may even change the pollen that the plants generate, resulting in smaller, fewer pollen grains. In the end, these consequences may result in the extinction of some plant species, which would reduce the diversity of the ecosystem and make it less able to support various pollinator species.

Studies on air pollution drew insights from different academic disciplines, such as forestry (Paoletti et al., 2010), agricultural sciences (Shannon & Mulchi, 1974), and plant physiology (Noorie et al., 2024). Nevertheless, there was a particular emphasis on agricultural environments, as evidenced by studies such as Drogoudi and Ashmore (2000) and Ryalls et al. (2022). This aligns with the predominant focus of current pollination research on the importance of the process for food production. Air pollution is more prevalent in urban areas, with the exception of ammonia (Defra, 2019), which was not included in the research findings. Although only three studies have been conducted in urban locations (Chauhan et al., 2004; Tommasi et al., 2022), it is worth noting that towns and cities can contribute to pollinator conservation efforts (Wenzel et al., 2020). Research has primarily focused on less contaminated landscapes, but it has still discovered the impact of air pollution which clearly indicates that pollination would be extra evident in highly polluted areas. Thus, if air pollution is not addressed, it is likely to worsen the problem of biodiversity by hindering plant reproductive success and disrupting plant-pollinator interactions, especially in urban areas.

8.3.5 Air Pollution Damages Pollinators' Bodies, Sometimes Killing Them

Like humans, pollinators suffer long-term physical injury from pollution exposure, which makes it harder for them to withstand additional dangers. It has been discovered by researchers that air pollution can impair bees' immune systems, making it more difficult for them to fend off diseases and recover from them. Similarly, pollution damages their circulatory systems, making breathing harder. The cognitive function of pollinators is also compromised. Polluted air may throw pollinators off-kilter and even impair their memory, which makes it harder for them to recognize and recall plant scents.

8.4 Pollinators and Global Food Security

Pollinators, both domesticated and wild, have dramatically decreased over the past few decades, endangering ecosystem resilience, species survival, and global food security. 87 of the world's principal food crops rely on animal pollination. All of them together make up 35% of the volume of food produced worldwide. For the human diet to include key micronutrients, pollinator-mediated crops are crucial. Insect pollinators are essential to the growth of many decorative plants as well as crops used for biofuels, lumber, textiles, feed, and phyto-pharmaceuticals.

8.4.1 Need of Pollinator Variety in Agricultural Productivity

According to studies, pollinator loss would result in malnutrition even though a small number of main courses are made from crops that do not need pollination and also have a high global output level, responsible for significant contribution from human nutritional requirements (Ellis et al., 2015; Potts et al., 2010). Pollinator diversity is critical to agricultural output in terms of both yield quantity and quality, according to mounting evidence (Cusser et al., 2016; Garibaldi et al., 2014; Vasiliev & Greenwood, 2020), despite some studies' contention that crop pollination alone is insufficient justification for protective conservation of wild pollinators (Richards, 2001).

The quality of agricultural produce may be determined by the diversity of pollinators. Coffee weight rises only in regions where both ants and flying insects offer pollination, according to pollinator exclusion studies conducted on a coffee plantation (Philpott et al., 2006). Like this, intercropping raises the fruit weight of mango (*Mangifera indica*) plantations by attracting high flower visitor richness (Rafique et al., 2016). Similarly, recent research has shown that the physiological and chemical characteristics of fruit production may be influenced by the richness of pollinator species. According to Tremlett et al. (2020), the removal of nectarivorous bats from pitaya (*Stenocereus queretaroensis*) flowers resulted in a 35% decrease in yield, a 16% reduction in fruit weight, and 13% less tasty fruit. Heavier cotton seeds were only discovered in portions that were pollinated by a variety of pollinators, according to a different study conducted on cotton, a significant crop used for fabric (Cusser et al., 2016). Besides, compared to locust (*Eriobotrya japonica*) fruits pollinated only by honeybees, locust fruits acquired by open pollination (visited by high species richness) are heavier, have greater sugar levels, and have stronger antioxidant characteristics (Ahmad et al., 2021).

The processes via which the diversity of pollinator species affects agricultural productivity and quality need to be investigated. Nonetheless, a number of scientists have hypothesized that functionally varied pollinators could increase genetic diversity and promote gene flow (Kumar et al., 2020). Research has demonstrated that cross-pollinated crops provide higher-quality crop yields than self-pollinated ones (Cusser et al., 2016). Due to their increased exposure to pollinator complementarity, these cross-pollinated crops are less susceptible to the detrimental impacts of a number of variables on pollination success, including temporal (day or night), environmental, and climatic conditions. Thus, a strategy for preserving excellent crop quality with the impact of climate change may involve taxonomically varied pollinators.

The variety of pollinators increases the amount of agricultural produce. Numerous crops, including coffee, almonds, pumpkin, and apples have shown that pollinator variety influences fruit set (Katumo et al., 2022). According to the findings of a recent meta-analysis conducted by Woodcock et al. (2019), oilseed rape (Brassica napus), which is a significant crop for the global economy, produced a greater amount of output when there was a high functional pollinator diversity. It has been established that the variety of wild bees has an effect on the growth of fruit in apple crop orchards located in rural areas (Mallinger & Gratton, 2015). This is true regardless of the number of honeybees that are regulated. According to Garibaldi et al. (2014), expansion in the number of wild pollinators offers a significant boost to the establishment of fruit under circumstances when honeybee flower visiting rates are high. Similar to this, a momentous rise in the variety of pollinator species in urban areas has a significant impact on the seed set of the jalapeño crop, which is a species of *Capsicum annuum* (Cohen et al., 2021).

In agricultural contexts with a high floral visitor richness, there is a correlation between pollinator complementarity, which is given by taxonomically diverse pollinators, and excellent crop productivity (Cusser et al., 2021). For example, unlike honeybees, wild bees had a lower favourability for floral density in an apple fruit orchard (Mallinger & Gratton, 2015). Honeybees have a higher preference for floral density. While wild bees visit flowers on all the trees in the orchard, indicating no preference for floral density, honeybees only visit flowers on apple trees that bloom profusely (Mallinger & Gratton, 2015). It is observed in the US Gulf Coast cotton agricultural environment that non-bee pollinators—especially flies and butterflies—provide an equivalent functional visitation space with bee pollinators (Cusser et al., 2021). Using a worldwide crop data collection, it is concluded that pollinator temporal complementarity was essential for creating inter-annual stability in global crop pollinator groups (Senapathi et al., 2021). Studies conducted in Germany provided evidence that bee activity generally increases on warmer, sunny days, although it increases more on colder days for bumblebees especially in an agroecosystem (Fründ et al., 2013). Improved agricultural productivity is a result of increased flower pollination during crop blooming seasons due to the complementarity between different pollinator guilds in both space and time. Understanding the complex interplay of stressors is crucial in determining the impact on biodiversity, including various forms of pollution. However, the understanding of how pollution affects the functioning of ecosystems is still not fully complete. Through a comprehensive analysis, the existing information on the impact of noise, light, and air pollution on the intricate process of pollination is brought together. It's worth mentioning that air pollution negatively affects pollination, resulting in reduced plant reproduction and disrupting the relationship between plants and pollinators. However, it has no direct impact on the pollinators themselves. However, there are still important gaps in our knowledge. In tropical, and middle- and low-income nations, where air pollution levels are high and the human population depends heavily on pollination-supported food production, there is a lack of available information. In addition, the impact of light and noise pollution is still not fully understood, as there is no concrete evidence of its effects on pollinators and plants, despite the potential risks it poses to their relationship. However, people can demonstrate the importance of addressing air pollution. This study emphasizes the need of decreasing air pollution levels to guarantee the optimal operation of pollination systems, while also enhancing the comprehension of the topic.

8.5 Regarding Pollinators and Their Value

Pollination is a crucial ecosystem function that plays a role in regulating, maintaining, and contributing to cultural aspects of the environment (Chagnon et al., 2015). The concept encompasses a cohesive network of connections that interrelate the plant life, animal life, and well-being of humans on Earth (Kevan & Menzel, 2012). Pollination is necessary for the fertilization of fruits and seeds in several crops and wild plants. Pollination may occur by passive self-pollination, anemophilous pollination, or zoophilous pollination, which involves invertebrates like bees, mammals like bats, and birds like hummingbirds. Self-pollination is a phenomenon that happens in some plant species; however, it is not often the primary method of pollination. Less than 10% of blooming plants rely on anemophilous (wind) pollination as their primary method of pollination. Nearly 94% of all flowering plants on Earth rely on zoophilous pollination for their reproduction and development (Ollerton et al., 2011; Vanbergen & the Insect Pollinators Initiative, 2013). According to Klein et al. (2007), animal pollination is essential for the growth of 87 of the world's main food crops. The crops mentioned include vegetables, fruits, nuts, edible oils, protein-rich crops, as well as spices and sauces (Maxim & Van der Sluijs, 2013). Insect pollination is crucial for the growth of both fiber and fodder crops. This means that the decline of insect pollinators may have an indirect impact on the animal agricultural industry as well. Animal pollination is necessary for the growth and production of biofuel crops, such as canola, as well as many trees that are cultivated or harvested for lumber. Most plants used for the development of phytopharmaceuticals rely on pollinators. This also applies to the majority of decorative plants. Our food crops which are enhanced by insect pollination include several important vegetable crops, fruit crops, nut crops, edible oil, and proteinaceous crops (Maxim & Van der Sluijs, 2013). These collectively constitute 35% of the global food production volume (Klein et al., 2007). Insect pollination is also necessary for many fiber crops (like cotton, flax, and hemp), fodder crops for cattle breeding (like alfalfa and soy), biofuel crops (like canola), timber crops (like eucalyptus), ornamental plants, and plants that produce phytopharmaceuticals (like Cinchona, which is used to make quinine, which is commonly used to treat malaria). In general, higher yields from animal pollination benefit around 75%

of all crop species. Additionally, pollinators can increase the genetic diversity of wild flowering plants and improve crop quality (especially fruit quality), shelf life, and commercial value (for example, poorly pollinated strawberries are malformed and look less attractive for consumers). Because of this, they have a large economic value (Van der Sluijs & Vaage, 2016). Gallai et al. (2009) calculated that the value of insect pollinators was €153 billion in 2005. According to their estimates, the average value of pollinator-dependent crops was 761 € per ton, whereas the average value of other crop types, such as cereals and roots and tubers, was 151 €. These crops significantly improve people's quality of life and food cultures worldwide by offering variety, nutritional value, and mouthwatering flavors (Chagnon et al., 2015). Pollinator-dependent agricultural production and consumption are increasing globally, although mostly in the rich world (16.7%, compared to 9.4% growth in the poor world between 1961 and 2006 (Aizen & Harder, 2009; Chagnon et al., 2015). About 40% of the world's population make their living from agriculture, in addition to providing food for everyone on the planet today. The decrease in pollinators may result in the disappearance or reduction of a substantial quantity of crops with significant cultural and economic worth. Furthermore, pollinator-mediated crops play a crucial role in supplying essential nutrients to the human food supply, even though wind-pollinated cereals account for the majority of calories in the human diet. They provide over 90% of vitamin C, 100% of lycopene, and nearly 100% of the antioxidants β-cryptoxanthin and β-tocopherol, as well as most lipids (74%), vitamin A (>70%) and related carotenoids (98%), calcium (58%), fluoride (62%), and a significant amount of folic acid (55%). Approximately 40% of the world's supply of nutrients for people comes from pollinator-mediated crops overall (Eilers et al., 2011). Hidden hunger, or a lack of certain micronutrients, affects an estimated 2 billion individuals worldwide (IFPRI, 2014; Nicole, 2015). Chronic vitamin and mineral shortages have substantial and long-lasting effects on health and wellness. People's capacity to live and develop is hampered by hidden hunger, which may sometimes have permanent negative health implications. This kind of malnourishment may happen to those who have no energy shortages and the ability to cope with fat. It mostly affects developing nations at the moment, holding them back in a vicious cycle of starvation, ill health, ongoing poverty, and decreased production (IFPRI, 2014). There is growing evidence that hidden hunger issues will only get worse if current trends in pollinator loss continue (Eilers et al., 2011; Nicole, 2015). A study by Ellis et al., (2015) found that up to 56% of populations in Mozambique could become newly at risk if pollinators were removed, and that the character of local diets is a crucial determinant of the susceptibility of concealed hunger to the decline of pollinators. Vitamin A is crucial in this context; fruits and vegetables, many of which are heavily dependent on pollinators, provide almost two thirds of the vitamin A in children's diets. The UN defines food security as meeting dietary choices and nutritional requirements in addition to a continuous supply of food. This is a worldwide concern, as the definition makes clear. Because many bird and animal species eat wild fruits (such as birds that eat a variety of wild berries in woods), wild nuts, and wild plant seeds, pollination is also necessary to maintain the food sources for wildlife. All species that rely on wild fruits, nuts, and seeds would suffer greatly from a sharp reduction in fruit, nut, and seed setting if all insect pollinators were eliminated. Food deprivation may also affect herbivores if pollinator-dependent plants are absent on which these herbivores rely, are no longer able to procreate. Many different types of animals, primarily insects, such as stingless bees, solitary bees, hover flies, butterflies, wasps, moths, beetles, midges, and other invertebrates, are known to pollinate; these include humans (hand pollination), bats, squirrels, parrots, hummingbirds, and some primates (Buchmann & Nabhan, 1997; Klein et al., 2007). Bees are the primary pollinators of agricultural crops (UNEP, 2010). In the past, domestic honeybees have received the most of the credit. Recent research, however, has shown that wild pollinators contribute more significantly to worldwide agricultural pollination than previously thought (Breeze et al., 2011). In the UK, estimates suggest that little more than one-third of crops are pollinated by controlled honeybees (*Apis mellifera*). Wild bees are an essential part of the uncontrolled pollinator population. There are around 25,000 species of bees in the world (Chagnon et al., 2015). But apart from bees, a variety of other insects are also effective pollinators, accounting for 39% of visits to agricultural flowers (Rader et al., 2016). The most efficient pollinators of fruit crops are thought to be wild insect species (Chagnon et al., 2015). According to Klein et al. (2007), three of the nine pollinator-dependent crops—vanilla, passion fruit, and atemoya—that did not show evidence of honeybee presence are now hand-pollinated in certain regions of the globe, partly because the necessary natural pollinators are less common there. Bees and humans have a long history together. Beekeeping has a long history that dates back to pre-agricultural periods (Dams, 1978; Pattinson, 2012). Large-scale beekeeping has existed since the dawn of agriculture (Bloch et al., 2010; Ebert, 2011). Because of their many applications, such as the medical use of propolis in combating infections (Banskota et al., 2001) and honey to speed up the healing of burn wounds (Jull et al., 2008), bees and apiary products have a significant extrinsic value. They also serve as an inspiration for scientific and technological research on robotic flight and navigation (Srinivasan, 2011). For centuries, honey has been a vital component of cuisine and continues to be so, for example, in traditional pastries (Ellis, 2014); aside from their function as pollinators, domestic bees as a species possess other significant attributes. Some of these qualities and values are undoubtedly inherent in and of themselves, even if the majority are extrinsic. Therefore, the present idea is that these species should be preserved for their own sake rather than only for their practical utility. There is aesthetic, nutritional, cultural, and economic benefits associated with other insects. Insect-shaped jewelry and decorations have been utilized for centuries worldwide; sayings, stories, and even religious respect have been influenced by the forms of butterflies, moths, and beetles (Southwood & Richard, 1977). It is easy to forget how important insects are to people and how important they are to natural systems when talking about pollinators and how they offer ecosystem services. When addressing the issue of pollinators' worldwide decrease, both the intrinsic and extrinsic worth of these creatures should be considered (Pearson, 2016). NGOs and the general public are generally not very concerned about the world's entomofauna. This is maybe partly because of the belief that all insects are pests—even unquestionably helpful ones like honeybees might harm you (Southwood & Richard, 1977). Also, partly due to the genetic gap that separates humans and insects on the tree of life, and the ensuing doubt about the applicability of ideas like well-being and suffering to insects. But as people become more conscious of how important pollinators are, local communities are implementing grassroots programs to save them, with bees serving as the most recognizable example.

8.6 Reasons for the Worldwide Pollinator Drop

Pollinators, both wild and domesticated, have dramatically decreased during the last several decades (Potts et al., 2010). Parallel decreases in pollinators (mostly bees, but also hoverflies) and insect-pollinated plants were documented in Britain and the Netherlands by Biesmeijer et al. (2006). There has been a notable and extensive decrease in bumblebee populations in the United States. Up to 96% of the relative profusions of four common species of bumblebees have decreased. Some of the examined geographic ranges have shrunk in the previous 20 years, ranging from 23 to 87% (Cameron et al., 2011). There has also been evidence of a drop in managed honeybee populations in the US (Allen-Wardel et al., 1998) and Europe (Potts et al., 2010). In contrast, the number of managed hives is growing faster than the number of colony collapses in other countries of the globe, including China and Argentina (Goulson et al., 2015). Since the 1990s, the loss of wild pollinators has compelled orchard owners in southwest China to use human hand-pollinators, which are significantly less efficient and unsustainable from an economic standpoint (Partap & Tang, 2012). However, there are some fruit varieties where hand pollination is advantageous, like certain pear varieties where male and female trees don't bloom at the same time. Strict quarantines are in place in Australia to prevent foreign species—such as viruses, parasites, or competing bees—from upsetting native honeybee populations. As of now, there have been no verified reports of a rise in colony failures. Beekeepers from Egypt have observed colony collapses in Africa (UNEP, 2010). On the other hand, not much is known about pollinators in these areas, especially when it comes to untamed species (Vanbergen & the Insect Pollinators Initiative, 2013). Over the last several decades, there have been observations of an increase in bee diseases, unusually high winter colony losses, and shortened bee queen lifespans (Maxim & Van der Sluijs, 2013, Pisa et al., 2015). Thus far, evidence suggests that there isn't a single factor responsible for the worldwide rise in winter colony losses. All viruses and other diseases associated with colony collapse are found year-round in healthy colonies as well, suggesting that these infections do not cause colony collapse on their own (Runckel et al., 2011). A confluence of mutually reinforcing factors seems to be more plausible (Goulson et al., 2015; Vanbergen & the Insect Pollinators Initiative, 2013). The widespread use of systemic pesticides, especially the so-called neonicotinoids, has become increasingly prevalent among them (EASAC, 2015; Van der Sluijs et al., 2015). According to van Lexmond et al. (2015), there has been a catastrophic worldwide drop in insects and arthropods in general, which includes the collapse of uncontrolled insect pollinators. Provincial studies (available largely for Europe and the US; see e.g., Chagnon et al., 2015) suggest that reduced pollination may lead to lower agricultural yields and species diversity, while worldwide evidence for this effect has not yet been established. The strengthening of agriculture, with its concomitant loss of natural habitats and foraging and nesting resources, large-scale use of agrochemicals like fertilizers, fungicides, and insecticides, nitrogen deposition, climate change, invasive species, pathogen spread, the multiplicity of increases in roads and motorized traffic, and the continent-wide nocturnal light pollution are the main causes of the ongoing collapse of Earth's pollinators (and entomofauna at large). In the broader scheme of things, pollinators are especially important since they support ecosystems and are essential to local and global food security. The important drivers are further discussed in the paragraphs that follow.

Bees and other pollinators feed on flowers that provide them with nectar (carbohydrates) and pollen (proteins and fats). Worker bees live for around four weeks, and honeybee colonies normally include forty thousand bees. This implies that during the whole foraging season, a continuous brood cycle from eggs to larvae to bees must be maintained. Additionally, food supplies necessary for the colony's overwintering must be gathered. Throughout the foraging season, fresh pollen of adequate nutritional quality, quantity, and variety must be accessible. This calls for an enormous and ongoing migration from both Europe (Potts et al., 2010) and the US (Allen-Wardel et al., 1998). In contrast, the number of managed hives is growing faster than the number of colonies collapses in other countries of the globe, including China and Argentina (Goulson et al., 2015). Since the 1990s, orchard farmers in southwest China have been forced by the decline of wild pollinators to use human hand-pollinators, which are significantly less effective and unsustainable from an economic standpoint (Partap & Tang, 2012). However, there are some fruit varieties where hand pollination is advantageous, like some pear varieties where the male and female trees do not flower at the same time. Strict quarantines are in place in Australia to prevent foreign species—such as viruses, parasites, or competing bees—from upsetting native honeybee populations. As of right now, there have been no verified reports of a rise in colony failures. Beekeepers from Egypt have reported colony collapses in Africa (UNEP, 2010). On the other hand, not much is known about pollinators in these areas, especially when it comes to untamed species (Vanbergen & the Insect Pollinators Initiative, 2013).

We can work toward cleaner air to help pollinators. They depend on us for protection, even though we can lessen the effects of air pollution by using air filters and respirators. Altogether, human beings are the most effective force for reducing air pollution by pressing for change. Environmental justice groups have a long history of addressing and fighting for clean air since many industrial facilities were and are purposefully located in areas that have been systemically disinvested in (Guenat & Dallimer, 2023).

8.7 The Actions Listed Below May Be Implemented to Improve the Air Quality for Pollinators and Our Communities

The current pollinator conservation policies are very disjointed, incomplete, and disorganized. Numerous policy areas, such as chemical control, agricultural policy, conservation policy, environmental policy, and spatial planning policy, are impacted by the problem of pollinator stewardship. It cuts across all levels of governance, from the local (e.g., what individuals can do to make their yard pollinator friendly) to the global (e.g., phasing out particular pesticides and mitigating climate change), in terms of agency and what can and should be done by whom. This presents significant obstacles to the concerted effort required to handle a problem with this level of complexity. There are some necessary steps to combat this situation as detailed here. First, people can join a community-based environmental or air quality advocacy organization. They should support, inspire or launch neighborhood air monitoring initiatives (Das et al., 2025). If they are planning to set up pollinator habitat next to an industrial sector or other sites where pollution has been happening already, they should test the soil and air for presence of toxins. They should also verify the Environmental Protection Agency database to check if any neighboring facilities are violating the Clean Air Act. They should appeal for more regulation of pollution sources by contacting their administrative and/or governing bodies. The most crucial thing to keep in mind is that the fight for human rights to safe and clean air is closely linked to the health of pollinators, who sustain the ecosystems on which we all rely.

8.8 Conclusion

Pollution has a negative impact on pollinators and is a key contributing factor to the loss of insects worldwide. Some types of air and soil contaminants have a significant impact on interactions between plant pollinators. In addition to contributing to the loss of biodiversity, low pollination rates have ecological and economic ramifications that should be discussed. Strict action is needed to protect these priceless animals from pollution and its effects on pollination, as well as to execute environmental regulations pertaining to conservation. To reduce the negative effects of air pollution on pollinators, it is recommended to use renewable fuels and clean energy production techniques. Integrated Pest Management (IPM) should be utilized to reduce the amount of chemicals used in gardens while introducing organic formation. To mitigate the effects of this loss, a worldwide agreement for the restoration of the environment supporting pollinators should be started. The current pollinator crisis jeopardizes local and global food security, exacerbates issues with hidden hunger, weakens ecosystem resilience, and has the potential to destabilize ecosystems that are essential to human survival. It is necessary to have an integrated strategy that tackles the main factors concurrently. This entails developing and restoring floral and nesting resources, ending the worldwide use of neonicotinoids and fipronil as preventative measures, enhancing test procedures for pesticide authorization, and preserving and restoring regulatory science's independence.

Acknowledgements The author wants to express her gratitude to Dr. Sanmitra Sarkar of IBM and Shriranjani Sarkar of Auxilium Convent School, Dum Dum for their sincere help in writing this chapter.

References

Agarwal, R., Awasthi, A., Mital, S. K., Singh, N., & Gupta, P. K. (2014). Statistical model to study the effect of agriculture crop residue burning on healthy subjects. *Mapan, 29*, 57–65.

Allen-Wardell, G., Bernhardt, P., Bitner, R., Burquez, A., Buchmann, S., Cane, J., ... & Walker, S. (1998). The potential consequences of pollinator declines on the conservation of biodiversity and stability of food crop yields. *Conservation biology*, 8–17.

Ahmad, S., Khalofah, A., Khan, S. A., et al. (2021). Effects of native pollinator communities on the physiological and chemical parameters of loquat tree (*Eriobotrya japonica*) under open field condition. *Saudi Journal of Biological Sciences, 28*, 3235–3241.

Aizen, M. A., & Harder, L. D. (2009). The global stock of domesticated honey bees is growing slower than agricultural demand for pollination. *Current Biology, 19*(11), 915–918.

Awasthi, A., Hothi, N., Kaur, P., Singh, N., Chakraborty, M., & Bansal, S. (2017). Elucidative analysis and sequencing of two respiratory health monitoring methods to study the impact of varying atmospheric composition on human health. *Atmospheric Environment, 171*, 32–37.

Banskota, A. H., Tezuka, Y., & Kadota, S. (2001). Recent progress in pharmacological research of propolis. *Phytotherapy Research, 15*(7), 561–571.

Bascompte, J., & Scheffer, M. (2023). The resilience of plant–pollinator networks. *Annual Review of Entomology, 68*(1), 363–380.

Biesmeijer, J. C., Roberts, S. P., Reemer, M., Ohlemuller, R., Edwards, M., Peeters, T., et al. (2006). Parallel declines in pollinators and insect-pollinated plants in Britain and the Netherlands. *Science, 313*(5785), 351–354.

Bloch, G., Francoy, T. M., Wachtel, I., Panitz-Cohen, N., Fuchs, S., & Mazar, A. (2010). Industrial apiculture in the Jordan valley during Biblical times with Anatolian honeybees. *Proceedings of the National Academy of Sciences, 107*(25), 11240–11244.

Breeze, T. D., Bailey, A. P., Balcombe, K. G., & Potts, S. G. (2011). Pollination services in the UK: How important are honeybees? *Agriculture, Ecosystems & Environment, 142*(3–4), 137–143.

Brook, R. D., Rajagopalan, S., Pope, C. A., et al. (2010). Particulate matter air pollution and cardiovascular disease: An update to the scientific statement from the American Heart Association. *Circulation, 121*(21), 2331–2378. https://doi.org/10.1161/CIR.0b013e3181dbece1

Buchmann, S. L., & Nabhan, G. P. (1997). *The forgotten pollinators*. Island Press.

Cameron, S. A., et al. (2011). Patterns of widespread decline in North American bumble bees. *Proceedings of the National Academy of Sciences of the United States of America*. https://doi.org/10.1073/pnas.1014743108

Chagnon, M., et al. (2015). Risks of large-scale use of systemic insecticides to ecosystem functioning and services. *Environmental Science and Pollution Research International*. https://doi.org/10.1007/s11356-014-3277-x

Chauhan, S. V. S., Chaurasia, B., & Rana, A. (2004). Impact of air pollution on floral morphology of *Cassia siamea* Lamk. *Journal of Environmental Biology, 25*(3), 291–297.

Choël, M., & Visez, N. (2019). Altérations du grain de pollen par la pollution atmosphérique Alterations of pollen grains due to air pollution.

Cohen, H., Philpott, S. M., Liere, H., et al. (2021). The relationship between pollinator community and pollination services is mediated by floral abundance in urban landscapes. *Urban Ecosystems, 24*(2), 275–290. https://doi.org/10.1007/s11252-020-01024-z
Cusser, S., Neff, J. L., & Jha, S. (2016). Natural land cover drives pollinator abundance and richness, leading to reductions in pollen limitation in cotton agroecosystems. *Agriculture, Ecosystems & Environment, 226*, 33–42. https://doi.org/10.1016/j.agee.2016.04.020
Cusser, S., Haddad, N. M., & Jha, S. (2021). Unexpected functional complementarity from non-bee pollinators enhances cotton yield. *Agriculture, Ecosystems & Environment, 314*, Article 107415. https://doi.org/10.1016/j.agee.2021.107415
Dams, L. R. (1978). Bees and honey-hunting scenes in the Mesolithic rock art of Eastern Spain. *Bee World*. https://doi.org/10.1080/0005772X.1978.11097692
Das, S., Roy, A., Masiwal, R., Mandal, M., Popek, R., Chakraborty, M., et al. (2023). Comprehensive analysis of PM1 composition in the Eastern Indo-Gangetic Basin: A three-year urban study. *Sustainability, 15*(20), 14894. https://doi.org/10.3390/su152014894
Das, S., Barman, C., Roy, A., et al. (2025). Air pollution may alter reproductive dynamics/efficacy in plants: Correlative evidences from an urban sprawl in Eastern Indo-Gangetic plain. *Aerobiologia, 41*, 35–53. https://doi.org/10.1007/s10453-024-09811-3
Defra. (2019). *Clean air strategy*. Defra. https://doi.org/10.5694/j.1326-5377.1962.tb59377.x
Doane, R. W. (1917). Effect of smelter gases on insects. *Science, 46*, 295–296.
Drogoudi, P. D., & Ashmore, M. R. (2000). Does elevated ozone have differing effects in flowering and deblossomed strawberry? *New Phytologist, 147*, 561–569.
Ebert, A. (2011). Nectar for the taking: The popularization of scientific bee culture in England, 1609–1809. *Agricultural History, 85*, 322–343.
Eilers, E. J., et al. (2011). Contribution of pollinator-mediated crops to nutrients in the human food supply. *Plos One*. https://doi.org/10.1371/journal.pone.0021363
Ellis, H. (2014). *Spoonfuls of honey. A complete guide to honey's flavours & culinary uses with over 80 recipes*. Pavilion Books.
Ellis, A. M., Myers, S. S., & Ricketts, T. H. (2015). Do pollinators contribute to nutritional health? *Plos One*. https://doi.org/10.1371/journal.pone.0114805
Erickson, E., Junker, R. R., Ali, J. G., McCartney, N., Patch, H. M., & Grozinger, C. M. (2022). Complex floral traits shape pollinator attraction to ornamental plants. *Annals of Botany, 130*(4), 561–577.
European Academies Science Advisory Council (EASAC). (2015). *Ecosystem services, agriculture and neonicotinoids*. European Academies Science Advisory Council.
Fisher, A., Tadei, R., Berenbaum, M., Nieh, J., Siviter, H., Crall, J., et al. (2023). Breaking the cycle: Reforming pesticide regulation to protect pollinators. *BioScience, 73*(11), 808–813.
Fründ, J., Dormann, C. F., Holzschuh, A., et al. (2013). Bee diversity effects on pollination depend on functional complementarity and niche shifts. *Ecology, 94*, 2042–2054. https://doi.org/10.1890/12-1620.1
Gallai, N., et al. (2009). Economic valuation of the vulnerability of world agriculture confronted with pollinator decline. *Ecological Economics*. https://doi.org/10.1016/j.ecolecon.2008.06.014
Garibaldi, L. A., Carvalheiro, L. G., Leonhardt, et al. (2014). From research to action: Enhancing crop yield through wild pollinators. *Frontiers in Ecology and the Environment, 12*, 439–447. https://doi.org/10.1890/130330.
Goulson, D., et al. (2015). Bee declines driven by combined stress from parasites, pesticides, and lack of flowers. *Science*. https://doi.org/10.1126/science.1255957
Green, F. B., Peterson, E. M., Emert, A. D., Subbiah, S., & Smith, P. N. (2023). Bee pollinator mortality due to pesticide-laden particulate matter from beef cattle feedyards. *Environmental Science & Technology, 57*(40), 14839–14848.
Guenat, S., & Dallimer, M. (2023). A global meta-analysis reveals contrasting impacts of air, light, and noise pollution on pollination. *Ecology and Evolution, 13*(4), Article e9990.

Harrison, R. M. (2020). Airborne particulate matter. *Philosophical Transactions of the Royal Society a: Mathematical, Physical and Engineering Sciences, 378*, 20190319. https://doi.org/10.1098/rsta.2019.0319

Harrison, R. M., Shi, J. P., Xi, S., et al. (2000). Measurement of number, mass and size distribution of particles in the atmosphere. *Philosophical Transactions of the Royal Society a: Mathematical, Physical and Engineering Sciences, 358*, 2567–2580. https://doi.org/10.1098/rsta.2000.0669

Hausmann, A., Ulrich, W., Segerer, A. H., Greifenstein, T., Knubben, J., Morinière, J., et al. (2022). Fluctuating insect diversity, abundance and biomass across agricultural landscapes. *Scientific Reports, 12*(1), 17706.

IFPRI. (2014). Global hunger index. International Food Policy Research Institute. Retrieved January 7, 2016, from http://www.ifpri.org/node/538

Jones, J., & Rader, R. (2022). Pollinator nutrition and its role in merging the dual objectives of pollinator health and optimal crop production. *Philosophical Transactions of the Royal Society B, 377*(1853), 20210170.

Juda-Rezler, K., Reizer, M., & Oudinet, J.-P. (2011). Determination and analysis of PM10 source apportionment during episodes of air pollution in Central Eastern European urban areas: The case of wintertime 2006. *Atmospheric Environment, 45*, 6557–6566. https://doi.org/10.1016/j.atmosenv.2011.08.020

Jull, A. B., Rodgers, A., & Walker, N. (2008). Honey as a topical treatment for wounds. *The Cochrane Database of Systematic Reviews*. https://doi.org/10.1002/14651858.CD005083.pub2

Jürgens, A., & Bischoff, M. (2017). Changing odour landscapes: The effect of anthropogenic volatile pollutants on plant–pollinator olfactory communication. *Functional Ecology, 31*, 56–64. https://doi.org/10.1111/1365-2435.12774

Katumo, D. M., Liang, H., Ochola, A. C., Lv, M., Wang, Q. F., & Yang, C. F. (2022). Pollinator diversity benefits natural and agricultural ecosystems, environmental health, and human welfare. *Plant Diversity, 44*(5), 429–435. https://doi.org/10.1016/j.pld.2022.01.005

Kelly, F. J., & Fussell, J. C. (2012). Size, source and chemical composition as determinants of toxicity attributable to ambient particulate matter. *Atmospheric Environment, 60*, 504–526. https://doi.org/10.1016/j.atmosenv.2012.06.039

Kevan, P. G., & Menzel, R. (2012). The plight of pollination and the interface of neurobiology, ecology and food security. *The Environmentalist*. https://doi.org/10.1007/s10669-012-9394-5

Kim, K. H., Kabir, E., & Kabir, S. (2015). A review on the human health impact of airborne particulate matter. *Environment International, 74*, 136–143. https://doi.org/10.1016/j.envint.2014.10.005

Klein, A.-M., Vaissière, B. E., Cane, J. H., Steffan-Dewenter, I., Cunningham, S. A., Kremen, C., & Tscharntke, T. (2007). Importance of pollinators in changing landscapes for world crops. *Proceedings. Biological Sciences/The Royal Society*. https://doi.org/10.1098/rspb.2006.3721

Kumar, P., Singh, G., & Singh, H. (2020). Impact of insect pollinators on quantitative and qualitative improvement in agricultural crops. *International Journal of Current Microbiology and Applied Sciences, 9*, 2359–2367. https://doi.org/10.20546/ijcmas.2020.909.295

Kumar, R., Nain, V., & Duhan, J. S. (2022). An ecological approach to control pathogens of Lycopersicon esculentum L. by slow release of mancozeb from biopolymeric conjugated nanoparticles. *Journal of Xenobiotics, 12*(4), 329–343.

Kumar, R., Goel, R., Singh, T., Priyadarshi, N., & Twala, B. (2024). Incentivizing green building technology: A financial perspective on sustainable development in India. *F1000Research, 13*, 924.

Mallinger, R. E., & Gratton, C. (2015). Species richness of wild bees, but not the use of managed honeybees, increases fruit set of a pollinator-dependent crop. *Journal of Applied Ecology, 52*, 323–330. https://doi.org/10.1111/1365-2664.12377

Mandal, M., Das, S., Roy, A., Rakwal, R., Jones, O. A., Popek, R., et al. (2023). Interactive relations between plants, the phyllosphere microbial community, and particulate matter pollution. *Science of the Total Environment, 890*, Article 164352. https://doi.org/10.1016/j.scitotenv.2023.164352

Manisalidis, I., Stavropoulou, E., Stavropoulos, A., & Bezirtzoglou, E. (2020). Environmental and health impacts of air pollution: A review. *Frontiers in Public Health, 8*, 1–13.

Maxim, L., & van der Sluijs, J. P. (2013). Seed-dressing systemic insecticides and honeybees. In European Environment Agency (Ed.), *Late lessons from early warnings: Science, precaution, innovation* (pp. 401–438). European Environment Agency (EEA) report 1/2013.

Nicole, W. (2015). Pollinator power: Nutrition security benefits of an ecosystem service. *Environmental Health Perspectives, 123*, A210–A215.

Noori, A., Hasanuzzaman, M., Roychowdhury, R., Sarraf, M., Afzal, S., Das, S., & Rastogi, A. (2024). Silver nanoparticles in plant health: Physiological response to phytotoxicity and oxidative stress. *Plant Physiology and Biochemistry,* 108538.

Ollerton, J., Winfree, R., & Tarrant, S. (2011). How many flowering plants are pollinated by animals? *Oikos*. https://doi.org/10.1111/j.1600-0706.2010.18644.x

Paoletti, E., Schaub, M., Matyssek, R., Wieser, G., Augustaitis, A., Bastrup-Birk, A. M., et al. (2010). Advances of air pollution science: From forest decline to multiple-stress effects on forest ecosystem services. *Environmental Pollution, 158*, 1986–1989. https://doi.org/10.1016/J.ENVPOL.2009.11.023

Partap, U., & Tang, Y. (2012). The human pollinators of fruit crops in Maoxian County, Sichuan, China: A case study of the failure of pollination services and farmers' adaptation strategies. Mountain Research and Development. https://doi.org/10.1659/MRD-JOURNAL-D-11-00108.1

Pattinson, D. (2012). Pre-modern beekeeping in China: A short history. *Agricultural History, 86*, 235–255.

Pearson, R. G. (2016). Reasons to conserve nature. *Trends in Ecology & Evolution*. https://doi.org/10.1016/j.tree.2016.02.005

Penn, H., Penn, J., Hagan, M., & Hu, W. (2020). The buzz about bee campuses: Student thoughts regarding pollinator conservation. *American Entomologist, 66*(4), 54–61.

Perera, F. (2018). Pollution from fossil-fuel combustion is the leading environmental threat to global pediatric health and equity: Solutions exist. *International Journal of Environmental Research and Public Health, 15*(1), 16.

Philpott, S. M., Uno, S., & Maldonado, J. (2006). The importance of ants and high–shade management to coffee pollination and fruit weight in Chiapas, Mexico. *Biodiversity & Conservation, 15*, 487–501. https://doi.org/10.1007/s10531-005-0602-1

Pisa, L. W., et al. (2015). Effects of neonicotinoids and fipronil on non-target invertebrates. *Environmental Science and Pollution Research*. https://doi.org/10.1007/s11356-014-3471-x

Potts, S. G., Biesmeijer, J. C., Kremen, C., Neumann, P., Schweiger, O., & Kunin, W. E. (2010). Global pollinator declines: Trends, impacts and drivers. *Trends in Ecology & Evolution*. https://doi.org/10.1016/j.tree.2010.01.007

Rader, R., Bartomeus, I., Garibaldi, L. A., Garratt, M. P., Howlett, B. G., Winfree, R., Cunningham, S. A., Mayfield, M. M., Arthur, A. D., Andersson, G. K., Bommarco, R., Brittain, C., Carvalheiro, L. G., Chacoff, N. P., Entling, M. H., Foully, B., Freitas, B. M., Gemmill-Herren, B., Ghazoul, J., Griffin, S. R., et al. (2016). Non-bee insects are important contributors to global crop pollination. *Proceedings of the National Academy of Sciences*. https://doi.org/10.1073/pnas.1517092112

Rafique, M. K., Quratulain, M. R., Stephen, E., & Ahmad, S. (2016). Pollination deficit in mango orchards at Multan, Pakistan. *Pakistan Journal of Zoology, 48*, 35–38.

Riaz, J., Razzaq, H., Amjad, T., Anjum, A., Afzal, T., Fatima, F., et al. (2020). Pollution and pollinators: A review. *Pure and Applied Biology, 9*(3), 2049–2058.

Richards, A. J. (2001). Does low biodiversity resulting from modern agricultural practice affect crop pollination and yield? *Annals of Botany, 88*(2), 165–172.

Roy, A., Mandal, M., Das, S., Popek, R., Rakwal, R., Agrawal, G. K., et al. (2024a). The cellular consequences of particulate matter pollutants in plants: Safeguarding the harmonious integration of structure and function. *Science of the Total Environment, 169763*. https://doi.org/10.1016/j.scitotenv.2023.169763

Roy, A., Das, S., Singh, P., Mandal, M., Kumar, M., Rajlaxmi, A., et al. (2024b). Summer-time monitoring and source apportionment study of both coarse, fine, and ultra-fine particulate pollution in Eastern Himalayan Darjeeling: A hint to health risk during peak tourist season. *Mapan, 39*(4), 995–1009. https://doi.org/10.1007/s12647-024-00776-9

Runckel, C., Flenniken, M. L., Engel, J. C., Ruby, J. G., Ganem, D., Andino, R., & DeRisi, J. L. (2011). Temporal analysis of the honey bee microbiome reveals four novel viruses and seasonal prevalence of known viruses, Nosema, and Crithidia. *Plos One*. https://doi.org/10.1371/journal.pone.0020656

Ryalls, J. M. W., Langford, B., Mullinger, N. J., Bromfield, L. M., Nemitz, E., Pfrang, C., & Girling, R. D. (2022). Anthropogenic air pollutants reduce insect-mediated pollination services. *Environmental Pollution, 297*, Article 118847. https://doi.org/10.1016/j.envpol.2022.118847

Senapathi, D., Fründ, J., Albrecht, M., et al. (2021). Wild insect diversity increases inter-annual stability in global crop pollinator communities. *Proceedings of the Royal Society B, 288*, 20210212. https://doi.org/10.1098/rspb.2021.0212

Shannon, J. G., & Mulchi, C. L. (1974). Ozone damage to wheat varieties at anthesis. *Crop Science, 14*, 335–337.

Siviter, H., Fisher, A., Baer, B., Brown, M. J. F., Camargo, I. F., Cole, J., et al. (2023). Protecting pollinators and our food supply: Understanding and managing threats to pollinator health. *Insectes Sociaux, 70*(1), 5–16.

Southwood, T., & Richard, E. (1977). Entomology and mankind: Insects over the ages have greatly affected man's health and food supply and have played an important role as religious and cultural symbols. *American Scientist, 65*(1), 30–39.

Srinivasan, M. V. (2011). Honeybees as a model for the study of visually guided flight, navigation, and biologically inspired robotics. *Physiological Reviews*. https://doi.org/10.1152/physrev.00005.2010

Thimmegowda, G. G., Mullen, S., Sottilare, K., Sharma, A., Mohanta, R., Brockmann, A., et al. (2020). A field-based quantitative analysis of sublethal effects of air pollution on pollinators. *Proceedings of the National Academy of Sciences, 117*(34), 20653–20661.

Tommasi, N., Pioltelli, E., Biella, P., Labra, M., Casiraghi, M., & Galimberti, A. (2022). Effect of urbanization and its environmental stressors on the intraspecific variation of flight functional traits in two bumblebee species. *Oecologia, 199*, 289–299.

Tremlett, C. J., Moore, M., Chapman, M. A., et al. (2020). Pollination by bats enhances both quality and yield of a major cash crop in Mexico. *Journal of Applied Ecology, 57*(2), 450–459.

UNEP. (2010). Global honey bee colony disorders and other threats to insect pollinators. UNEP emerging issues report, United Nations Environmental Program, Nairobi.

Van der Sluijs, J. P., Amaral-Rogers, V., Belzunces, L. P., Bijleveld Van Lexmond, M. F. I. J., Bonmatin, J. M., Chagnon, M., Downs, C. A., Furlan, L., Gibbons, D. W., Giorio, C., Girolami, V., Goulson, D., Kreutzweiser, D. P., Krupke, C., Liess, M., Long, E., McField, M., Mineau, P., Mitchell, E. A. D., Morrissey, C. A., et al. (2015). Conclusions of the worldwide integrated assessment on the risks of neonicotinoids and fipronil to biodiversity and ecosystem functioning. *Environmental Science and Pollution Research, 22*(1), 148–154. https://doi.org/10.1007/s11356-014-3229-5

Van der Sluijs, J. P., & Vaage, N. S. (2016). Pollinators and global food security: The need for holistic global stewardship. *Food Ethics, 1*, 75–91. https://doi.org/10.1007/s41055-016-0003-z

Van Lexmond, M. B., Bonmatin, J. M., Goulson, D., & Noome, D. A. (2015). Worldwide integrated assessment on systemic pesticides: Global collapse of the entomofauna: Exploring the role of systemic insecticides. *Environmental Science and Pollution Research, 22*, 1–4.

Vanbergen, A. J., & the Insect Pollinators Initiative. (2013). Threats to an ecosystem service: Pressures on pollinators. *Frontiers in Ecology and the Environment, 11*(5), 251–259. https://doi.org/10.1890/120126

Vasiliev, D., & Greenwood, S. (2020). Pollinator biodiversity and crop pollination in temperate ecosystems, implications for national pollinator conservation strategies. *Science of the Total Environment, 744*, Article 140880.

Wenzel, A., Grass, I., Belavadi, V. V., & Tscharntke, T. (2020). How urbanization is driving pollinator diversity and pollination—A systematic review. *Biological Conservation, 241*, Article 108321.
Woodcock, B., Garratt, M. P. D., Powney, G. D., et al. (2019). Meta-analysis reveals that pollinator functional diversity and abundance enhance crop pollination and yield. *Nature Communications, 10*, 1–10. https://doi.org/10.1038/s41467-019-09393-6
World Health Organization. Regional Office for Europe. (2013). Health effects of particulate matter: policy implications for countries in Eastern Europe, Caucasus and central Asia. World Health Organization. Regional Office for Europe.

Chapter 9
Cold-Chain Aerosols as Microbial Vectors: Listeria–Particulate Interactions in Food Storage Air

Swati Ray and Mukesh Singh

Abstract The pathogen(s) *Listeria sp* are a group of prominent microorganisms that has a history of causing encephalitis in livestock animals and may act like an abortifacient, causing miscarriage or immature termination of pregnancy in domesticated mammals. In 1981, it was first reported to cause foodborne illness from contaminated animal feed, fodder and silage, where the main responsible species were *Listeria monocytogenes* and *Listeria ivanovii.* Since the outbreaks often lead to fatality, food handlers have to maintain stringent control and implement regulatory practices to eradicate this entity, especially maintaining the air quality of the food handling areas. It can travel through the water droplets present in air particulates when they flow over soil, water bodies or any contaminated as well as healthy animal or human which may unknowingly carry them in their skin or clothes. Through air, it spreads very quickly onto food-contact or non-contact surfaces, deli utensils or equipment, hands, skin surfaces and ultimately to foods. Refrigerated foods may also harbour this microorganism in viable condition, and it remains alive for a very long period. Thus, food processing, cooking or serving areas may colonize Listeria and contaminate food, causing Listeriosis, which is a deadly disease affecting people with compromised immune systems, such as pregnant women or the elderly. This chapter gives narrative overviews about airborne microbe transmission in cold storage leading to food poisoning with particular reference to *Listeria sp* on illustrations based on previously published constructive literature reviews.

Keywords Contaminated food · Food borne · Food contact surface · *Listeria monocytogenes* · Listeriosis · Outbreak

S. Ray (✉)
Department of Food Technology, Haldia Institute of Technology, Haldia, West Bengal, India
e-mail: rayswati279@gmail.com

M. Singh
Kabi Nazrul College, Birbhum, West Bengal, India

A. Sarkar et al. (eds.), *From Sky to Soil*, https://doi.org/10.1007/978-3-032-31991-3_9

9.1 Introduction

Air pollution is a matter of great concern in the world today, and the data from the World Health Organization (WHO) confirms the fact about 99% of the population inhales polluted air, which is much beyond the WHO guidelines for air quality index (AQI). Air contains high levels of pollutants, and exposure assessment shows that moderate to low-income countries are the worst affected ones. Air pollution can be defined as "any contamination of the indoor or outdoor environment by any chemical, physical or biological agent that modifies the natural characteristics of the atmosphere". Automobiles, combustion devices, factories emitting aerosols or effluents and/or forest fires are the birthplace of air pollution that includes the emission of pollutants such as carbon monoxide, particulate matter, nitrogen dioxide, sulphur dioxide and ozone. These pollutants can saturate indoor and outdoor air, causing pulmonary infections or diseases that are morbid and may lead to even fatality (World Health Organization, 2024).

Air pollution is one of the top three global agendas, which is not only detrimental to health but also hinders economic progression. A 2016 data revealing 4.2 million deaths through particulate matter (PM) of the outdoor air was attributed to $PM_{2.5}$ ($\leq$ 2.5 μm in diameter). These particles may travel great distances with considerable changes in their physical and chemical properties over time. The Data Integration Task Force of WHO developed the Data Integration Model for Air Quality (DIMAQ) to estimate $PM_{2.5}$ exposure worldwide. They provide air quality profiles of different countries through satellite investigation of the aerosol optical depth and chemical transport models that follow a consistent trend (Shaddick et al., 2020).

Microbes such as bacteria and viruses may build up in different sections of the environment, and they often use air as their vehicle for transmission. This aerial transmission via particulate matter leads to hazardous effects on human health (Fujiyoshi et al., 2017; Mandal et al., 2024). Inhalation of these particulates carrying harmful pathogens may lead to problems in the upper and lower respiratory tract. The source of these microorganisms can be the environment, animals or other human beings, which can spread contagiously to any healthy human, causing diseases. Viral infections like flu, influenza, acute respiratory syndrome, Middle East respiratory syndrome, enteric infections, measles, hantaviral disease and varicella are pretty common. Some fungal diseases like coccidiomycosis, histoplasmosis, cryptococcosis, aspergillosis and blastomycosis may also occur through airborne fungi. The bacterial population is the most commonly available organism among all microbes. Exposure to pathogenic gram-positive or gram-negative bacteria, their endo and exotoxins, actinomycetes etc., may enter through inhalation. Bacterial infections like tuberculosis, Listeriosis, Legionnaires' disease, and non-tubercule mycobacterial disease are considered airborne and have become a threat. These microbes often pass to the food substances through the air, causing foodborne outbreaks (Gonzalez-Martin, 2019). There are a certain number of studies that reported Listeria infection in the past (Allerberger & Wagner, 2009) including contamination reported in food production areas (Osek et al., 2022), milk and dairy products (Shamloo et al., 2018), through soil and water (Bongiovanni et al., 2023) and meat, seafood and fish products (Matle et al., 2020; Peratikos et al., 2023) etc. Though the natural habitat of listeria is principally soil, it may get transmitted to water, and via airborne water droplets and particulate matter, it contaminates food. Through air transmission, Listeria can contaminate food in different production and storage regions, as reported earlier (Berrang et al., 2013; Dobeic et al., 2011; Yong et al., 2024; Zhang et al., 2007). A few isolated studies reported Listeria survival in lower temperatures and cold chains transmitted through contaminated raw material (Guan et al., 2021; Liu et al., 2016; Possas et al., 2022), but there are no significant works published on Listeria transmission inside cold storage areas through air particulate matter transmission.

This chapter starts with a general overview of particulate matter as a pollutant and its pollution cycle in the environment. It describes the details of Listeria transmission with various outbreak studies around the world based on the investigations published by distinguished authors in the past. Lastly, it recounts a comprehensive study of Listeria propagation inside cold storage, vehicles associated with its transmission and preventive options to eliminate the microbe from food handling locations.

9.2 Particulate Matter as a Major Air Pollutant

Particulate matter (PM), or ambient particulate matter, is defined as a "complex mixture of solid and liquid particles that remains suspended in air". The level of pollutants present in particulate matter depends on the source of those pollutants, which may combine with other reactive species of the atmosphere to form secondary particulates. Other factors, such as the presence of gaseous mixtures, meteorology, geography or seasonal character, may also affect the nature of the PM generated (Adams et al., 2015; Mandal et al., 2023).

The first broad classification of PM is based on the source of the pollutants suspended in particulate matter. Primary PM is formed due to industrial pollution, traffic, dust, seawater, blowing wind, etc., which principally contain various organic compounds along with carbon, metals and their oxides and ions. Secondary PM originates from the chemical transformation of the gaseous constituents like the conversion of sulphur dioxide, nitrogen dioxide, ammonia and volatile organic components to sulphates, nitrates, ammonium ions and secondary organic aerosols, respectively (Das et al., 2023). Alternatively, PM can also be classified into four major types depending on their aerodynamic diameter (AD) and nature of distribution in the atmosphere as shown in Fig. 9.1.

The TSP and PM_{10} are usually coarse materials of principally geological origin, pollens or sea salts, whereas the $PM_{2.5}$ are mainly accumulated particles deposited through condensation or droplets. They are primarily composed of fine geological materials, sulphates, nitrates, ammonium ions, organic or elemental carbon, heavy metals, etc. Ultrafine PM are usually superfine particulates of condensed metal vapours, carbon fines, sulphuric acid, etc. (Johnson et al., 2011).

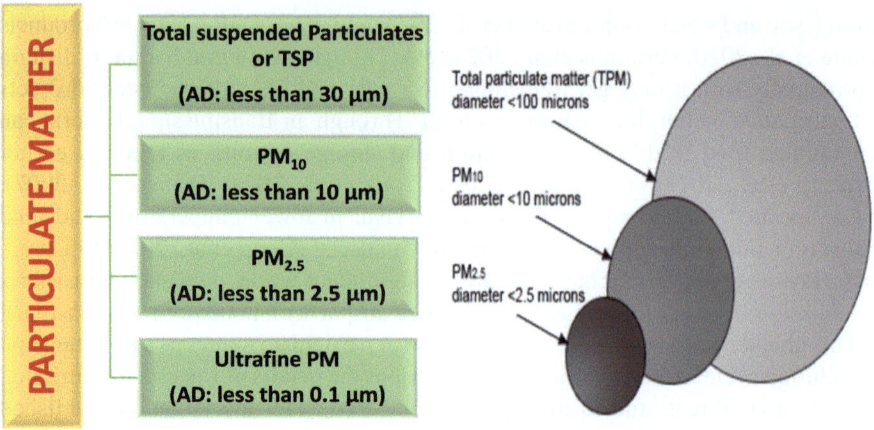

Fig. 9.1 Types and relative aerodynamic diameter of the particulate matters available in the atmosphere

The amount of particulates and other pollutants can be assessed and monitored using the air quality index (AQI), also known as the Air Pollution Index (API) or Pollutant Standards Index (PSI). It usually denotes a numeric value that reflects the air quality daily with respect to human health. The value is evaluated based on the concentration of various pollutants with National Ambient Air Quality Standards (NAAQS) (Suman, 2021). The colour-schemed indicator for AQI is used to specify air quality, which covers environmental pollutants regulated under the Clean Air Act, particle pollution (PM_{10} and $PM_{2.5}$), ozone, carbon monoxide, nitrogen dioxide, and sulphur dioxide. It educates the mass about the condition of air exposure daily and suggests preventive measures to resist health issues related to inferior air quality (Abulude et al., 2022; Roy et al., 2024a). Figure 9.2 illustrates the colour-coded air quality indices and the significance of the values indicative of day-to-day air quality of any particular region.

9.3 Background Information

The impact of particulate matter on human health is repeatedly reported and proven to be hazardous. In developing countries, about 28% of fatalities are caused by inappropriate indoor air quality. In 2014, WHO reported about 7 million deaths from exposure to unhealthy indoor or outdoor air quality, which is complemented by the reports from Delhi, India, which sees approximately 10,000 to 30,00 deaths every year. The inhalation of particulate matter PM_{10} and $PM_{2.5}$ caused acute and chronic respiratory problems, including damage of organs, asthma, cardiovascular and lung diseases and finally resulted in hospitalization. Children and old people are the most affected groups depending on the duration of exposure. Since 85% of the people

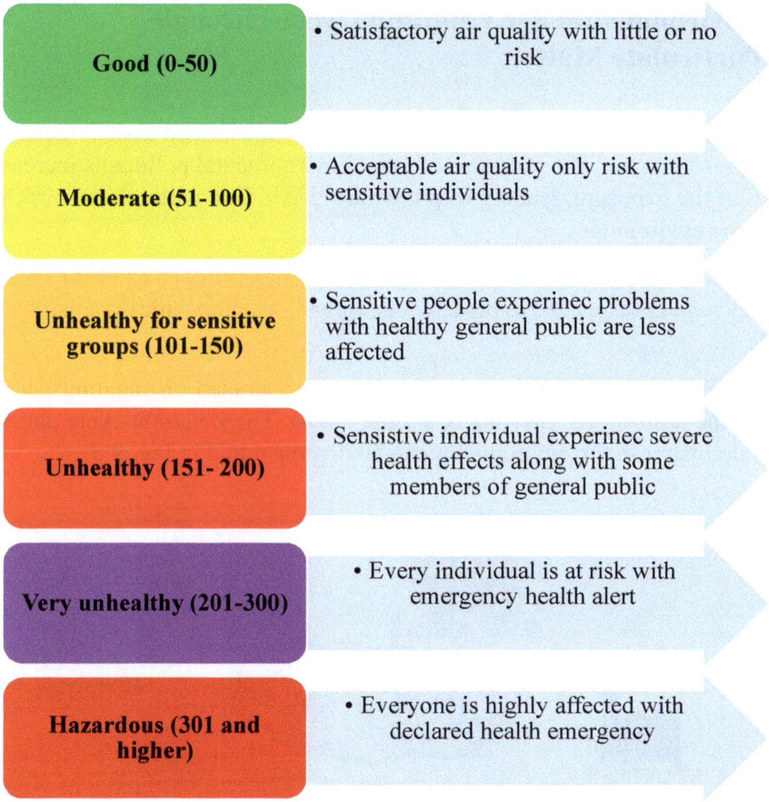

Fig. 9.2 Types of coloured indicators starting from green, yellow, orange, red, purple and maroon indicating respective air quality indices (Abulude et al., 2022)

in the world live in a space of unhealthy air, the number of premature deaths can soar as high as 5.5 million. Besides, the PM also is responsible for poor visibility, corrosion to structures, irreparable damage to vegetation and topsoil, disturbing the atmospheric radiation balance and altering cloud formation (Abulude, 2016).

Kuzma et al. (2020) carried out a study based on the mortality rate of the inhabitants of North Eastern Poland on ten-year data from 2008 to 2017 and found that irrespective of differences in air quality, PM forms the primary cause of mortality, pulmonary diseases and cardiovascular diseases, which typically shows symptoms one or two days after exposure. Studies also proved the association of the Covid-19 outbreak with exposure to particulate matter, and they played an important role as vehicles for the transmission of the disease (Bianconi et al., 2020).

9.4 Environmental Air Pollution Cycle Through Particulate Matter

Airborne bacterial, viral or fungal pathogens can be transmitted easily via particulate matters present in air. The presence of other environmental pollutants increases the severity of the exposure, leading to premature death. The particulates work in two ways in the environment:

(a) They enter the respiratory system and cause respiratory or heart diseases
(b) They remain suspended as bioaerosols, spreading pathogenic organisms through the air.

The vulnerability of infection and their spread worsen on inhaling particulate matter in the atmosphere (Myung & Joung, 2024). The pollution cycle pathway of particulate matter in the environment can be demonstrated in Fig. 9.3.

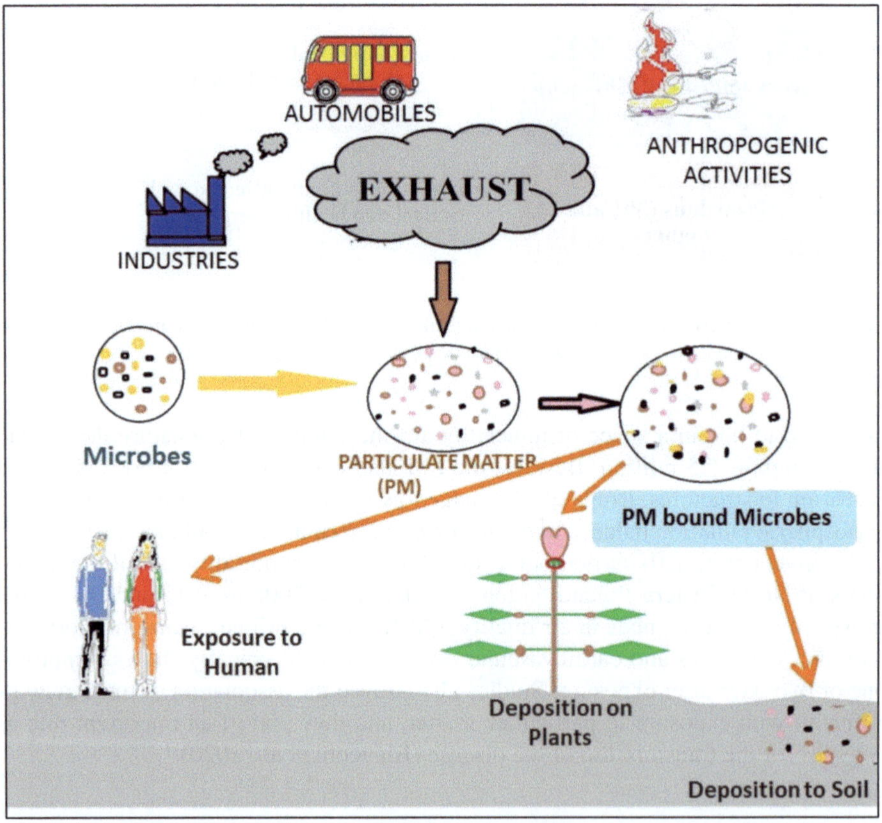

Fig. 9.3 Transmission and pathway of environmental distribution of particulate matter

The fumes from automobiles, flue gases from industries and various anthropogenic activities release particulate matter into air through exhaust which when comes in contact with aerially suspended microbes from different sources forms PM-bound microorganisms (Roy et al., 2024b). Various pathogens may also be present in these varieties of suspended microbes which may get transmitted to humans, plants or animals through their food. Some part of it is also deposited over top soil which may again become airborne and get mixed with particulate matter when air flows over these regions. Thus, the cycle continues in the environment where particulate matter forms a vehicle for carrying pathogenic microorganisms to humans and animals.

9.5 Significance of *Listeria* Sp. In Disease Outbreak

Nearly one fourth of the total world population is under threat from disease by consumption of contaminated food, making it one of the major public health concerns worldwide. Contamination can occur from biological agents like bacteria, viruses, and other parasites which find their way to invade food through natural vehicles such as air, water, soil etc. The most prevalent ones are the bacterial contamination of which Listeria spp pose a serious threat to the health of humans and animals. This intrinsic pathogen is a non-spore-forming Gram-positive microaerophilic bacillus belonging to the family of *Corynebacteriaceae.* The most pathogenic variety *Listeria monocytogenes* can be isolated from water, soil, the body of humans and animals from which they transmit via air to the food products. *Listeria monocytogenes* causes Listeriosis that disseminates through consumption of contaminated food and water which may lead to hospitalisation and lethality (Shamloo et al., 2018).

Listeria monocytogenes have 13 serotypes of which serotype 4b is mostly the cause of outbreaks (Abdi Moghadam et al., 2015). Beside soil, water and air, these serotypes are chiefly distributed in food surfaces, processed foods, meat, fish, vegetables and dairy products. They are also isolated from hospitals, clinical facilities and healthcare zones. They also behave as a highly salt and draught tolerant psychrophile which allows them to survive in refrigeration conditions as well (Mahdavi et al., 2012). It is of great concern since it can tolerate high acidity, osmolality, oxygen scarcity, bile salts and peptides present in gastrointestinal tract, causing chronic infection (Shamloo et al., 2018). Along with Listeriosis it can cause severe effects like pregnancy termination, premature birth, stillbirth, sepsis, and meningitis to weaker-immuned patients, like aged people, or those suffering from cancer, leukaemia, Hodgkin lymphoma, AIDS, etc. Among the mild symptoms it manifests fever, diarrhoea, nausea and vomiting and is also present in the stool of healthy individuals who are asymptomatic. The seasonal mortality trend shows its peak during July to October where about 206 deaths among 1600 infected people were reported earlier in USA (Sauders et al., 2003; Thomas, 2016). Listeria monocytogenes also affect vertebrate species like birds, other than human beings, whereas the other pathogenic species Listeria ivanovii mainly affects ruminants and rodents and has a lower pathogenicity towards humans. The nature of damage is found to be similar to monocytogenes with very few cases reported so far. Some documented cases—one of visceral parasitic gastroenteritis in weaned lambs from five different farms of United Kingdom (Dunnett et al., 2020) and another of abortion of fully matured foetus of 390 sheep of Argentina over a one-month period were reported (Della rosa et al., 2019). In both cases colonies of *Listeria ivanovii* were isolated from the infected animal and found that the animals mainly got affected by this organism through contaminated feed which when stopped has led to no further propagation. Reports confirming human intervention of this species are also found in very rare cases of which only two are reported so far. They mainly include mild gastroenteritis, bacteremia in weak-immunity population and in some cases premature loss of foetus in pregnant women (Rossi et al., 2022).

9.5.1 Mode of Action **of Listeria**

Listeria can enter the body through any of its sources mainly passing through the intestine. Liver is the principal organ of target where they colonize and multiply rapidly. This colonization induces cell-mediated immune response of anti-listeria memory T cells. The presence of this memory T cell is due to regular subclinical exposure of any healthy individual to listerial antigens (non-invasive Listeriosis). In immune-deficient or sensitive individuals, this uncontrolled cell proliferation in liver leads to mild bacteremia targeting secondary organs like brain or uterus opening pathways for clinical condition (Invasive Listeriosis). Both the pathogenic Listeria Spp can survive inside macrophages and ravage the neighbouring epithelial, endothelial cells and hepatocytes completing their 4 stage intercellular life cycle. The four stages of listeria invasion inside the cell involves-

I. Prompt break out from phagocytic vacuole
II. Rapid multiplication in cytoplasm
III. Actin-based motility
IV. Spreading to neighbouring cells.

They can restart the cycle every time they gain actin-based motility to transfer themselves to the neighbouring cells sheltered inside the macrophages. The actin-mediated motility gained by the bacteria which helps their translocation from one cell to another involves host cell actin polymerization. So irrespective of its techniques for entry into the host cell, it can penetrate the intestinal wall by basolateral distribution causing enteritis (Vázquez-Boland et al., 2001).

Summatively, the basic transmission stages in the environment as apparent from outside includes- contamination of food from airborne *Listeria sp*-bound particulate matter, consumption of this contaminated food by humans, *Listeria sp* colonizing in GI tract and gets transmitted to liver following circulation in the body through lymphatic systems.

Liver forms the main organ of target where Listeria can colonize and multiply rapidly in number. Then it spreads to different organs and cells causing targeted cell

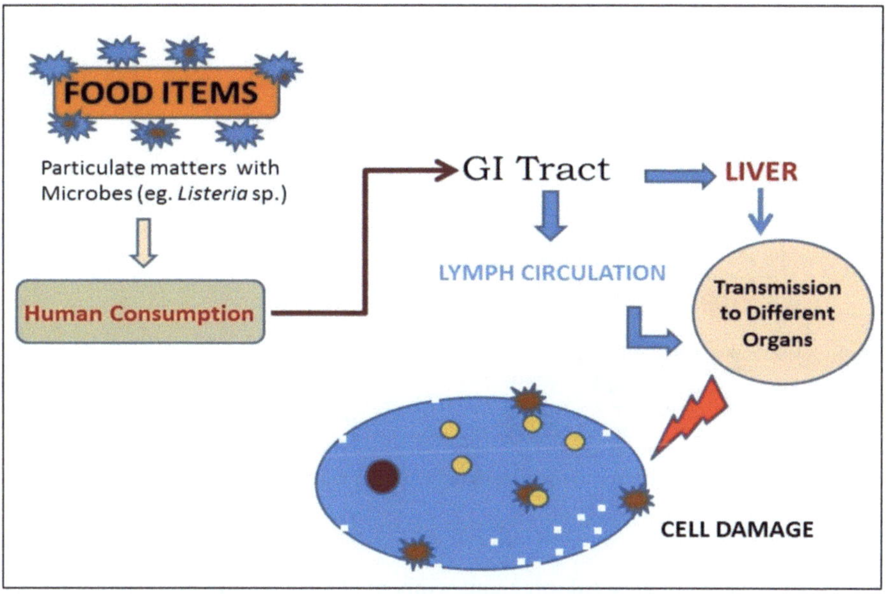

Fig. 9.4 Mode of transmission of Listeria in human body and its environmental life-cycle

damage and Listeriosis. The basis of Listeria transmission in the environment to the human body and the nature of its propagation inside the body can be illustrated collectively in Fig. 9.4. After it enters the blood stream, it targets the vascular degeneration and lysis. They multiply inside the cytosol and engage themselves in actin-bound polymerization and propulsion to the neighbouring cells. The mechanism of Listeria invasion in the human body can be shown in Fig. 9.5 (Sibanda & Buys, 2022).

This chain-wise cell-to-cell transmission causes the spread of this organism in the body's organs and systems, causing different diseases and outbreaks.

9.5.1.1 Case Studies of Listeria Outbreak

Although Listeriosis is a deadly and serious disease, its frequency of occurrence is quite rare with up to 10 person affected per 1 million people annually depending on the region of the world. The limited ingestion of this microbe usually does not affect individuals extremely but it can be severe if taken in large numbers. The high death rate of the infected individuals makes it a global concern in disease outbreak study. With an increase in awareness, the number of Listeria outbreaks has significantly decreased. Between the years 1990–2003, USA had a decrease from 7.7 to 3.3 cases per million and UK recorded a decrease from 4.5 to 3.4 cases per million (Clark 2015–2017). According to the data published by Centre for Disease Control and Prevention (CDC), about 206 deaths were reported so far of 1600 infected people each year of which 99% of the patients needed hospitalization. The worst affected

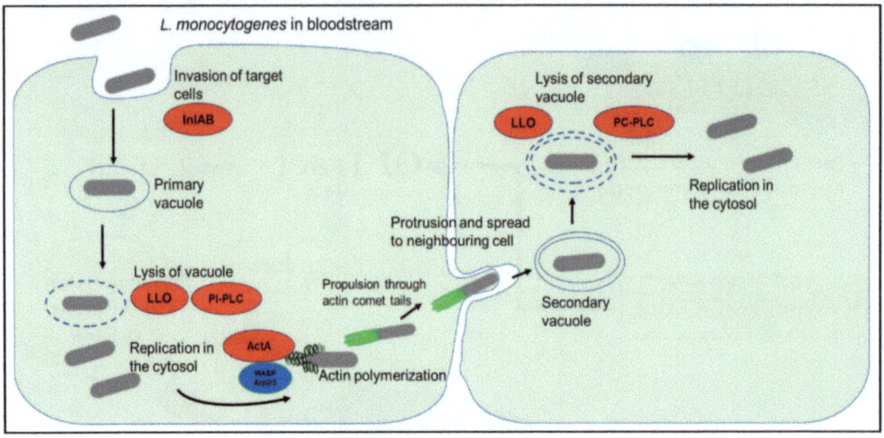

Fig. 9.5 Mechanism of Listeria invasion inside affected human body causing Listeriosis (Sibanda & Buys, 2022)

groups were new-borns, weak pregnant women, elderly people aged above 65 and immune-deficient individuals (Centre for Disease Control and Prevention, 2024).

In 2001, Listeriosis was promoted to the list of nationally notifiable diseases by CDC. National Molecular Subtyping Network for Foodborne Disease Surveillance instructed all public health labs to collect every available isolate of Listeria for subtyping. The healthcare professional was directed to report any cases of Listeriosis with detailed interviews of the infected persons about the sources of their food intake (Shamloo et al., 2018). After these regulations were implemented, in 2006, a single outbreak was reported by CDC, followed by 122 outbreaks in 2007 and 158 outbreaks in 2008. The outbreaks, when studied in detail, are found to be associated with some food products such as dairy products, contaminated fruits and vegetables, ice cream, shellfish, crustaceans, molluscs, meat, juice and Chinese cuisines (Buchanan, 2017; Zhu et al., 2017; Wu et al., 2015). Some of the most recent Listeria outbreak incidents are listed in Table 9.1.

In India, however, no such robust data is available except a few published literature data on clinical Listeriosis by variant *Listeria monocytogenes*. In most cases, the source of contamination is unknown, and patients mainly belong to the vulnerable populations of pregnant mothers, infants and old age people. Table 9.2 enlists data of Listeriosis exposure in Indian population.

Listeriosis is both acute and chronic, reported from different parts of the world in epidemic or sporadic forms. However, in most Asian countries, including India, it remained under-reported and un-documented due to a lack of proper monitoring and negligence. The disease also remains undiagnosed in many cases due to a lack of a reliable and easy-to-use diagnostic kit or a lack of awareness about symptoms indicating Listeriosis. As a result, the data available is not satisfactory or ample to evaluate its magnitude in humans and animals (Barbuddhe et al., 2012).

Table 9.1 Summary of outbreaks caused by Listeria spp. in the year 2014–2024 in the United States (*Data Source* Centers for Disease Control and Prevention, National Center for Emerging and Zoonotic Infectious Diseases (NCEZID), Division of Foodborne, Waterborne, and Environmental Diseases (DFWED)

Year of occurrence	Associated element	Total Infected	Deaths reported
February, 2024	Queso Fresco and Cotija cheese	26	2
October, 2023	Ice cream and the serving cups	2	0
June, 2023	Leafy greens	18	0
January, 2023	Peaches, Nectarines, and Plums	11	1
November, 2022	Enoki Mushrooms	5	0
September, 2022	Reheat deli meat and deli-sliced cheese	13	1
August, 2022	Brie and Camembert Cheese	5	0
August, 2022	Ice cream	27	1
December, 2021	Packaged Salads	10	1
October, 2021	Packaged Salads	18	3
September, 2021	Cooked Chicken	3	1
May, 2021	Queso Fresco cheese	13	1
October- August, 2020	Deli Meat	12	1
June, 2020	Enoki Mushrooms	36	4
December, 2019	Hard-boiled Eggs	8	1
November, 2019	Unidentified source	24	2
September, 2019	Deli-Sliced Meats and Cheeses	10	1
November, 2018	Pork Products	4	0
October, 2018	Deli Ham	4	1
March, 2017	Soft raw milk cheese	8	2
December, 2016	Raw Milk	2	1
July, 2016	Frozen Vegetables	9	3
March, 2016	Packaged Salads	19	1
October, 2015	Soft Cheeses	30	3
June, 2015	Ice cream, frozen yogurt, sherbet, and frozen snacks	10	3
December, 2014	Pre-packaged Caramel Apples	35	7
October 2014	Bean Sprouts	5	2
August, 2014	Cheese	5	1
April, 2014	Dairy Products	8	1

Table 9.2 Summary of outbreaks caused by Listeria monocytogenes in the year 1966–2011 in India (*Data Source* Barbuddhe et al., 2012; Tirumalai, 2013)

Year	Clinical condition	Number of cases
1966	Maternal Listeriosis in the cervix with poor obstetric history	16
1972	Maternal Listeriosis with abortion and premature birth	3
1974	Perinatal Listeriosis with bad obstetric history	9
1977	Maternal Listeriosis with abortion history	4
1988	Septicaemia	1
1980	Neonatal meningitis	1
1981	Meningitis	1
1995	Pericarditis	1
1997	Neonatal with congenital heart disease and digestive failure	3
1997	Septicaemia	2
1998	Perinephricbscess in malnourished child	1
2002	Abortion in second trimester	1
2003	Maternal Listeriosis	1
2005	Neonatal Listeriosis	1
2006	Meningitis	1
2006	Maternal Listeriosis	1
2009	Meningoencephalitis in immunocompetent child	1
2009	Neonatal Meningitis	1
2011	Neonatal Meningitis	1

9.6 Transmission of Listeria in Cold Storage

9.6.1 Cold Chain Conditions Promoting Listeria Attack

Cold storage or cold chain indicates a large storage area where the required optimum storage temperature is maintained. Maintaining this temperature is critical for preventing damage to food or pharmaceutical products that will be released into the market. Along with temperature, other factors like humidity, air pressure, flow rate, air quality, light conditions and shock or impact resistance capacity must be maintained inside the storage area (Cold Chain Technologies, 2022). The various zones of temperature that cold storage can maintain specifically or all at once can be described in Fig. 9.5. It is a proven fact that *Listeria sp* can adapt itself to different low and freezing temperature conditions due to the presence of more stress regulatory as well as transcriptional or translational proteins in their body (Santos et al., 2019). This fact increases the chances of Listeria survival in cold chain or storage areas.

Some reports about the cold, frozen, refrigerated, cold chain or retail stored products are discussed here. (Guan et al., 2021) pointed out the frequency of cold storage apples affected by Listeria and investigated the effectiveness of gaseous intervention in eliminating it. It can grow on vegetable products like freshly cut and refrigerated salad (Kaur & Yemmireddy, 2022), refrigerated tomato juice (Diakogiannis et al., 2017) and cut mushrooms, where they principally grow by cross-contamination. Dairy products are considered the major food group and the principal target of Listeria. Ultra-high temperature (UHT) treated dairy products in cold storage (Lobacz & Kowalik, 2015), concentrated and fermented milk products like Cheese (Possas et al., 2022), and kefir (Kalamaki & Angelidis, 2023) reported Listeria contamination causing outbreaks. In certain samples of mango-flavoured fermented milk and ice cream, a toxigenic strain Listeria monocytogenes CFR 1302 has been isolated, which survives and grows (Jayanth & Varadaraj, 2017; Kevenk & Aras, 2023). Ready-to-eat foods and ready-to-cook canned products in cold storage, like cold smoked salmon and yellowfin tuna, can allow conditions for the growth of Listeria and temperature regulation cannot be certainly used for control of this agent (Huang et al., 2023; Liu et al., 2016). Besides food, it has been identified in plant equipment, processing plant areas (Kara & Aslan, 2021), food-contact surfaces, bodies of workers, retail store areas, etc., chiefly due to accumulated condensation in and around these places. The magnitude of the problem increases since it was found that prolonged storage of *Listeria sp* may help them adapt and emerge as a cold-tolerant and more potent variant (Hingston et al., 2019).

9.6.2 Aerial Transmission Mechanism of Listeria in Cold Chain

By definition, any "Airborne" bacteria are those attached to dust particles in the air, whereas the "Aerosolized" bacteria remain suspended in liquid droplets in the air. Listeria can get transmitted to different food substances, both raw and processed, through the dust air and aerosols present in the air current. Since air is not Listeria's natural habitat, these species' presence in the air is attributed to the suspended particulate matter or air current flowing over different contaminated surface areas. (Berrang & Frank, 2012) showed how Listeria-laden drain water may contaminate storage area air during high-speed spray hose washing, which cross-contaminates foods in the warehouse. Dry air is hostile to any microbial survival. Investigation shows that air with higher relative humidity helps Listeria survive over a long period (Zoz et al., 2016). Zhang et al. (2007) showed how droplets of diameter 1–2 μm helped in the propagation and survival of Listeria spreads in the aerosol. Figure 9.6 shows an example of transmission of Listeria inside a cold chain or storage system storing fruits and vegetables through air particulate matter or droplets (Fig. 9.7).

The contaminated freezing or cooling tunnels, the unclean refrigerators, evaporators as well as cooling ducts form the main breeding grounds for Listeria contamination inside cold storage. To date, adequate clean air filters are not mounted in most of the cold storage areas, and the ones having filters are not cleaned regularly, leading to more and more contamination in circulating air. Aerial decontamination of cold

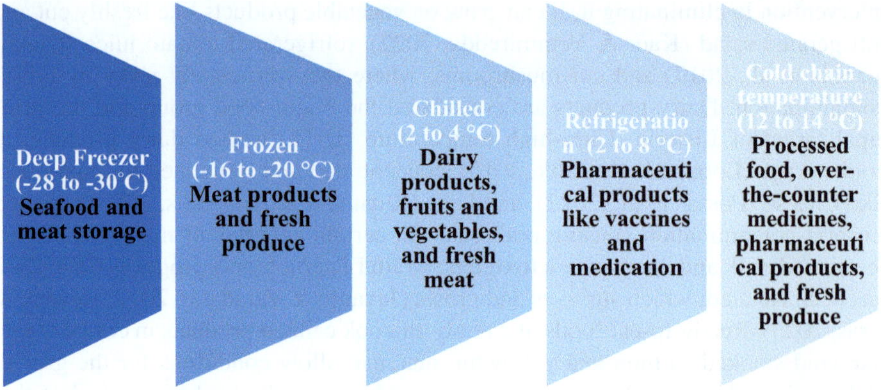

Fig. 9.6 Temperature profile and targeted storage products of different zones of a cold storage line or cold chain (Cold Chain Technologies, 2022)

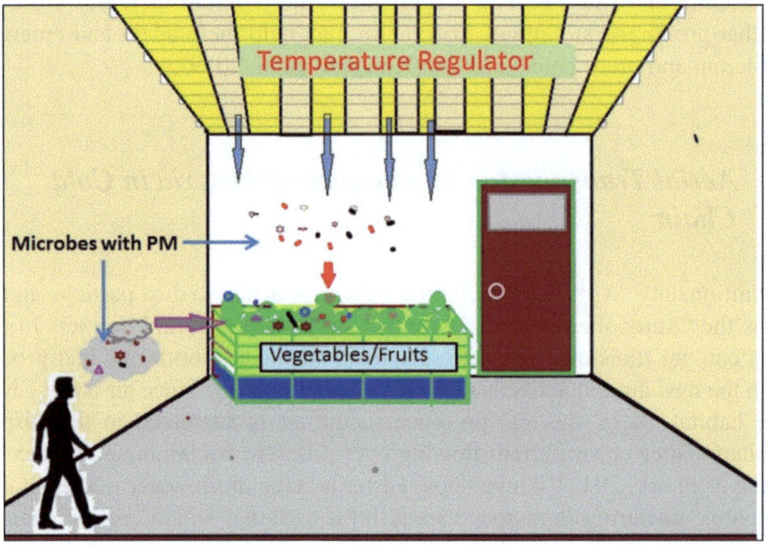

Fig. 9.7 Aerial particulate or droplet transmission cycle of Listeria sp inside cold storage

storage facilities for maintaining food quality and preventing cross-contamination of Listeria is an absolute necessity to prevent their distribution and spread (Oliveira et al., 2020). Due to the unclean filters, Listeria-bound particulate matter enters cold storage systems and contaminates the foods inside. Since Listeria can survive long term in various degrees of chilled and cooling temperatures, it may cause Listeriosis to the person consuming this contaminated food.

9.7 Preventive Measures to Mitigate Listeria Contamination

For the preservation and storage of food, a certain degree of hygiene and sterility is necessary to maintain in the area. Though retaining a total asepsis is absurd in practical cases—the best way seems to be frequent cleaning and disinfection of the zone. Many methods are available for this purpose, which may include physical processes such as irradiation, ionisation, ozonolysis and chemical methods like chemical fogging using vapours of chlorine dioxide, hydrogen peroxide, etc. (Oliveira et al., 2020).

To prevent Listeria, the first step is identifying and characterizing the microbe, which involves simple tests such as agar-plate culture techniques, monoclonal antibodies specific immunoassay and PCR-based techniques (Tabit, 2018). In the second step mass awareness programs, social media platforms, electronic or any other public communication media must be availed to spread awareness to ordinary people about Listeria contamination, spread and its preventive measures. The World Health Organization repeatedly publishes several guidelines for controlling Listeria on a ground level at source. The general Guidelines for Listeria prevention are listed as below:

1. Adequate cleaning provision must include constancy in cleaning water supply, proper drainage of contaminated water, clean toilet facilities and installation of hand wash basins.
2. A proper system-based storage facility with a sensitive temperature and humidity controller must be present.
3. Proper filtered air flow with adequate flow rate must be maintained.
4. Clean and sanitized refrigerated food-contact surfaces must be operated below 5 °C.
5. Processing kitchen tools like knives, graters, cutting and chopping boards must be washed before and after use.
6. Separate storage for raw and cooked food and/or temperature-variant foods must be indicated.
7. Enforcement of food safety hazards and hygiene practices through HACCP must be implemented.
8. A system for maintaining sanitary conditions in production, packaging and storage areas must be established.
9. Effectively monitored cleaning programs, pest control, or waste management systems must be employed.
10. Adequate personal hygiene, which includes the personal cleanliness behaviour of staff along with their daily record of health status, must be maintained.
11. Removal of faulty transport facilities with proper temperature and humidity control must be undertaken.
12. Refresher and advanced courses on hygiene practices to maintain the food facility must be conducted (Bolívar & Pérez-Rodríguez, 2023).

Table 9.3 Some eradication methods of Listeria species from contaminated food or food-contact surfaces

Type of method	Name of the method	Specification of dosage	Reference
Physical method	UV-C Irradiation	330 m J/cm^2 UV-C	Brugnini et al. (2021)
	Ionizing radiation	<10 KGy	Munir and Federighi (2020)
	Ozonolysis	50 ppm, 6 h exposure	Panebianco et al. (2021)
	Saturated steam	100 °C steam for 6 s reduce 2–3 log cycle	Hua et al. (2021)
Chemical method	Commercial alkaline, acidic and enzymatic cleaning agent	2–8% for 5 log reduction for 7 day old biofilm	Fagerlund et al. (2020)
	Quaternary ammonium compound (QAC)	5-min treatments of 400 ppm QAC	Hua et al. (2019)
	Per-chloro acetic acid	Chlorine and 80 ppm PAA	Wang et al. (2023)
	Chlorine di oxide vapours	3.5 log reduction at 20 ppmv for 15 min	Guan et al. (2021)
	Hydrogen peroxide mists	12–20 °C, at concentrations of 35–80 ppm for 1–2 h	Møretrø et al. (2019)
	Lactic Acid	2.6% Lactic Acid	Brugnini et al. (2021)

Besides undertaking preventive action, the decontamination of identified listeria, as discussed above, can be described in detail in Table 9.3.

After scientific application of these techniques in different food handling and storage zones, Listeria sp can be eliminated, causing a depletion in the number of annual Listeriosis cases worldwide.

9.8 Conclusion

The greatly ubiquitous, facultative psychrotrophic organism listeria can sustain and propagate in normal to harsh environments, which gets transmitted to food, causing foodborne outbreaks. The primary sources can be air, water, soil or any food-contact surfaces through which diseases spread to the human system. The species Listeria is also zoonotic, causing illness to livestock animals and birds. Particulate matter, aerosols and moisture droplets in air may carry these organisms and contaminate any storage area. The cold storage area, having a different set of low-temperature zones, is more susceptible towards Listeria contamination since it continues to proliferate under the cooler conditions of the atmosphere. Only rigorous implementation of hygienic practices, along with frequent cleaning and disinfection procedures, can eliminate and destroy listeria completely.

9.9 Knowledge Gap to Date

Through a literature review, we found that there are studies reporting listeria survival and contamination in food systems under low-temperature conditions and refrigeration. A few isolated reports also suggested listeria transmission using air as a vehicle. The study of the spread of listeria in cold storage or cold chain systems through aerial transmission was almost nil.

9.10 Future Prospects

The content of this chapter holds an enormous prospect, which is concerned with the spread of pathogenic *Listeria* sp in food chains and cold storage. It will make the food handlers in the cold chain aware of the probable cross-contamination, which can introduce Listeria inside the storage. It will also attract the attention of the stakeholders and policymakers for incorporating safety protocols about air quality in cold stores that is useful for the mitigation of Listeria and preventing the occurrence of foodborne outbreaks.

References

Abdi Moghadam, Z., Shamloo, E., & Atefi, M. (2015). Frequency of Listeria species in raw milk and traditional dairy products in Isfahan, Iran. *Iranian Journal of Nutrition Sciences & Food Technology, 10*, 101–107.

Abulude, F. O., Arifalo, K. M., Kenni, A. M., Akinnusotu, A., Oluwagbayide, S. D., & Sunday, A. (2022). Air quality index levels of particulate matter (pm2.5) in Venogua, Nigeria. *Jurnal Geografi GEA, 22*(2), 95–105. https://doi.org/10.17509/gea.v22i2.46307

Abulude, F. O. (2016). Particulate Matter: An Approach to Air Pollution. Preprints, Basel, Switzerland, pp. 1–14.

Adams, K., Greenbaum, D. S., Shaikh, R., Van Erp, A. M., & Russell, A. G. (2015). Particulate matter components, sources, and health: Systematic approaches to testing effects. *Journal of the Air and Waste Management Association, 65*(5), 544–558. https://doi.org/10.1080/10962247.2014.1001884

Allerberger, F., & Wagner, M. (2009). Listeriosis: A resurgent foodborne infection. *Clinical Microbiology & Infection, 16*(1), 16–23. https://doi.org/10.1111/j.1469-0691.2009.03109.x

Barbuddhe, S. B., Malik, S. V. S., Kumar, J. A., Kalorey, D. R., & Chakraborty, T. (2012). Epidemiology and risk management of listeriosis in India. *International Journal of Food Microbiology, 154*(3), 113–118. https://doi.org/10.1016/j.ijfoodmicro.2011.08.030

Berrang, M. E., & Frank, J. F. (2012). Generation of airborne Listeria innocua from model floor drains. *Journal of Food Protection, 75*(7), 1328–1331. https://doi.org/10.4315/0362-028X

Berrang, M., Frank, J., & Meinersmann, R. (2013). Contamination of raw poultry meat by airborne Listeria originating from a floor drain. *Journal of Applied Poultry Research, 22*(1), 132–136. https://doi.org/10.3382/japr.2012-00676

Bianconi, V., Bronzo, P., Banach, M., Sahebkar, A., Mannarino, M. R., & Pirro, M. (2020). Particulate matter pollution and the COVID-19 outbreak: Results from Italian regions and provinces. *Archives of Medical Science, 16*(5), 985–992. https://doi.org/10.5114/aoms.2020.95336

Bolívar, A., & Pérez-Rodríguez, F. (2023). Listeria in food: Prevalence and control. *Foods, 12*(7), 1378. https://doi.org/10.3390/foods12071378

Bongiovanni, M., Cavallo, C., Barda, B., Strulak, L., Bernasconi, E., & Cardia, A. (2023). Clinical findings of Listeria monocytogenes infections with a special focus on bone localizations. *Microorganisms, 12*(1), 178. https://doi.org/10.3390/microorganisms12010178

Brugnini, G., Rodríguez, S., Rodríguez, J., & Rufo, C. (2021). Effect of UV-C irradiation and lactic acid application on the inactivation of Listeria monocytogenes and lactic acid bacteria in vacuum-packaged beef. *Foods, 10*(6), 1217. https://doi.org/10.3390/foods10061217

Buchanan, R. L., Gorris, L. G., Hayman, M. M., Jackson, T. C., & Whiting, R. C. (2017). A review of Listeria monocytogenes: An update on outbreaks, virulence, dose-response, ecology, and risk assessments. *Food Control, 75*, 1–13. https://doi.org/10.1016/j.foodcont.2016.12.016

Centers for Disease Control and Prevention. (2024). National Center for Emerging and Zoonotic Infectious Diseases (NCEZID), Division of Foodborne, Waterborne, and Environmental Diseases (DFWED). https://www.cdc.gov/listeria/outbreaks/index.html

Centre for Disease Control and Prevention. (2024). https://www.cdc.gov/listeria/index.html

Clark, M. (2015–2017). About Listeria. http://www.about-listeria.com/listeria_incidence/#.Wjaly1wbIU

Cold Chain Technologies. (2022). What is Storage Temperature? https://www.coldchaintech.com/blog/uncategorized/what-is-storage-temperature#:~:text=Make%20sure%20you%20know%20these,and%20vegetables%2C%20and%20fresh%20meat

Das, S., Roy, A., Masiwal, R., Mandal, M., Popek, R., Chakraborty, M., Prasad, D., Chyliński, F., Awasthi, A., & Sarkar, A. (2023). Comprehensive analysis of PM1 composition in the Eastern Indo-Gangetic Basin: a three-year urban study. *Sustainability, 15*(20), 14894. https://doi.org/10.3390/su152014894

Della Rosa, P., Colque Caro, L. A., Cantón, G. J., Morrell, E. L., Hecker, Y. P., Paolicchi, F. A., & Fiorentino, M. A. (2019). Aborto ovino asociado a Listeria ivanovii; Proceedings of the XV Congreso Argentino de Microbiología; Buenos Aires, Argentina, pp. 25–27.

Diakogiannis, L., Proestos, C., Varzakas, T., & Markaki, P. (2017). Survival of Listeria Monocytogenes in Tomato Juice at 5 and 30°C Storage. *Current Research in Nutrition and Food Science, 5*(1). https://doi.org/10.12944/CRNFSJ.5.1.01

Dobeic, M., Kenda, E., Mičunovič, J., & Zdovc, I. (2011). Airborne Listeria spp. in the red meat processing industry. *Czech Journal of Food Sciences, 29*(4), 441–447. https://doi.org/10.17221/88/2010-CJFS

Dunnett, E., Florea, L., Thurston, L., Floyd, T., Collins, R., & Otter, A. (2020). Deaths of weaned lambs with visceral Listeria ivanovii infections. *Veterinary Record Case Reports, 8*, Article e001254. https://doi.org/10.1136/vetreccr-2020-001254

Fagerlund, A., Heir, E., Møretrø, T., & Langsrud, S. (2020). Listeria monocytogenes biofilm removal using different commercial cleaning agents. *Molecules, 25*(4), 792. https://doi.org/10.3390/molecules25040792

Fujiyoshi, S., Tanaka, D., & Maruyama, F. (2017). Transmission of airborne bacteria across built environments and its measurement standards: A review. *Frontiers in Microbiology, 8*, Article 291562. https://doi.org/10.3389/fmicb.2017.02336

Gonzalez-Martin, C. (2019). Airborne Infectious Microorganisms. *Encyclopedia of Microbiology* 52–60. https://doi.org/10.1016/B978-0-12-809633-8.13002-X

Guan, J., Lacombe, A., Rane, B., Tang, J., Sablani, S., & Wu, V. C. H. (2021). A review: Gaseous interventions for listeria monocytogenes control in fresh apple cold storage. *Frontiers in Microbiology, 12*, Article 782934. https://doi.org/10.3389/fmicb.2021.782934

Hingston, P. A., Truelstrup, H. L., Pombert, J., & Wang, S. (2019). Characterization of Listeria monocytogenes enhanced cold-tolerance variants isolated during prolonged cold storage. *International Journal of Food Microbiology, 306*, Article 108262. https://doi.org/10.1016/j.ijfoodmicro.2019.108262

Hua, Z., Korany, A. M., El-Shinawy, S. H., & Zhu, J. (2019). Comparative evaluation of different sanitizers against listeria monocytogenes biofilms on major food-contact surfaces. *Frontiers in Microbiology, 10*, 2462. https://doi.org/10.3389/fmicb.2019.02462

Hua, Z., Younce, F., Tang, J., Ryu, D., Rasco, B., Hanrahan, I., & Zhu, M. (2021). Efficacy of saturated steam against Listeria innocua biofilm on common food-contact surfaces. *Food Control, 125*, Article 107988. https://doi.org/10.1016/j.foodcont.2021.107988

Huang, L., Hwang, C., & Sheen, S. (2023). Shelf-life boundaries of Listeria monocytogenes in cold smoked salmon during refrigerated storage and temperature abuse. *International Food Research Journal, 173*, Article 113362. https://doi.org/10.1016/j.foodres.2023.113362

Jayanth, H. S., & Varadaraj, M. C. (2017). Survival of a native toxigenic isolate of Listeria monocytogenes CFR 1302 during storage of milk-based foods can be a potential cause of health risk. *Food Science and Technology International, 23*(5), 426–436. https://doi.org/10.1177/1082013217698329

Johnson, T. M., Guttikunda, S., Wells, G. J., Artaxo, P., Bond, T. C., Russell, A. G., Watson, J. C., & West, J. (2011).Tools for improving air quality management. A review of Top-down source apportionment techniques and their application in developing countries. ESMAP Formal report. World Bank: Washington, DC, USA.

Kalamaki, M. S., & Angelidis, A. S. (2023). Growth and survival of Listeria monocytogenes during the manufacture and storage of artisanal kefir. *Food Science and Technology International, 29*(8), 789–798. https://doi.org/10.1177/10820132221117462

Kara, R., & Aslan, S. (2021). Investigation of Listeria monocytogenes in workers, equipment and environments at Kaymak processing plants. *Food Science and Technology, 41*(2), 449–452. https://doi.org/10.1590/fst.02620

Kaur, A., & Yemmireddy, V. (2022). Effect of different pre-growth temperatures on the survival kinetics of Salmonella enterica and listeria monocytogenes in fresh-cut salad during refrigerated storage. *Foods, 12*(23), 4287. https://doi.org/10.3390/foods12234287

Kevenk, T. O., & Aras, Z. (2023). Occurrence and characterization of listeria monocytogenes. https://doi.org/10.2139/ssrn.4656518

Kuźma, Ł, Dąbrowski, E. J., Kurasz, A., Bachórzewska-Gajewska, H., & Dobrzycki, S. (2020). The 10-year study of the impact of particulate matters on mortality in two transit cities in North-Eastern Poland (PL-PARTICLES). *Journal of Clinical Medicine, 9*(11), 3445. https://doi.org/10.3390/jcm9113445

Liu, C., Jing, M., & Yi-Cheng, S. (2016). Behavior of Salmonella and listeria monocytogenes in Raw Yellowfin Tuna during cold storage. *Foods, 5*(1), 16. https://doi.org/10.3390/foods5010016

Lobacz, A., & Kowalik, J. (2015). A predictive model for listeria monocytogenes in UHT dairy products with various fat content during cold storage. *Journal of Food Safety, 35*(1), 119–127. https://doi.org/10.1111/jfs.12163

Mahdavi, M., Jalali, M., Safaei, H. G., & Shamloo, E. (2012). Microbial quality and prevalence of Salmonella and Listeria in eggs. *International Journal of Environmental Health Engineering, 1*(6), 16–20.

Mandal, M., Roy, A., Popek, R., & Sarkar, A. (2024). Micro-and nano-plastic degradation by bacterial enzymes: A solution to'white pollution. *The Microbe, 3*, Article 100072. https://doi.org/10.1016/j.microb.2024.100072

Mandal, M., Das, S., Roy, A., Rakwal, R., Jones, O. A., Popek, R., Agarwal, G. K., & Sarkar, A. (2023). Interactive relations between plants, the phyllosphere microbial community, and particulate matter pollution. *Science of the Total Environment, 890*, 164352. https://doi.org/10.1016/j.scitotenv.2023.164352

Matle, I., Mbatha, K. R., & Madoroba, E. (2020). A review of Listeria monocytogenes from meat and meat products: Epidemiology, virulence factors, antimicrobial resistance and diagnosis. *Onderstepoort Journal of Veterinary Research, 87*(1). https://doi.org/10.4102/ojvr.v87i1.1869

Møretrø, T., Fanebust, H., Fagerlund, A., & Langsrud, S. (2019). Whole room disinfection with hydrogen peroxide mist to control Listeria monocytogenes in food industry related environments. *International Journal of Food Microbiology, 292*, 118–125. https://doi.org/10.1016/j.ijfoodmicro.2018.12.015

Munir, M. T., & Federighi, M. (2020). Control of foodborne biological hazards by ionizing radiations. *Foods, 9*(7), 878. https://doi.org/10.3390/foods9070878

Myung, H., & Joung, Y. S. (2024). Contribution of particulates to airborne disease transmission and severity: A review. *Environmental Science and Technology, 58*(16), 6846–6867. https://doi.org/10.1021/acs.est.3c08835

Oliveira, M., Tiwari, B. K., & Duffy, G. (2020). Emerging technologies for aerial decontamination of food storage environments to eliminate microbial cross-contamination. *Foods, 9*(12), 1779. https://doi.org/10.3390/foods9121779

Osek, J., Lachtara, B., & Wieczorek, K. (2022). Listeria monocytogenes—how this pathogen survives in food-production environments? *Frontiers in Microbiology, 13*, Article 866462. https://doi.org/10.3389/fmicb.2022.866462

Panebianco, F., Rubiola, S., Chiesa, F., Civera, T., & Di Ciccio, P. A. (2021). Effect of gaseous ozone on listeria monocytogenes planktonic cells and biofilm: An in vitro study. *Foods, 10*(7), 1484. https://doi.org/10.3390/foods10071484

Peratikos, P., Tsitsos, A., Damianos, A., Kyritsi, M. A., Hadjichristodoulou, C., Soultos, N., & Economou, V. (2023). Listeria monocytogenes from Marine fish and the seafood market environment in Northern Greece: Prevalence, molecular characterization, and antibiotic resistance. *Applied Science, 14*(7), 2725. https://doi.org/10.3390/app14072725

Possas, A., Hernández, M., Esteban-Carbonero, O., Valero, A., & Rodríguez-Lázaro, D. (2022). Listeria monocytogenes survives better at lower storage temperatures in regular and low-salt soft and cured cheeses. *Food Microbiology, 104*, Article 103979. https://doi.org/10.1016/j.fm.2022.103979

Rossi, F., Giaccone, V., Colavita, G., Amadoro, C., Pomilio, F., & Catellani, P. (2022). Virulence characteristics and distribution of the Pathogen Listeria ivanovii in the environment and in food. *Microorganisms, 10*(8), 1679. https://doi.org/10.3390/microorganisms10081679

Roy, A., Das, S., Singh, P., Mandal, M., Kumar, M., Rajlaxmi, A., Vijayan, N., Awasthi, A., Chhetri, H., Roy, S., & Popek, R. (2024a). Summer-time monitoring and source apportionment study of both coarse, fine, and ultra-fine particulate pollution in Eastern Himalayan Darjeeling: A hint to health risk during peak tourist season. *MAPAN, 39*(4), 995–1009. https://doi.org/10.1007/s12647-024-00776-9

Roy, A., Mandal, M., Das, S., Kumar, M., Popek, R., Awasthi, A., Giri, B. S., Mondal, K. C., & Sarkar, A. (2024b). Non-exhaust particulate pollution in Asian countries: A comprehensive review of sources, composition, and health effects. *Environmental Engineering Research, 29*(3), 44–68. https://doi.org/10.4491/eer.2023.384

Santos, T., Viala, D., Chambon, C., Esbelin, J., & Hébraud, M. (2019). Listeria monocytogenes biofilm adaptation to different temperatures seen through shotgun proteomics. *Frontiers in Nutrition, 6*, Article 439004. https://doi.org/10.3389/fnut.2019.00089

Sauders, B. D., Fortes, E. D., Morse, D. L., Dumas, N., Kiehlbauch, J. A., Schukken, Y., Hibbs, J. R., & Wiedmann, M. (2003). Molecular subtyping to detect human listeriosis clusters. *Emerging Infectious Diseases, 9*, 672.

Shaddick, G., Thomas, M. L., Mudu, P., Ruggeri, G., & Gumy, S. (2020). Half the world's population are exposed to increasing air pollution. *Nature Partner Journals, 3*(1), 1–5. https://doi.org/10.1038/s41612-020-0124-2

Shamloo, E., Hosseini, H., Moghadam, Z. A., Larsen, M. H., Haslberger, A., & Alebouyeh, M. (2018). Importance of Listeria monocytogenes in food safety: A review of its prevalence, detection, and antibiotic resistance. *Iranian Journal of Veterinary Research, 20*(4), 241–254.

Sibanda, T., & Buys, E. M. (2022). Listeria monocytogenes pathogenesis: The role of stress adaptation. *Microorganisms., 10*(8), 1522. https://doi.org/10.3390/microorganisms10081522

Suman. (2021). Air quality indices: A review of methods to interpret air quality status. *Materials Today: Proceedings, 34*(3), 863–868. https://doi.org/10.1016/j.matpr.2020.07.141

Tabit, F. T. (2018). Contamination, prevention and control of listeria monocytogenes in food processing and food service environments. In *Listeria Monocytogenes.* In Techopen. https://doi.org/10.5772/intechopen.76132.

Thomas, A. (2016). Listeria monocytogenes in fresh fruits and vegetables. Doctoral dissertation, Virginia Tech.

Tirumalai, P. S. (2013). Listeriosis and Listeria monocytogenes in India. *Wudpecker Journal of Food Technology, 1*(6), 98–103.

Vázquez-Boland, J. A., Kuhn, M., Berche, P., Chakraborty, T., Domínguez-Bernal, G., Goebel, W., González-Zorn, B., Wehland, J., & Kreft, J. (2001). Listeria pathogenesis and molecular virulence determinants. *Clinical Microbiology Reviews, 14*(3), 584–640. https://doi.org/10.1128/CMR.14.3.584-640.2001

Wang, R. Y., Shen, X., Su, Y., Critzer, F., & Zhu, M. (2023). Chlorine and peroxyacetic acid inactivation of Listeria monocytogenes in simulated apple dump tank water. *Food Control, 144*, Article 109314. https://doi.org/10.1016/j.foodcont.2022.109314

World health organisation. Air Pollution. https://www.who.int/health-topics/air-pollution#tab=tab_1

Wu, S., Wu, Q., Zhang, J., Chen, M., & Hu, H. (2015). Listeria monocytogenes prevalence and characteristics in retail raw foods in China. *PLoS One, 10*(8), e0136682. https://doi.org/10.1371/journal.pone.0136682

Yong, S. S., Lee, J. I., & Kang, D. H. (2024). Airborne survival and stress response in Listeria monocytogenes across different growth temperatures. *Journal of Hazardous Materials, 468*, Article 133706. https://doi.org/10.1016/j.jhazmat.2024.133706

Zhang, G., Ma, L., Oyarzabal, O. A., & Doyle, M. P. (2007). Aerosol Studies with Listeria innocua and Listeria monocytogenes. *Journal of Food Protection, 70*(8), 1857–1865. https://doi.org/10.4315/0362-028x-70.8.1857

Zhu, Q., Gooneratne, R., & Hussain, M. A. (2017). Listeria monocytogenes in fresh produce: Outbreaks, prevalence and contamination levels. *Foods, 6*(3), 21. https://doi.org/10.3390/foods6030021

Zoz, F., Iaconelli, C., Lang, E., Iddir, H., Guyot, S., Grandvalet, C., Gervais, P., & Beney, L. (2016). Control of relative air humidity as a potential means to improve hygiene on surfaces: A preliminary approach with listeria monocytogenes. *PLoS ONE, 11*(2), Article e0148418. https://doi.org/10.1371/journal.pone.0148418

Chapter 10
Assessing Human Exposure to Potentially Toxic Trace Elements in Urban Soils and Road Dust: A Case Study from Southern Africa

Sutapa Adhikari, Ricart Barend Boneschans, and Madeleen Struwig

Abstract Human exposure to particle-borne potentially toxic trace elements (TEs) in public spaces adjacent to busy roads remains underexplored in smaller cities, particularly in southern Africa. In this case study, total concentrations of Ba, Cr, Fe, Mn, Ni, Pb, Sr, Ti, V and Zn in roadside soils and road dust from diverse urban facilities in Mahikeng, a relatively small South African city, were measured to assess traffic influence and evaluate health risks. The findings reveal distinct TE concentration trends between the two media, with significantly ($p < 0.05$) higher mean levels of Ni and Sr in road dust. Children face significantly elevated health risks from both media at most sites, with non-carcinogenic risks driven by dermal exposure to Cr and from all TEs via both pathways, and cancer risks from Cr and Ni ingestion. For adults, cancer risk from ingestion of Ni in both media is concerning. Both exposure pathways present potential health consequences, with road dust exhibiting greater risks. Traffic origin for most TEs in both media is evident, including the key health risk

Supplementary Information The online version contains supplementary material available at https://doi.org/10.1007/978-3-032-31991-3_10.

S. Adhikari
South African Environmental Observation Network, Ndlovu Node, WITS Rural Campus, Acornhoek 1360, South Africa

S. Adhikari (✉) · R. B. Boneschans · M. Struwig
Unit for Environmental Sciences and Management, North-West University, Potchefstroom 2531, South Africa
e-mail: Sutapa.Adhikari@nwu.ac.za

R. B. Boneschans
e-mail: Ricart.Boneschans@nwu.ac.za

M. Struwig
e-mail: Madeleen.Struwig@nwu.ac.za

M. Struwig
Botany Department, Faculty of Natural and Agricultural Sciences, North-West University, Mmabatho 2745, South Africa

A. Sarkar et al. (eds.), *From Sky to Soil*, https://doi.org/10.1007/978-3-032-31991-3_10

drivers, Cr and Ni. Hence, systematic monitoring and mitigation of TEs in roadside soil and road dust are essential in public spaces, even in smaller cities.

10.1 Introduction

Human health consequences of exposure to particle-borne potentially toxic trace elements (TEs) are becoming increasingly severe due to expanding industrialisation and urbanisation worldwide. The expansion of road networks and increasing traffic volumes in urban areas are major contributors to aerial TE loads, resulting in the contamination of other environmental compartments (Kabir et al., 2021; Najmeddin et al., 2018). Traffic linked emissions typically contain a mixture of TEs, including barium (Ba), chromium (Cd), iron (Fe), manganese (Mn), nickel (Ni), lead (Pb), strontium (Sr), titanium (Ti), vanadium (V) and zinc (Zn) (Fussell et al., 2022; Raza Ahmad et al., 2019; Wang et al., 2022; Yang et al., 2017). Disintegration of road construction materials, such as asphalt, fly ash and concrete, and pavement wear and corrosion of roadside metal barriers further contribute to the elemental content in road dust (Duong & Lee, 2011; Fiala & Hwang, 2021; Pagotto et al., 2001). Hence, various transportation-related factors play a critical role in determining the element composition of urban road dust (Duong & Lee, 2011; Wang et al., 2022). Parallel to this, aerial deposition and runoff elevate concentrations of TEs in roadside soil, specifically at the surface horizon (Adimalla et al., 2020; Qiao et al., 2014; Sager, 2020; Sparks, 2005). Due to contributions from multiple sources, road dust and roadside soil typically contain particles of various size fractions, enriched with an array of potentially toxic TEs at varied concentrations (Fortoul et al., 2015).

Human exposure to particle-borne TEs near road networks occurs through multiple pathways, leading to adverse health effects, particularly among pedestrians and individuals waiting for transport. Therefore, TEs, generally found in excess quantities in roadside environments, have garnered extensive attention worldwide. Among these, the introduction of relatively more toxic Pb into the human body, even at low concentrations, may be harmful (Alhassan et al., 2012; Al-Shidi et al., 2020; Cai & Zhang, 2021; Fiala & Hwang, 2021). The International Agency for Research on Cancer classifies Pb as 'probably carcinogenic to humans' and is placed in Group 2A (IARC, 2025). Despite bans and reductions in the use of Pb in fuels, its presence in other commodities, including various automobile and road infrastructure components, continues to contribute substantially to environmental pollution (Carver & Gallicchio, 2017). In addition, carcinogenic properties of Cr and Ni have been discussed widely (Kamunda et al., 2016; Shahzad et al., 2018). Other comparatively less hazardous TEs emitted by traffic-related factors may further elevate potential health risks for exposed populations (Apeagyei et al., 2011; Chang et al., 2021; Han et al., 2023; Jose & Srimuruganandam, 2020). The United States Environmental Protection Agency included Cr, Ni, Pb and Zn in the Priority Pollutant List (US EPA, 2014), and hazardous levels of Zn in the atmosphere have been recommended.

Although both soil and road dust have long been employed worldwide to assess urban pollution, most studies have focused on only one medium, leaving the comprehensive understanding of these closely related media largely unexplored (Chang et al., 2009; Jonidi Jafari et al., 2018; Pagotto et al., 2001; Qiao et al., 2014). In this regard, limited information is available on smaller cities worldwide, specifically in southern African nations (Semerjian et al., 2021), and South Africa is no exception. In mineral-rich countries like South Africa, where soil elemental concentrations are typically higher than those in many other nations (Herselman et al., 2005), anthropogenic activities further elevate these levels considerably in the environment. While substantial data exist on the influence of anthropogenic activities on hazards associated with airborne particles in larger cities and mining hotspots of southern Africa (Adhikari et al., 2022; Batwa-Ismail et al., 2021; Kamunda et al., 2016; Mugudamani et al., 2022; Mvovo & Magagula, 2021; Novela et al., 2020; Van der Westhuizen, 2024; Williams et al., 2021), comparatively limited information is available on TE concentrations in soils and road dust, as well as the associated health risks, in smaller urban regions, particularly in public spaces. Public spaces are frequently intersected by busy roads, exhibit higher foot traffic and accommodate gatherings throughout the year. Pollution data for these localities are essential for urban planning and for developing cost-effective pollution management and sustainable remediation strategies, including the use of vegetation to enhance urban greenery, promote cleaner environments and support urban ecosystem services.

We hypothesise that direct human exposure to potentially toxic TE-containing roadside soil particles and road dust in public spaces adjacent to busy roads poses significant health threats in even smaller cities or towns, such as Mahikeng, the capital of the North West Province of South Africa, where urban foot traffic is relatively high. Therefore, the objectives of this investigation were to (1) quantify the total concentrations of an array of potentially toxic TEs, namely Ba, Cr, Fe, Mn, Ni, Pb, Sr, Ti, V and Zn, in the surface soil and road dust from various public spaces in Mahikeng, and compare soil element concentrations with global averages and national and international permissible standards, (2) identify sources of TEs in both media and (3) evaluate potential human health risks associated with the exposure to the selected TEs in these media. The preliminary findings of this study may encourage further research to address existing knowledge gaps and support the development of sustainable strategies for mitigating particle-borne TE pollution at national and regional scales.

10.2 Materials and Methods

10.2.1 Study Locality

The city of Mahikeng falls under the Mahikeng Local Municipality, the smallest of the five municipalities within the Ngaka Modiri Molema District, one of four districts in the North West Province of South Africa (Profile: Ngaka Modiri Molema District, 2023). The province is a prominent mining region in the country, and alongside agriculture, mining activities are major drivers of the municipality's economy (Council for Geoscience, South Africa, CGS, 2022; Profile: Ngaka Modiri Molema District, 2023). Despite being the smallest among the five municipalities, Mahikeng Municipality boasts the second-highest population growth rate (1.5%) and the highest population density (78 people per km^2) (Annual Report, 2024–2025). Accordingly, this relatively small southern African city was selected as the study region for this investigation (Fig. 10.1a).

Three shopping centres (S1, S2 and S3) and the main gate of North-West University (S4) were designated for investigation, based on four key criteria: (1) distance from mining activities, ensuring that they were not situated in the immediate vicinity of such operations; (2) proximity to busy roads, prioritising localities near high traffic roads and streets (Fig. 10.1b); (3) transportation hubs, where sites intersected with both public and private transportation systems and (4) pedestrian activity, focusing on areas with high foot traffic.

10.2.2 Sampling

Sampling was conducted in July 2023, during the winter season following a dry period of fifteen days. This season was selected based on the meteorological factors influencing dust fall-out rates in South Africa (Dudu et al., 2018): (1) low rainfall, which increases dust fall-out rates, and (2) prevailing dry wind, which enhances particle resuspension and deposition. Both soil and road dust samples were collected within 5 m of the nearest road or street at each site, and at depths ≤ 0.1 m for soils at all four sites. This is based on the general observation that surface soil within 7 m of traffic generally contains higher levels of pollutants in urban regions (Sager, 2020; Sun et al., 2017; Timofeev et al., 2019). At each of the four sites, two sampling points, located within 100 m of each other and < 5 m from the nearest road or street, were selected to quantify elemental levels in the two examined matrices. At each sampling point, samples were collected from five randomly selected positions and subsequently combined to form a composite sample of soil and road dust for each of the eight sampling points, resulting in a total of eight composite samples per medium across the four sites. In this paper, soil (SL) and road dust (RD) samples are referred to as SL1, SL2, RD1, RD2 (Crossing Shopping Centre, S1); SL3, SL4, RD3, RD4 (Megacity Mall, S2); SL5, SL6, RD5, RD6 (Morena Mall, S3) and SL7, SL8, RD7,

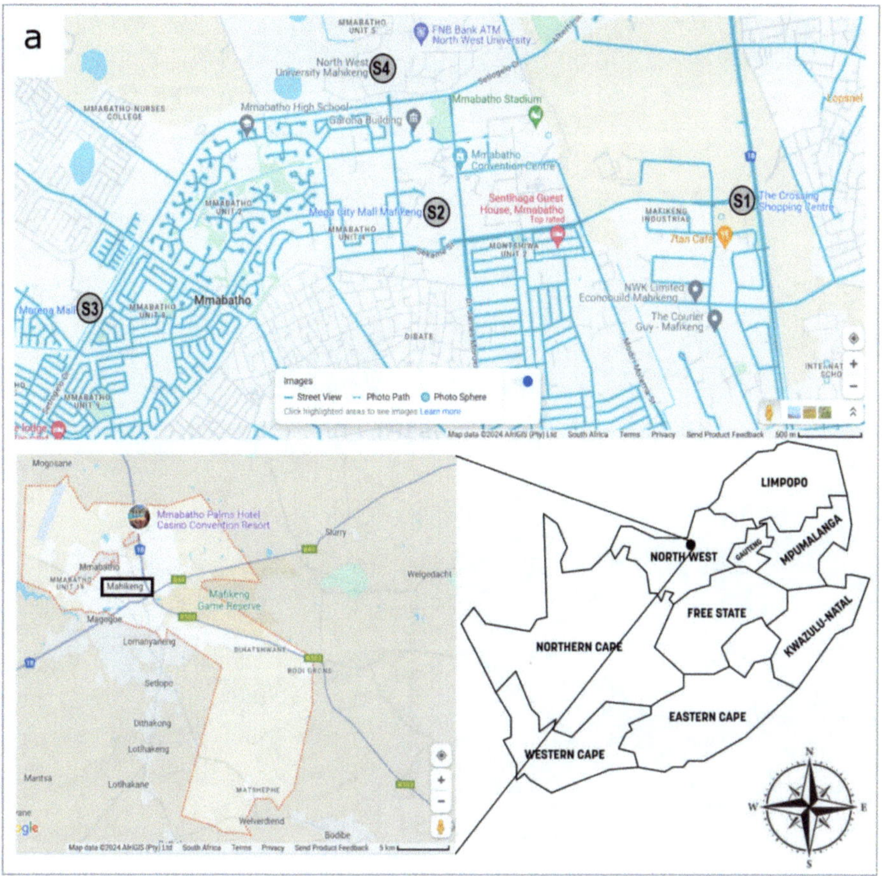

Fig. 10.1 **a** Location of the four sampling sites in Mahikeng, North West Province, South Africa, and **b** proximity of the sampling sites to roadways

RD8 (NWU main gate, S4). At each sampling point, road dust was collected using a new plastic brush (Raza Ahmad et al., 2019), and soil with a stainless steel garden spade, which was cleaned before sampling at each site. Collected soil and road dust samples were transported to the Botany Department, Mahikeng Campus, North-West University. Subsequently, the samples were thoroughly air-dried and stored in sealed, numbered plastic containers at room temperature.

10.2.3 Preparation and Analysis of Samples

Portable X-ray fluorescence analysis was performed at the School of Geo- and Spatial Sciences at the North-West University, Potchefstroom Campus, South Africa.

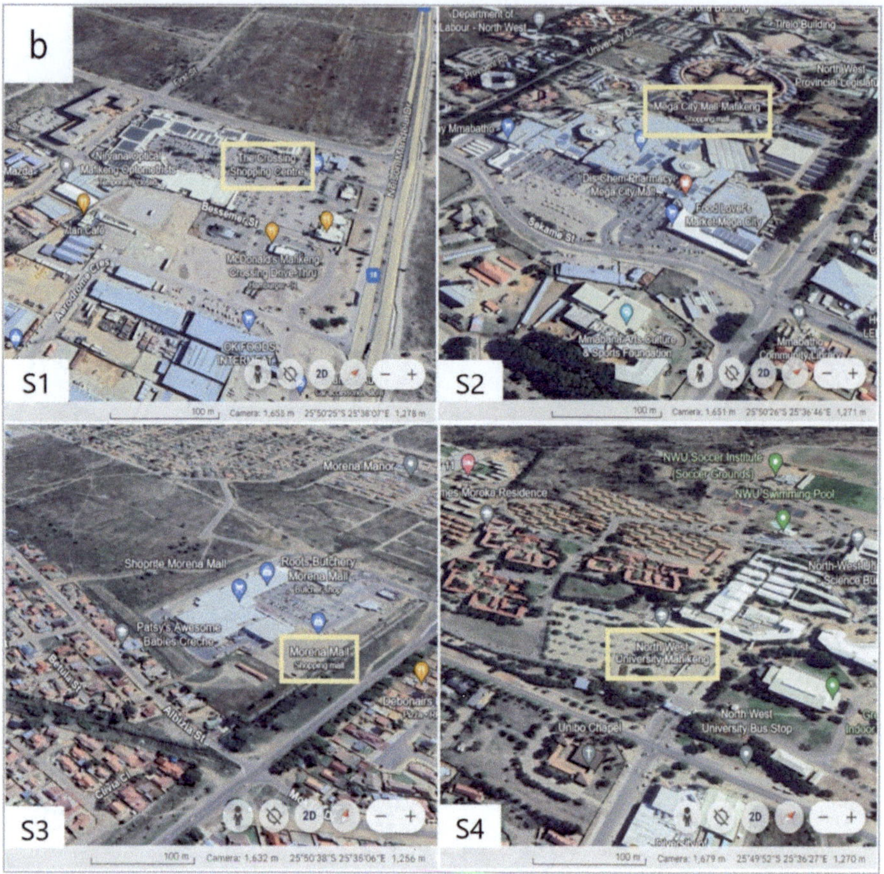

Fig. 10.1 (continued)

Samples of air-dried soil and road dust were passed through a 2-mm sieve to remove larger-sized particles, plant remains and inorganic debris. Thereafter, samples were ground using a porcelain mortar and pestle, which ensured that the samples were thoroughly homogenised. To avoid cross-contamination, all surfaces of the mortar and pestle were carefully cleaned with 90% ethanol before each sample was powdered. Non-destructive handheld X-ray fluorescence was utilised to measure the total concentrations of TEs in soil and road dust samples (Qiao et al., 2014). The analysis was conducted using a Thermo Scientific Niton XL3t GOLDD+ handheld XRF analyser (Thermo Fisher Scientific, Waltham, MA, USA). Prior to analysis, the built-in system check function was performed to ensure the instrument operated within specified parameters. An XRF sample cup filled with homogenised sample was covered with polypropylene XRF film and scanned for a total of 120 s, applying the TestAll Geo analysis mode. This mode employs both the Compton Normalisation calibration (Soil Mode) and the Fundamental Parameters calibration (Mining

Mode) to evaluate the validity of the soil calibration mode and to determine whether the total metal concentration exceeds the threshold permissible for reliable use of the Compton mode. To ensure data validity and reliability, each soil and road dust sample was analysed in triplicate and averaged, following recommended protocols and equipment safety measures.

10.2.4 Human Health Risk Assessment

This study applied health risk assessment models recommended by the US EPA (2002) to protect human health in formal residential sites. These models have been adopted in the Framework for the Management of Contaminated Land by the Department of Environmental Affairs, South Africa, with values adjusted to meet national criteria (DEA, 2010). In this study, non-carcinogenic and cancer risks were assessed for two target population groups, adults and children, via ingestion and dermal contact—the primary exposure routes in formal residential environments (US EPA, 2002).

10.2.4.1 Non-carcinogenic Risk

The potential non-carcinogenic risk of TEs for the exposed population was assessed by calculating the following indices: (1) Hazard Quotient (HQ), which indicates hazard related to each TE through ingestion and dermal exposure, and (2) Hazard Index (HI), representing the total non-carcinogenic hazard of all assessed TEs through each exposure route. First, the Average Daily Dose (ADD) of each element was calculated for ingestion (Eq. (10.1)) and dermal contact pathway (Eq. (10.2)). Then, the ADD value was divided by the reference dose (RfD, this represents the level of an exposed contaminant at which no negative effects are expected) of each element for oral exposure (RfD_O) or dermal contact pathway (RfD_{ABS}, dermally adjusted reference dose) to obtain the HQ (Eq. (10.3)). Subsequently, the HI was calculated by summing the HQ values of all eight elements for each of the two exposure pathways (Eq. (10.4)). Due to the unavailability of RfD_{ABS} for Ti, this element was excluded from the dermal exposure assessments.

$$ADD_{ing} = \frac{C_{SL/RD} \times IngR \times EF \times ED}{BW \times AT} \times CF \tag{10.1}$$

$$ADD_{derm} = \frac{C_{SL/RD} \times SA \times AF \times ABS \times EF \times ED}{BW \times AT} \times CF \tag{10.2}$$

$$HQ = \frac{ADD_{ing/derm}}{RfD_{O/ABS}} \tag{10.3}$$

$$HI = \sum HQ \tag{10.4}$$

In Eqs. (10.1) and (10.2), ADD_{ing} and ADD_{derm} (both in mg/kg-day) present the ADD through direct ingestion and dermal contact, respectively. $C_{SL/RD}$ represents the concentration (mg/kg) of each TE in soil or road dust. In Eq. (10.1), IngR indicates the rate of ingestion (mg/day). In Eqs. (10.1) and (10.2), EF indicates exposure frequency (day/year), ED is the exposure duration (years), AT is the average exposure time (days), BW presents the average body weight of a child or adult (kg) and CF is the conversion factor (kg/mg). In Eq. (10.2), SA presents exposed skin or dermal surface area (cm^2), ABS is the dermal absorption efficiency for each element (unitless) and AF is the skin adherence factor (mg/cm^2). Table S1 (Supplementary Information) summarises the values of RfD of evaluated TEs for ingestion and dermal exposure. Nationally adjusted reference values for the parameters applied in Eqs. (10.1) and (10.2) for the two age groups are summarised in Table S2 (DEA, 2010). A value of HQ and HI higher than 1 indicates an exceedance of non-carcinogenic health guidelines, suggesting a lifetime risk associated with exposure to non-carcinogens (US EPA, 2002).

10.2.4.2 Cancer Risk

The Cancer Risk (CR) for the exposed population represents the potential lifelong likelihood of developing cancer from ingestion of Cr, Ni and Pb (Eq. (10.5)). These TEs were selected based on global reports identifying them as potential carcinogens, primarily through the ingestion pathway (Miletić et al., 2023). In Eq. (10.5), CR_{ing} represents the increased risk of cancer associated with TEs via ingestion. SF_O stands for the cancer Slope Factor of an element (mg/kg-day) through ingestion. CR values lower than 1×10^{-6} could be interpreted as insignificant lifelong cancer development risk to exposed individuals. Values between 1×10^{-6} and 1×10^{-4} are considered acceptable, and exceeding 1×10^{-4} should be considered unacceptable and require further attention (US EPA, 2020). Table S1 summarises the SF values for the evaluated TEs through ingestion.

$$CR_{ing} = ADD_{ing} \times SF_O \tag{10.5}$$

10.2.5 Data Analysis

Soil and road dust data showed normal distribution (Shapiro–Wilk test, $p > 0.05$). Independent samples t-tests were conducted to assess if the mean concentrations of TEs differed significantly between the two media. The same tests were applied to estimate if the means of HQ, HI and CR differed between the two age groups, pathways and media. Hierarchical cluster analysis, using Euclidean distance and Word linkage, was performed to reveal associations among sites based on similar TE contents in soil and road dust. Pearson correlation analysis was conducted to investigate relationships among TEs in both matrices. Additionally, hierarchical cluster analysis revealed associations among TEs combining both matrices and sites. The correlation matrix and the cluster analysis were used to identify the potential sources of TEs in the two matrices. Statistical significance was considered at $p < 0.05$. Data was analysed using SPSS software (IBM, v. 29.0, Armonk, NY, USA).

10.3 Results and Discussion

10.3.1 *Trace Element Concentrations in Roadside Soil and Road Dust*

In general, TE concentration means exhibited the following descending order in roadside soils (Fe > Ti > Mn > Ba > Cr > Sr > Zn > V > Ni > Pb) and road dust (Fe > Ti > Mn > Ba > Sr > Cr > Ni > Zn > V > Pb) (Table 10.1). While half of the assessed elements (Fe, Ti, Mn, V and Zn) showed higher mean concentrations in soil, the other half of TEs, consisting of Ba, Cr, Sr, Ni and Pb, exhibited higher means in road dust. Among the sites, soils at S2 recorded the highest mean concentrations of most of the evaluated TEs, including Ba, Cr, Ni, Ti and V (Table 10.1). Regarding road dust, S3 showed the highest mean concentrations of a maximum number of TEs (Ba, Fe, Mn, Ni, Pb and V) (Table 10.1). Combining all sites, significant differences between the two media were observed for three elements: Fe ($p = 0.039$, soil > road dust), Ni ($p = 0.005$, road dust > soil) and Sr ($p = 0.013$, road dust > soil).

The cluster analysis dendrograms further illustrated variations among sites regarding TE levels in the two matrices (Fig. S1). For soil, S2 and S3 were grouped in one cluster, reflecting higher overall TE loads, whereas S1 and S4, which exhibited relatively lower TE concentrations, formed a separate cluster (Fig. S1a). In the case of road dust, S3, with noticeably higher TE levels, constituted an independent cluster, while the remaining three sites, showing similar elemental contents, were grouped in the other cluster (Fig. S1b).

Compared to the TE levels determined in soils from Mahikeng, lower (Cr, Fe, Ni, Sr), higher (Ba, Mn, Pb, Zn) and similar (V) concentrations of elements were reported in urban and specifically roadside soils worldwide (Sun et al., 2017; Timofeev et al., 2019; Wang et al., 2022). Similarly, with respect to road dust, investigations in urban traffic environments elsewhere have documented comparable, as well as both higher and lower, TE concentrations relative to those observed in this study (Al-Shidi et al., 2020; Duong & Lee, 2011; Moradi et al., 2020; Pagotto et al., 2001). Studies from other regions of South Africa have similarly reported varying TE concentration patterns in road dust (Mugudamani et al., 2022; Mvovo & Magagula, 2021).

Table 10.1 Concentrations (mg/kg) of potentially toxic trace elements in roadside soil and road dust from Mahikeng

Sampling sites	Sample No.	Ba	Cr	Fe	Mn	Ni	Pb	Sr	Ti	V	Zn
S1	SL1	141.12	79.70	10,727.50	144.77	49.47	6.47	73.51	1688.27	33.14	44.67
	SL2	174.46	72.26	23,440.26	344.46	59.48	13.03	64.93	2711.53	54.34	88.68
S2	SL3	411.79	76.20	24,512.52	393.57	65.51	31.04	87.42	3613.96	72.21	63.84
	SL4	191.16	82.95	22,798.18	318.09	48.41	12.41	76.25	3114.30	50.96	52.34
S3	SL5	127.80	66.55	24,718.98	366.53	38.09	6.46	36.63	3096.76	59.91	26.15
	SL6	168.04	66.87	21,698.44	438.48	48.58	18.26	49.39	3093.49	65.30	20.94
S4	SL7	157.82	76.82	17,154.07	342.28	29.65	40.95	105.19	1842.52	38.47	67.70
	SL8	168.41	59.46	16,823.27	247.37	61.52	25.45	84.79	2106.18	43.67	61.52
Min		127.80	59.46	10,727.50	144.77	29.65	6.46	36.63	1688.27	33.14	20.94
Max		411.79	82.95	24,718.98	438.48	65.51	40.95	105.19	3613.96	72.21	88.68
Mean		192.58	72.60	20,234.15	324.44	50.09	19.26	72.26	2658.38	52.25	53.23
SD		90.73	7.84	4912.27	91.44	12.09	12.32	21.83	698.67	13.43	22.35
S1	RD1	132.95	92.97	10,541.90	199.94	58.27	13.03	70.45	2475.67	41.76	34.04
	RD2	222.58	101.59	15,317.94	231.13	72.12	6.58	133.97	2619.55	38.37	46.37
S2	RD3	222.02	51.41	13,386.85	165.62	64.08	8.57	144.73	2342.31	35.31	56.60
	RD4	241.47	79.51	14,346.09	203.32	64.55	11.04	118.56	1610.42	35.84	51.52
S3	RD5	442.69	65.43	23,247.46	498.40	85.43	94.13	95.82	2453.24	65.10	16.41
	RD6	205.72	66.93	16,398.02	340.16	55.87	31.50	116.70	2183.12	44.30	96.55
S4	RD7	186.53	106.59	13,835.07	187.18	69.67	33.64	93.41	2042.18	39.52	44.51
	RD8	268.77	121.09	15,229.64	274.48	81.75	37.05	78.61	2486.73	44.12	63.73
Min		132.95	51.41	10,541.90	165.62	55.87	6.58	70.45	1610.42	35.31	16.41

(continued)

Table 10.1 (continued)

Sampling sites	Sample No.	Ba	Cr	Fe	Mn	Ni	Pb	Sr	Ti	V	Zn
Max		442.69	121.09	23,247.46	498.40	85.43	94.13	144.73	2619.55	65.10	96.55
Mean		240.34	85.69	15,287.87	262.53	68.97	29.44	106.53	2276.65	43.04	51.22
SD		91.05	23.82	3659.19	110.38	10.52	28.86	26.26	326.10	9.54	23.35

Min, minimum; Max, maximum; SD, standard deviation (n = 8). SL1-SL8 and RD1-RD8 represent soil and road dust samples, respectively; S1–S4, sampled sites

Comparable and contrasting TE concentrations in roadside soils and road dust have also been documented in Nigeria (Akpan & William, 2014; Odediran et al., 2021).

Relative to soil, higher concentrations of Ba, Cr, Ni, Sr and Pb in road dust, particularly the significantly elevated levels of Ni and Sr, are consistent with previous studies that frequently reported enrichment of these TEs in road dust (Ali et al., 2017; Pagotto et al., 2001). A similar comparative study by Chang et al. (2009) reported notably higher elemental concentrations in road dust than in soils, along with a comparable gradient in elemental concentrations across both matrices in industrial areas of Taiwan, trends that were partly evident in this Mahikeng case study as well.

Road dust at S3 exhibited distinctly higher TE levels, likely due to the relatively greater traffic load, which may also have contributed to elevated TE concentrations in the roadside soil at this site through deposition and runoff (Popek et al., 2025; Werkenthin et al., 2014). Overall, similarities in TE concentration patterns between the two media may reflect the increased potential for mixing during the drier winter season (Mandal et al., 2025). The limited sample size constrained our ability to test whether TE concentration differences among sites were statistically significant, which would be necessary to draw definitive conclusions. Consistent with the conclusions of Pagotto et al. (2001), the observed similarities and differences in potentially toxic TE levels across both media, within this study and among reviewed reports, are likely influenced by variations in multiple local and regional factors, including traffic dynamics, urban pollution intensity, soil characteristics, geology, geography, meteorological conditions, soil management as well as sampling design and analytical procedures for sample preparation and element quantification.

10.3.2 Comparison of Soil Trace Element Concentrations with Permissible Limits

Levels of TEs in roadside soils from the study area were compared with world average values reported by Kabata-Pendias (2011), and other available national and international permissible levels reported in the literature (Table S3). Some TEs, namely Cr, Ni, Pb and Zn, exhibited concentration values higher than some of the corresponding world reference values (Fig. 10.2). Concentrations of Ba at all sites were only over the minimum limit recommended by the US EPA, while remained lower than other available standards (Fig. 10.2a). Likewise, Cr levels were slightly higher than the global average and the lower limit of Trigger Action Value (TAV) at all sites, but were far below the Maximum Allowable Concentration (MAC) (Fig. 10.2b). Conversely, Ni levels exceeded the world average and lower limit of MAC at all sampled sites, and the highest level of MAC at S2 and S4, but showed concentrations below the TAV at all four sites (Fig. 10.2c). Pb concentrations were higher than global mean and MAC at only S2 and S4 (Fig. 10.2d). Whereas Zn level was over the world average at only S1 (Fig. 10.2e). All concentrations however remained below the available South African Soil Screening Value 2 (SSV2) for residential areas (DEA,

2013; Roy et al., 2025a). The rest of the elements showed concentrations far below the available national and international standards (Table S3).

Roadside soils sampled in Mahikeng generally contained TE concentrations below their respective reference values. However, compared to global mean levels, higher concentrations of Cr and Ni at all sampling points and Pb and Zn at a few sampling points highlight the need for continued monitoring to prevent soil degradation over time. A recent investigation by Adhikari and Struwig (2024) documented multi-element contamination in roadside edible ornamental plants at the same sites examined here and further identified aerial deposition as a primary source of potentially toxic TEs. These results collectively emphasise the importance of future assessments of potential food chain contamination across broader areas of the city, as well as the evaluation of TE concentrations in different soil particle-size fractions. Altogether, these findings indicate spatially heterogeneous TE distribution patterns in roadside soils (Jonidi Jafari et al., 2018; Roy et al., 2025b). Soil properties and wind patterns merit additional attention, as they could provide valuable insights into this matter (Pagotto et al., 2001; Werkenthin et al., 2014).

10.3.3 Source Identification

Pearson correlation analysis revealed significant relationships between multiple TEs in the two media (Table 10.2). Significant positive correlations between Fe, Mn, Ti and V in soil with Sr in road dust suggest potential common sources. In contrast, negative correlations between Sr in soil and Fe, Mn and V in road dust, as well as between Ti and V in soil and Cr in road dust, indicate diverse sources. Furthermore, various significant positive correlations among Fe, Mn, Ti and V in soils, and among Ba, Fe, Mn, Pb and V in road dust point to shared sources in respective media.

The dendrogram derived from the cluster analysis classified most TEs in both media within the top cluster, whereas Ti in soil and road dust were assigned to the middle cluster, and Fe in both media formed the bottom cluster (Fig. 10.3). This pattern indicates differing predominant sources for the elements represented in each cluster. However, Ti in both matrices showed a closer association with the top cluster, suggesting common sources. Overall, the correlation matrix together with the cluster analysis indicates the presence of both shared and distinct TE sources across the two matrices at the investigated sites.

Strong positive correlations between different pairs of Fe, Mn, Ti, Sr and V in roadside soils indicate that their origin could be primarily associated with parent material (Apeagyei et al., 2011; Fussell et al., 2022; Kabata-Pendias, 2011; Sun et al., 2017; Yang et al., 2017). The closeness of clusters bearing Fe and Ti in both media further supports the assumption that Fe adsorption onto Ti-bearing minerals is common in soil environments (Kabata-Pendias, 2011). In addition, in both matrices, traffic could represent a common source of Ti and the TEs in the top cluster due to the presence of these elements in various transportation-derived emissions (Fussell et al., 2022; Popek et al., 2024). Positive correlations between Sr in road dust and

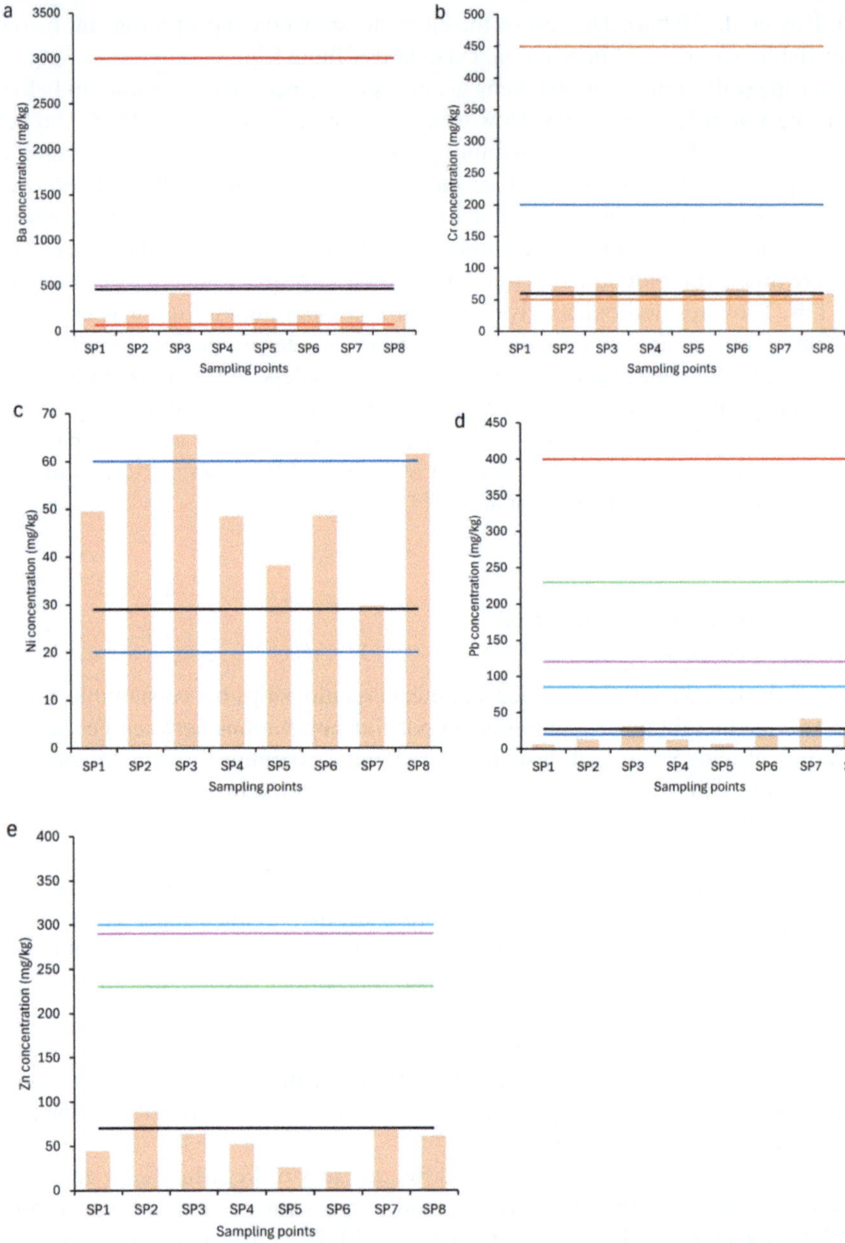

Fig. 10.2 Soil concentrations higher than various reference values for **a** barium (Ba), **b** chromium (Cr), **c** nickel (Ni), **d** lead (Pb) and **e** zinc (Zn). Colour code for reference threshold levels: red line, US EPA standard; black line, world average; purple line, Canadian reference values and Canadian environmental quality guidelines; orange line, trigger action value; sky blue line, World Health Organisation target value; dark blue line, MAC and green line, South African SSV2. SP1–SP8, sampling points

Table 10.2 Pearson correlation matrix of potentially toxic trace elements in roadside soil and road dust. Significant relationships are highlighted in bold

	Ba-SL	Cr-SL	Fe-SL	Mn-SL	Ni-SL	Pb-SL	Sr-SL	Ti-SL	V-SL	Zn-SL	Ba-RD	Cr-RD	Fe-RD	Mn-RD	Ni-RD	Pb-RD	Sr-RD	Ti-RD	V-RD
Cr-SL	0.225																		
Fe-SL	0.394	−0.112																	
Mn-SL	0.340	−0.183	**0.832***																
Ni-SL	0.585	−0.188	0.158	−0.071															
Pb-SL	0.424	−0.035	−0.029	0.290	−0.072														
Sr-SL	0.347	0.357	−0.397	−0.274	0.048	**0.771***													
Ti-SL	0.587	−0.002	**0.901****	**0.748***	0.330	−0.115	−0.418												
V-SL	0.609	−0.196	**0.857****	**0.825***	0.364	0.011	−0.426	**0.946****											
Zn-SL	0.280	0.210	−0.004	−0.155	0.357	0.396	0.636	−0.205	−0.214										
Ba-RD	−0.160	−0.487	0.579	0.327	−0.173	−0.291	−0.583	0.422	0.379	−0.338									
Cr-RD	−0.527	−0.220	−0.591	−0.568	−0.062	0.193	0.448	**−0.819***	**−0.792***	0.508	−0.264								
Fe-RD	−0.285	−0.548	0.609	0.499	−0.295	−0.286	**−0.709***	0.422	0.436	−0.438	**0.940****	−0.291							
Mn-RD	−0.460	−0.618	0.340	0.295	−0.313	−0.462	**−0.834***	0.244	0.292	−0.662	**0.830***	−0.253	**0.923****						
Ni-RD	−0.247	−0.571	0.245	−0.015	−0.087	−0.015	−0.144	−0.044	−0.046	0.151	**0.789***	0.312	0.665	0.526					
Pb-RD	−0.429	−0.543	0.202	0.196	−0.521	−0.185	−0.535	0.057	0.100	−0.569	**0.845****	−0.134	**0.872****	**0.899****	0.671				
Sr-RD	0.677	0.218	**0.806***	**0.727***	0.410	0.145	−0.063	**0.812***	**0.796***	0.296	0.006	−0.559	0.052	−0.202	−0.229	−0.376			
Ti-RD	−0.035	−0.539	−0.139	−0.209	0.378	−0.188	−0.264	−0.191	0.021	0.163	0.135	0.174	0.143	0.256	0.343	0.177	−0.149		
V-RD	−0.478	−0.534	0.163	0.094	−0.412	−0.425	**−0.707***	0.060	0.105	−0.610	**0.821***	−0.176	**0.857****	**0.947****	0.607	**0.957****	−0.383	0.334	
Zn-RD	0.217	−0.265	0.075	0.400	0.342	0.320	0.066	0.216	0.335	−0.132	−0.407	−0.059	−0.245	−0.185	−0.506	−0.385	0.296	−0.208	−0.432

SL, represents soil content; RD, represents road dust content. Significance level: * $p < 0.05$; ** $p < 0.01$

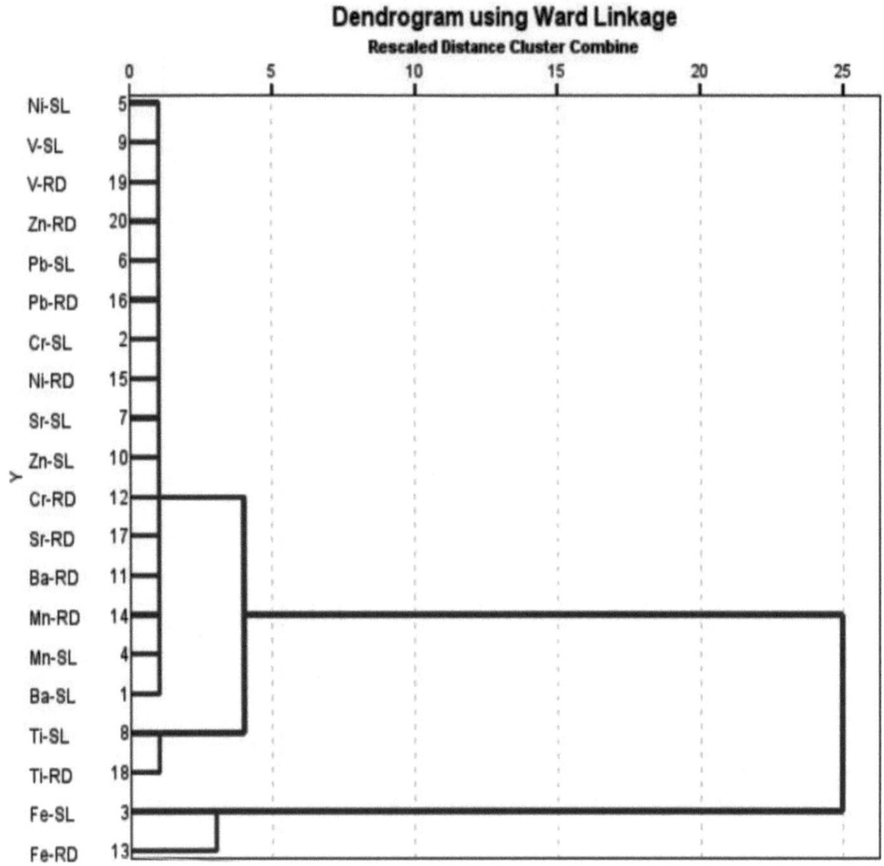

Fig. 10.3 Hierarchical cluster analysis dendrogram of potentially toxic trace element concentrations combining soil and road dust across the sampling sites

Fe and Mn in soil suggest that a portion of Sr in road dust may have originated from soil, as Sr is the fifteenth most abundant element in the Earth's crust, with mineral dust being one of its dominant natural sources (Peng et al., 2021).

In road dust, strong positive correlations among Ba, Fe, Mn, Ni, Pb and V indicate anthropogenic influence, most likely from traffic-related factors (Alhassan et al., 2012; Jose & Srimuruganandam, 2020; Miazgowicz et al., 2020; Mielke et al., 2010). In particular, the correlation between Ba and Pb may be associated with vehicle brake wear and fuel combustion, as reported by the Agency for Toxic Substances and Disease Registry and documented in relevant studies (Apeagyei et al., 2011; ATSDR, 2007; Miazgowicz et al., 2020). Excluding only Ti and Fe, grouping of other TEs in both roadside soil and road dust under one cluster further suggests common origins, pointing to traffic as a key contributor (Fussell et al., 2020).

Additionally, the mineral-enriched geology and valuable element prospecting in neighbouring regions may have played a determining role, as reported for other mining regions in South Africa (Adhikari et al., 2022; CGS, 2022; Kamunda et al., 2016). Furthermore, Pb and Sr levels in both media could be influenced by coal combustion, a major energy source at industrial and domestic levels nationally (Widory et al., 2010). Therefore, the elemental composition of urban road dust and soil particles in Mahikeng is likely affected by factors beyond traffic, subsequently resulting in mixing of particles within each medium. These combined effects may account for some of the inconsistencies identified in the correlation and cluster analyses, needing additional research (Moradi et al., 2020).

10.3.4 Human Health Risk

For adults, all HQ and HI values for both exposure pathways and media remained below the safe threshold of 1 (Tables S4 and S5). Likewise, for children, HQ for all TEs through ingestion remained < 1 for both media (Tables S4 and S5, Fig. 10.4a, b). However, for this age group, one sampling point at S1 and S3, and both sampling points at S2, showed HI higher than 1 for soil. Similarly, for road dust, one sampling point at S3 and both sampling points at S4 exhibited HI > 1 for children (Fig. 10.4a, b). Furthermore, the HQ of Cr in soil exceeded the safe threshold of 1 at one sampling point each at S1 and S4, and at both sampling points at S2 via dermal contact for children (Tables S4 and S5, Fig. 10.4c, d). The corresponding HI > 1 at all four sites followed the trend S2 > S3 > S1 > S4. Likewise, for dermal contact with road dust, HQ of Cr surpassed the safe limit of 1 for children at identical sampling locations recorded for soil (Tables S4 and S5, Fig. 10.4c, d). The resulting HI values exceeded 1 across all sites, following the descending order S4 > S3 > S1 > S2. In all the above cases for children, HQ values remained between 1 and 2, while HI values ranged from 1 to 2.5 for both media. Through ingestion, Cr, Fe, V and Pb were the largest contributors to HI for both age groups and media (Tables S4 and S5, Fig. 10.4). For dermal contact, Cr, V, Fe and Mn, in that order, contributed most to HI in both media and for both age groups.

Independent samples t-tests showed that the mean HQ for all assessed elements and the mean HI were significantly higher ($p < 0.01$) for children compared with adults in both media. Although road dust exhibited higher HQ for Cr for both adults and children, the mean did not differ significantly between the two media. However, for both age groups, comparison of the two exposure routes revealed significantly higher ($p < 0.01$) mean HQ for Cr via dermal contact than ingestion of both matrices. Additionally, for both adults and children, dermal contact resulted in a significantly higher mean HI than ingestion across both media.

Cancer risk values followed the sequence Ni > Cr > Pb across both media and age groups (Table S6, Fig. 10.5). For adults, CR values for Cr ranged between 1×10^{-6} and 1×10^{-4} via both media (Fig. 10.5a), while substantially higher values, mostly between 1×10^{-4} and 1×10^{-3}, were observed for Ni (Fig. 10.5a). For

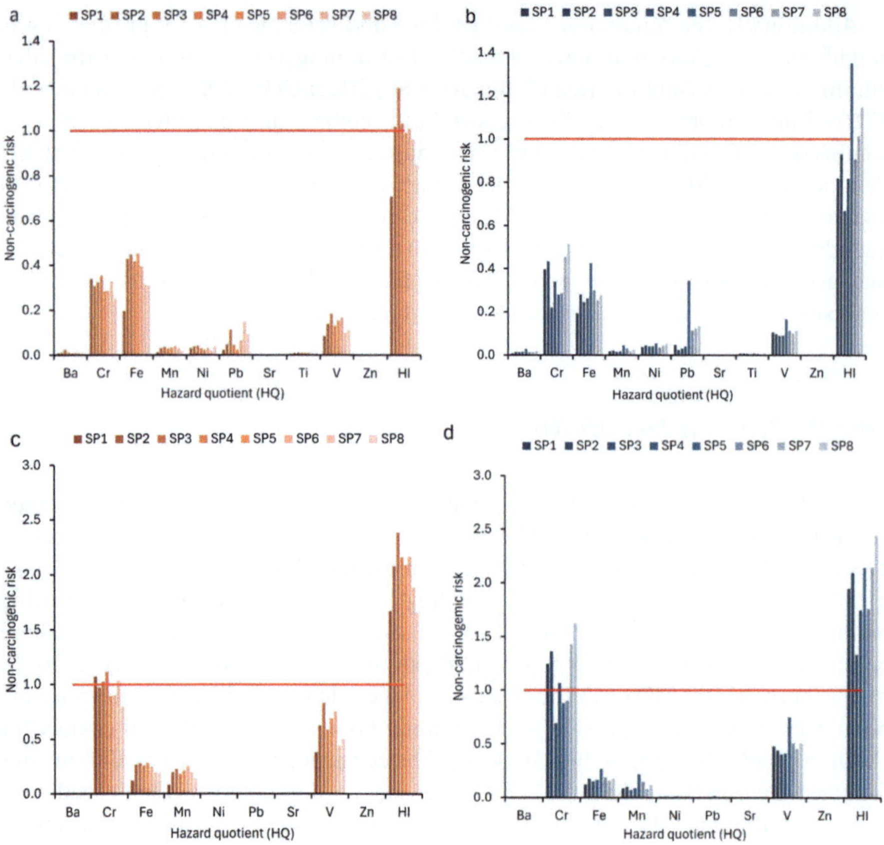

Fig. 10.4 Hazard quotient (HQ) for potentially toxic trace elements and corresponding hazard index (HI) for children through **a** ingestion of soil, **b** ingestion of road dust, **c**. dermal exposure to soil and **d** dermal exposure to road dust. Red line indicates the safe threshold. SP1–SP8, sampling points

children, CR values for Cr ranged between 1×10^{-4} and 1×10^{-3}, whereas those for Ni were mostly above 1×10^{-3} through both media (Fig. 10.5b). CR values for Pb generally remained below 1×10^{-6} for adults and mostly stayed between 1×10^{-6} and 1×10^{-4} for children through both soil and road dust (Fig. 10.5c). Independent samples t-tests revealed significantly higher CR ($p < 0.05$) from all three assessed TEs for children compared with adults across the two matrices. However, no significant differences in estimated CR were observed between the two media.

Among the two age groups, non-carcinogenic risk from potential ingestion of the combined TEs in both media appears to be a concern for only children at selected sites, particularly for the soil at the three shopping centres, and at Morena Mall and the university gate for road dust. Additionally, dermal exposure to soil- and road dust-borne Cr, as well as the cumulative TE load, poses potential hazards to children

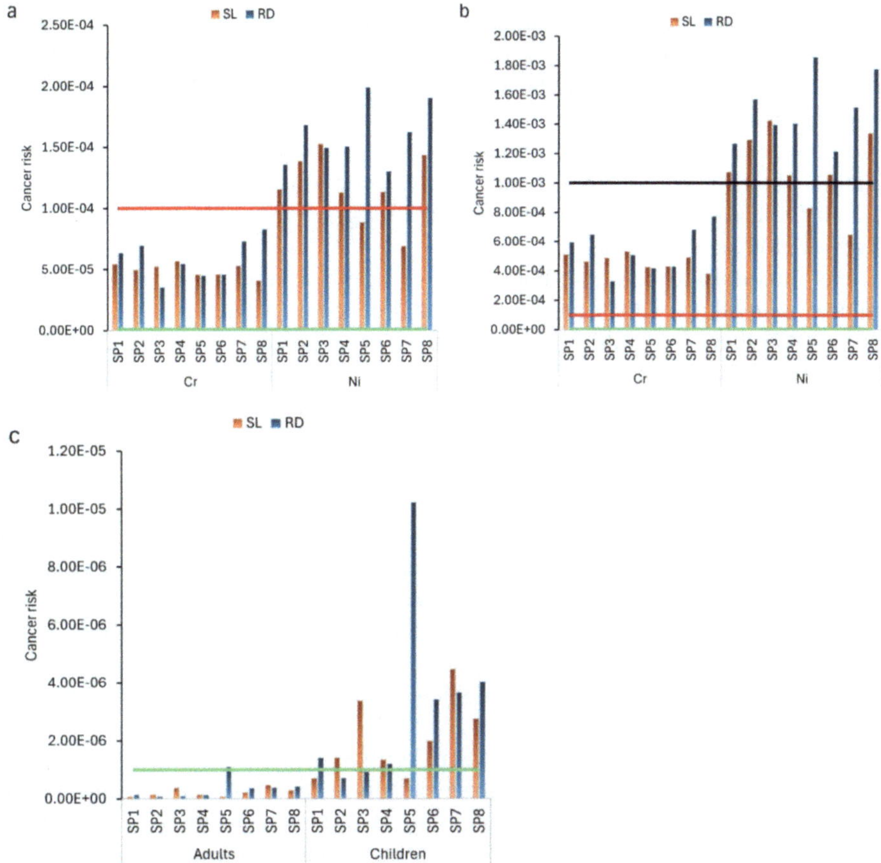

Fig. 10.5 Cancer risk from ingestion of **a** Cr and Ni for adults, **b** Cr and Ni for children and **c** Pb for both age groups. Green, red and black lines represent cancer risk of 1×10^{-6}, 1×10^{-4} and 1×10^{-3}, respectively. SP1–SP8, sampling points

at all four sites. Simultaneous ingestion-related cancer risks from both Cr and Ni for children and from Ni for adults may be significant via both media at all four locations, further enhancing cumulative health consequences for children. Identification of Cr as the principal health risk driver in Mahikeng is a significant finding, as this potentially toxic TE typically exhibits high levels in particles derived from anthropogenic sources in South Africa (Dudu et al., 2018). Overall, unacceptable non-carcinogenic risks and cancer risks observed in this study, particularly the relatively greater cancer risk from Ni for both age groups, and the non-carcinogenic risk and cancer risk potential of Cr for children exclusively, merit further in-depth toxicological assessments (ATSDR, 2022).

Several assessments of traffic-impacted soils and road dust in urban regions worldwide have identified these potentially toxic TEs as notable health threats to human populations (Jose & Srimuruganandam, 2020; Moradi et al., 2020; Timofeev et al., 2019; Yang et al., 2017). In the African context, investigations from South Africa (Mugudamani et al., 2022) and Nigeria (Odediran et al., 2021; Taiwo et al., 2017) have reported non-carcinogenic and cancer risk values exceeding safe thresholds for TE-enriched road dust through various exposure pathways. Compared to this investigation, other studies from South Africa have documented varying CR levels from Ni and Cr in road dust across different urban regions (Mugudamani et al., 2022; Mvovo & Magagula, 2021).

It is important to note that both types of health risks could be far higher for children than adults, which has been a global concern (Adimalla et al., 2020; Ali et al., 2017; Jiang et al., 2021; Mielke et al., 2010; Mugudamani et al., 2022; Raza Ahmad et al., 2019; Taiwo et al., 2017). Also, this study indicated sufficient health risks associated with both exposure pathways. Past reports have indicated both or either as dominant human exposure routes; for instance, Raza Ahmad et al. (2019) identified dermal contact as the main exposure pathway for road dust in Pakistan. Whereas elevated health consequences associated with ingestion of these matrices have also been reported frequently (Adimalla et al., 2020; Albuja et al., 2024; Ali et al., 2017). A study by Mvovo and Magagula (2021) reported higher risks for children and adults due to ingestion and dermal contact, respectively, from road dust in other parts of South Africa. The observed variations in health risk levels associated with both media among cities again point to traffic and other dominant factors, influencing a city's pollution gradients and patterns, as discussed in previous studies (Moradi et al., 2020; Pagotto et al., 2001).

Environmental exposure-driven health risks associated with potentially toxic TEs, including Cr and Ni, have been extensively discussed. Non-carcinogenic health consequences of prolonged Cr exposure may include various illnesses in humans, such as dermal rash, diseases affecting the liver, kidneys and respiratory system, as well as gastric problems, such as ulcers and upset stomach (Kamunda et al., 2016; Khan et al., 2020; Sun et al., 2017). Existing literature has also linked potential cancer risks to the ingestion of Cr and Ni (Kamunda et al., 2016; Sun et al., 2017). In particular, Ni exposure beyond accepted levels has been associated with oral and intestinal cancers (Kamunda et al., 2016). Exposure to Ni may induce diverse carcinogenic mechanisms, and the generation of free radicals may also trigger carcinogenic processes (Carver & Gallicchio, 2017).

We acknowledge that this study is based on limited data and assessed only a selected group of potentially toxic TEs for estimating health risks. Consequently, the risk levels could be far more concerning if other hazardous and carcinogenic elements present in roadside soil particles and road dust were included. Building on these findings, ongoing research is underway to assess soil pollution status, ecological risks and the phytoremediation potential of ornamental plant species in the studied public spaces in Mahikeng.

10.4 Conclusions

This study evaluated human health risks associated with exposure to potentially toxic TE-containing roadside soil and road dust at selected public spaces in Mahikeng, a small urban centre in South Africa. Documented soil concentrations of Ni and Cr, and to a lesser extent Pb and Zn, exceeded world soil averages, requiring further assessment in relation to local background standards to elucidate potential soil degradation. In addition to natural contributions, traffic was identified as a major source of most TEs in both matrices, including those exceeding permissible levels and posing potential health risks, particularly Ni and Cr, which also exhibited increased concentrations in road dust. Children appeared more vulnerable, as both ingestion and dermal contact with Cr in soil and road dust could result in non-carcinogenic hazards. Moreover, cumulative hazards associated with the combined presence of all assessed potentially toxic TEs in both media through both exposure routes should not be underestimated. Cancer risks related to the ingestion of Ni and Cr in both matrices could further increase potential health concerns for this age group. Adults may also face cancer risks from the ingestion of Ni in these media. Overall, unacceptable levels of both types of health risk from TE loads in both matrices, particularly elevated risks associated with road dust, confirm these media as potential health threats in these public spaces. Although sites and matrices exhibited distinct differences in TE concentrations and associated health risk levels, larger surveys are needed to confirm these statistically. Further investigations of potentially toxic TE enrichment in different particle-size fractions of roadside soil and road dust, as well as assessments of element contamination in other environmental matrices and the associated ecological risks, are required in the city. Taken together, the findings of this case study underline the importance of continuous monitoring of both matrices in roadside environments to prevent soil degradation and support the timely integration of sustainable remediation strategies into urban planning for managing particle-borne potentially toxic TE pollution, thereby ensuring safer environments in relatively smaller cities.

Acknowledgements The authors sincerely thank North-West University, South Africa, for financial support. The School of Geo- and Spatial Sciences at the North-West University, Potchefstroom Campus, South Africa, is gratefully acknowledged for providing the handheld XRF used in the soil analysis.

Funding This work was supported by the North-West University, South Africa (NWU PDRF Fund NW.1G01487, received by Dr Sutapa Adhikari).

References

Adhikari, S., & Struwig, M. (2024). Profile of potentially toxic elements in leaves of edible ornamental plants cultivated in roadside soils in Mahikeng, South Africa. *South African Journal of Botany, 174*, 75–84. https://doi.org/10.1016/j.sajb.2024.09.001

Adhikari, S., Jordaan, A., Beukes, J. P., & Siebert, S. J. (2022). Anthropogenic sources dominate foliar chromium dust deposition in a mining-based urban region of South Africa. *International Journal of Environmental Research and Public Health, 19*(4), 2072. https://doi.org/10.3390/ijerph19042072

Adimalla, N., Qian, H., Nandan, M. J., & Hursthouse, A. S. (2020). Potentially toxic elements (PTEs) pollution in soils in a typical urban region of South India: An application of health risk assessment and distribution pattern. *Ecotoxicology and Environmental Safety, 203*, 111055. https://doi.org/10.1016/j.ecoenv.2020.11105

Akpan, I. O., & William, E. S. (2014). Assessment of elemental concentrations of roadside soils in relation to traffic density in Calabar, Nigeria. *International Journal of Scientific and Technology Research, 3*, 1–8.

Albuja, M., Jeong, H., & Ra, K. (2024). Distribution of potentially toxic elements and health risk assessment of road dust in a steel industrial area. *International Journal of Environmental Science and Technology, 21*, 4967–4978. https://doi.org/10.1007/s13762-023-05327-4

Alhassan, A. J., Sule, M. S., Atiku, M. K., Wudil, A. M., Dangambo, M. A., Mashi, J. A., & Ibrahim, N. A. (2012). Study of correlation between heavy metal concentration, street dust and level of traffic in major roads of Kano Metropolis, Nigeria. *Nigerian Journal of Basic and Applied Science, 20*(2), 161–168.

Ali, M. U., Liu, G., Yousaf, B., Abbas, Q., Ullah, H., Munir, M. A. M., & Fu, B. (2017). Pollution characteristics and human health risks of potentially (eco)toxic elements (PTEs) in road dust from metropolitan area of Hefei, China. *Chemosphere, 181*, 111–121. https://doi.org/10.1016/j.chemosphere.2017.04.061

Al-Shidi, H. K., Al-Reasi, H. A., & Sulaiman, H. (2020). Heavy metals levels in road dust from Muscat, Oman: Relationship with traffic volumes, and ecological and health risk assessments. *International Journal of Environmental Health Research, 32*(2), 264–276. https://doi.org/10.1080/09603123.2020.1751806

Annual Report. (2024–2025). Mahikeng Local Municipality, Republic of South Africa. https://www.mahikeng.gov.za/download/annual-report-2024-2025/?wpdmdl=22809&refresh=69a1f4626686a1772221538

Apeagyei, E., Bank, M. S., & Spengler, J. D. (2011). Distribution of heavy metals in road dust along an urban-rural gradient in Massachusetts. *Atmospheric Environment, 45*(13), 2310–2323. https://doi.org/10.1016/j.atmosenv.2010.11.015

ATSDR. (2007). *Agency for toxic substances and disease registry, toxicological profile for barium and barium compounds: Potential for human exposure*. U.S. Department of Health and Human Services, Public Health Service. https://www.atsdr.cdc.gov/toxprofiles/tp24.pdf

ATSDR. (2022). *Agency for toxic substances and disease registry, public health assessment guidance manual*. US Department of Health and Human Services. https://www.atsdr.cdc.gov/pha-guidance/index.html

Batwa-Ismail, M. Z., Moodley, R., & Mutanga, O. (2021). Elemental analysis of soils along the South African National Road (N3)—A combined approach including statistics, pollution indicators, and geographic information system (GIS). *Environmental Monitoring and Assessment, 193*, 559. https://doi.org/10.1007/s10661-021-09374-7

Cai, Y., & Zhang, J. (2021). Pollution and health risk assessment of arsenic in street dust and soil from an industrial zone. In *IOP conference series: Earth and environmental science, Volume 791, 3rd international conference on advances in civil engineering, energy resources and environment engineering, 28–30 May, Qingdao, China*. https://doi.org/10.1088/1755-1315/791/1/012184

Carver, A., & Gallicchio, V. S. (2017). Heavy metals and cancer. In F. Atroshi (Ed.), *Cancer causing substances* (pp. 1–19). InTech Open Access. https://www.intechopen.com/chapters/56652

CGS. (2022). *North West provincial mining and energy investment conference*. Council for Geoscience, 25–26 January. https://www.dmre.gov.za/Portals/0/Energy_Website/files/Mining/MiningInvestmentConference/Mining-and-Energy-Investment-Conference-Council-for-Geoscience.pdf

Chang, S. H., Wang, K. S., Chang, H. F., Ni, W. W., Wu, B. J., Wong, R. H., & Lee, H. S. (2009). Comparison of source identification of metals in road-dust and soil. *Soil and Sediment Contamination, 18*(5), 669–683. https://doi.org/10.1080/15320380903085691

Chang, X., Yu, Y., & Li, Y. (2021). Response of antimony distribution in street dust to urban road traffic conditions. *Journal of Environmental Management, 296*, 113219. https://doi.org/10.1016/j.jenvman.2021.113219

DEA. (2010). The framework for the management of contaminated land, South Africa. https://cer.org.za/wp-content/uploads/2012/06/Framework-for-the-Management-of-Contaminated-Land-May-2010.pdf

DEA. (2013). National environmental management: Waste Act, 2008 (Act no. 59 of 2008), South Africa. In Draft national norms and standards for the remediation of contaminated land and soil quality. Government Gazette, 10 May 2013. https://cer.org.za/wp-content/uploads/2014/02/natl-norms-and-stds-for-remediation-of-contaminated-land-and-soil-quality.pdf

Dudu, V. P., Mathuthu, M., & Manjoro, M. (2018). Assessment of heavy metals and radionuclides in dust fallout in the West Rand mining area of South Africa. *Clean Air Journal, 28*, 42–52. https://doi.org/10.17159/2410-972x/2018/v28n2a17

Duong, T. T., & Lee, B. (2011). Determining contamination level of heavy metals in road dust from busy traffic areas with different characteristics. *Journal of Environmental Management, 92*(3), 554–562. https://doi.org/10.1016/j.jenvman.2010.09.010

Fiala, M., & Hwang, H. M. (2021). Influence of highway pavement on metals in road dust: A case study in Houston, Texas. *Water Air Soil Pollution, 232*, 185. https://doi.org/10.1007/s11270-021-05139-7

Fortoul, T. I., Rodriguez-Lara, V., Gonzalez-Villalva, A., Rojas-Lemus, M., Colin-Barenque, L., Bizarro-Nevares, P., García-Peláez, I., Ustarroz-Cano, M., López-Zepeda, S., Cervantes-Yépez, S., López-Valdez, N., Meléndez-García, N., Espinosa-Zurutuza, M., Cano-Gutierrez, G., & Cano-Rodríguez, M. C. (2015). Health effects of metals in particulate matter. In F. Nejadkoorki (Ed.), *Current air quality issues* (pp. 571−608). InTech Open Access. https://doi.org/10.5772/59749. https://www.intechopen.com/chapters/48145

Fussell, J. C., Franklin, M., Green, D. C., Gustafsson, M., Harrison, R. M., Hicks, W., Kelly, F. J., Kishta, F., Miller, M. R., Mudway, I. S., Oroumiyeh, F., Selley, L., Wang, M., & Zhu, Y. (2022). A review of road traffic-derived non-exhaust particles: Emissions, physicochemical characteristics, health risks, and mitigation measures. *Environmental Science and Technology, 56*(11), 6813−6835. https://doi.org/10.1021/acs.est.2c01072

Han, Q., Wang, M., Xu, X., Li, M., Liu, Y., Zhang, C., Li, S., & Wang, M. (2023). Health risk assessment of heavy metals in road dust from the fourth-tier industrial city in central China based on Monte Carlo simulation and bioaccessibility. *Ecotoxicology and Environmental Safety, 252*, 114627. https://doi.org/10.1016/j.ecoenv.2023.114627

Herselman, J. E., Steyn, C. E., & Fey, M. V. (2005). Baseline concentration of Cd, Co, Cr, Cu, Pb, Ni and Zn in soils of South Africa. *South African Journal of Science, 101*, 509–512. https://doi.org/10.1007/s13762-020-02667-3

IARC. (2025). *International Agency for Research on Cancer, Agents classified by the IARC monographs* (Vols. 1–139). IARC. Retrieved March 26, 2021, from https://monographs.iarc.who.int/agents-classified-by-the-iarc/

Jiang, H.-H., Cai, L.-M., Hu, G.-C., Wen, H.-H., Luo, J., Xu, H.-Q., & Chen, L.-G. (2021). An integrated exploration on health risk assessment quantification of potentially hazardous elements in soils from the perspective of sources. *Ecotoxicology and Environmental Safety, 208*, 111489. https://doi.org/10.1016/j.ecoenv.2020.111489

Jonidi Jafari, A., Kermani, M., Kalantary, R. R., & Arfaeinia, H. (2018). The effect of traffic on levels, distribution and chemical partitioning of harmful metals in the street dust and soil from urban areas of Tehran, Iran. *Environmental Earth Sciences, 77*(2). https://doi.org/10.1007/s12665-018-7226-8

Jose, J., & Srimuruganandam, B. (2020). Investigation of road dust characteristics and its associated health risks from an urban environment. *Environmental Geochemistry and Health, 42*, 2819–2840. https://doi.org/10.1007/s10653-020-00521-6

Kabata-Pendias, A. (2011). *Trace elements in soils and plants*. CRC Press.

Kabir, M. H., Kormoker, T., Islam, M. S., Khan, R., Shammi, R. S., Tusher, T. R., Proshad, R., Islam, M. S., & Idris, A. M. (2021). Potentially toxic elements in street dust from an urban city of a developing country: ecological and probabilistic health risks assessment. *Environmental Science and Pollution Research, 28*(40), 57126–57148. https://doi.org/10.1007/s11356-021-14581-3

Kamunda, C., Mathuthu, M., & Madhuku, M. (2016). Health risk assessment of heavy metals in soils from Witwatersrand Gold Mining Basin, South Africa. *International Journal of Environmental Research and Public Health, 13*(7), 663. https://doi.org/10.3390/ijerph13070663

Khan, M. A., Majeed, R., Fatima, S. U., Khan, M. A., & Shaukat, S. S. (2020). Occurrence, distribution and health effects of heavy metals in commercially available vegetables in Karachi. *International Journal of Biology and Biotechnology, 17*, 319–328.

Mandal, M., Roy, A., Ghosh, S., Mondal, A., Przybysz, A., Popek, R., Ghosh, T., Dash, S. K., Agrawal, G. K., Rakwal, R., & Sarkar, A. (2025). Assessing the influences of leaf functional traits on plant performances under dust deposition and microplastic retention. *Atmosphere, 16*(7), 861. https://doi.org/10.3390/atmos16070861

Miazgowicz, A., Krennhuber, K., & Lanzerstorfer, C. (2020). Metals concentrations in road dust from high traffic and low traffic area: A size dependent comparison. *International Journal of Environmental Science and Technology, 17*, 3365–3372.

Mielke, H. W., Laidlaw, M. A. S., & Gonzales, C. (2010). Lead (Pb) legacy from vehicle traffic in eight California urbanized areas: Continuing influence of lead dust on children's health. *Science of the Total Environment, 408*(19), 3965–3975. https://doi.org/10.1016/j.scitotenv.2010.05.017

Miletić, A., Lučić, M., & Onjia, A. (2023). Exposure factors in health risk assessment of heavy metal(loid)s in soil and sediment. *Metals, 13*(7), 1266. https://doi.org/10.3390/met13071266

Moradi, Q., Mirzaei, R., Alipour, M., Bay, A., Ghaderpoori, M., Asadi, A., Fakhri, Y., Sorooshian, A., & Mousavi Khaneghah, A. (2020). The concentration, characteristics, and probabilistic health risk assessment of potentially toxic elements (PTEs) in street dust: A case study of Kashan, Iran. *Toxin Reviews, 40*(4), 1421–1430. https://doi.org/10.1080/15569543.2020.1728336

Mugudamani, I., Oke, S. A., & Gumede, T. P. (2022). Influence of urban informal settlements on trace element accumulation in road dust and their possible health implications in Ekurhuleni Metropolitan Municipality, South Africa. *Toxics, 10*(5), 253. https://doi.org/10.3390/toxics10050253

Mvovo, I., & Magagula, H. B. (2021). Health risks of heavy metal contamination in road surface dusts from selected major roads in East London, South Africa. *International Journal of Environmental Health Research, 32*(11), 2425–2434. https://doi.org/10.1080/09603123.2021.1969340

Najmeddin, A., Keshavarzi, B., Moore, F., & Lahijanzadeh, A. (2018). Source apportionment and health risk assessment of potentially toxic elements in road dust from urban industrial areas of Ahvaz megacity, Iran. *Environmental Geochemistry and Health, 40*(4), 1187–1208. https://doi.org/10.1007/s10653-017-0035-2

Novela, R. J., Gitari, W. M., Chikoore, H., Molnar, P., Mudzielwana, R., & Wichmann, J. (2020). Chemical characterization of fine particulate matter, source apportionment and long-range transport clusters in Thohoyandou, South Africa. Clean Air Journal, *30*(2), 1−12. https://doi.org/10.17159/caj/2020/30/2.8735

Odediran, E. T., Adeniran, J. A., Yusuf, R. O., Abdulraheem, K. A., Adesina, O. A., Sonibare, J. A., & Du, M. (2021). Contamination levels, health risks and source apportionment of potentially toxic elements in road dusts of a densely populated African City. *Environmental Nanotechnology, Monitoring & Management, 15*, 100445. https://doi.org/10.1016/j.enmm.2021.100445

Pagotto, C., Remy, N., Legret, M., & Le Cloirec, P. (2001). Heavy metal pollution of road dust and roadside soil near a major rural highway. *Environmental Technology, 22*(3), 307–319. https://doi.org/10.1080/09593332208618280

Peng, H., Yao, F., Xiong, S., Wu, Z., Niu, G., & Lu, T. (2021). Strontium in public drinking water and associated public health risks in Chinese cities. *Environmental Science and Pollution Research, 28*(18), 23048–23059. https://doi.org/10.1007/s11356-021-12378-y

Popek, R., Łukowski, A., Roy, A., Mandal, M., Przybysz, A., Zając, Z., & Sarkar, A. (2025). Combined efficiency of tree and shrub vegetation barriers in mitigating PM, TEs, and PAHs along urban roadways. *International Journal of Phytoremediation*, 1–9. https://doi.org/10.1080/15226514.2025.2555595

Popek, R., Roy, A., Mandal, M., Przybysz, A., Drążkiewicz, K., Romanowska, P., & Sarkar, A. (2024). Enhancing urban sustainability: How spatial and height variability of roadside plants improves pollution capture for greener cities. *Sustainability, 16*(24), 11131. https://doi.org/10.3390/su162411131

Profile: Ngaka Modiri Molema District. (2023). *Profile and analysis—District development model*. Department of Cooperative Governance and Traditional Affairs, Republic of South Africa. https://www.cogta.gov.za/cgta_2016/wp-content/uploads/2023/11/DistrictProfile_NgakaModiriM11072020.pdf

Qiao, X., Schmidt, A. H., Tang, Y., Xu, Y., & Zhang, C. (2014). Demonstrating urban pollution using toxic metals of road dust and roadside soil in Chengdu, southwestern China. *Stochastic Environmental Research and Risk Assessment, 28*, 911–919. https://doi.org/10.1007/s00477-013-0790-2

Raza Ahmad, H., Mehmood, K., Sardar, M. F., Maqsood, M. A., Ur Rehman, M. Z., Zhu, C., & Li, H. (2019). Integrated risk assessment of potentially toxic elements and particle pollution in urban road dust of megacity of Pakistan. *Human and Ecological Risk Assessment: An International Journal*, 1–22. https://doi.org/10.1080/10807039.2019.1611415

Roy, A., Mandal, M., Binha, S. K., Prasad, D., Popek, R., Przybysz, A., & Sarkar, A. (2025a). Festive pollution: A global concern—A comparative study of Diwali in India and New Year's Eve in Poland. *Atmosphere, 16*(4), 442. https://doi.org/10.3390/atmos16040442

Roy, A., Mandal, M., Popek, R., Przybysz, A., & Sarkar, A. (2025b). Decoding leaf micro- and macro-morphology: A path to effective particulate matter phytoremediation. *International Journal of Phytoremediation*, 1–18. https://doi.org/10.1080/15226514.2025.2523938

Sager, M. (2020). Urban soils and road dust—Civilization effects and metal pollution—A review. *Environments, 7*(11), 98. https://doi.org/10.3390/environments7110098

Semerjian, L., Okaiyeto, K., Ojemaye, M. O., Ekundayo, T. C., Igwaran, A., & Okoh, A. I. (2021). Global systematic mapping of road dust research from 1906 to 2020: Research gaps and future direction. *Sustainability, 13*(20), 11516. https://doi.org/10.3390/su132011516

Shahzad, B., Tanveer, M., Rehman, A., Cheema, S. A., Fahad, S., Rehman, S., & Sharma, A. (2018). Nickel; whether toxic or essential for plants and environment—A review. *Plant Physiology and Biochemistry, 132*, 641–651. https://doi.org/10.1016/j.plaphy.2018.10.014

Sparks, D. L. (2005). Toxic metals in the environment: The role of surfaces. *Elements, 1*(4), 193–197. https://doi.org/10.2113/gselements.1.4.193

Sun, J., Hu, G., Yu, R., Lin, C., Wang, X., & Huang, Y. (2017). Human health risk assessment and source analysis of metals in soils along the G324 Roadside, China, by Pb and Sr isotopic tracing. *Geoderma, 305*, 293–304. https://doi.org/10.1016/j.geoderma.2017.06.017

Taiwo, A. M., Awomeso, J. A., Taiwo, O. T., Oremodu, B. D., Akintunde, O. O., Ojo, N. O., Elegbede, O. O., Olanrewaju, H. H., & Arowolo, T. A. (2017). Assessment of health risks associated with road dusts in major traffic hotspots in Abeokuta metropolis, Ogun state, southwestern Nigeria. *Stochastic Environmental Research and Risk Assessment, 31*, 431–447. https://doi.org/10.1007/s00477-016-1302-y

Timofeev, I., Kosheleva, N., & Kasimov, N. (2019). Health risk assessment based on the contents of potentially toxic elements in urban soils of Darkhan, Mongolia. *Journal of Environmental Management, 242*, 279–289. https://doi.org/10.1016/j.jenvman.2019.04.090

US EPA. (2002). *Supplemental guidance for developing soil screening levels for superfund sites, solid waste and emergency response*. United States Environmental Protection Agency. https://nepis.epa.gov/Exe/ZyPDF.cgi/91003IJK.PDF?Dockey=91003IJK.PDF

US EPA. (2014). *Priority pollutant list* (Document ID: EPA-HQ-OPPT-2019-0080-0173). United States Environmental Protection Agency. https://www.epa.gov/sites/production/files/2015-09/documents/priority-pollutant-list-epa.pdf

US EPA. (2020). *RCRA Delisting technical support document, Chapter 4: Risk and hazard assessment*. https://www.epa.gov/sites/default/files/2020-09/documents/chap4.pdf

Van der Westhuizen, D., Howlett-Downing, C., Molnár, P., Boman, J., Wichmann, J., & Von Eschwege, K. G. (2024). Source apportionment of fine atmospheric particles in Bloemfontein, South Africa, using positive matrix factorization. *Environmental Monitoring and Assessment, 196*, 188. https://doi.org/10.1007/s10661-023-12293-4

Wang, X., Birch, G. F., & Liu, E. (2022). Traffic emission dominates the spatial variations of metal contamination and ecological-health risks in urban park soil. *Chemosphere, 297*, 134155. https://doi.org/10.1016/j.chemosphere.2022.134155

Werkenthin, M., Kluge, B., & Wessolek, G. (2014). Metals in European roadside soils and soil solution—A review. *Environmental Pollution, 189*, 98–110. https://doi.org/10.1016/j.envpol.2014.02.025

Widory, D., Liu, X., & Dong, S. (2010). Isotopes as tracers of sources of lead and strontium in aerosols (TSP & $PM_{2.5}$) in Beijing. *Atmospheric Environment*, *44*(30), 3679–3687. https://doi.org/10.1016/j.atmosenv.2010.06.036

Williams, J., Petrik, L., & Wichmann, J. (2021). $PM_{2.5}$ chemical composition and geographical origin of air masses in Cape Town, South Africa. *Air Quality, Atmosphere and Health, 14*, 431–442. https://doi.org/10.1007/s11869-020-00947-y

Yang, J., Teng, Y., Wu, J., Chen, H., Wang, G., Song, L., Yue, W., Zuo, R., & Zhai, Y. (2017). Current status and associated human health risk of vanadium in soil in China. *Chemosphere, 171*, 635–643. https://doi.org/10.1016/j.chemosphere.2016.12.058

Chapter 11
Harnessing Plants for Cleaner Air: Phytoremediation as a Green Strategy Against Particulate Matter Pollution

Debajyoti Saha, Kakan Ball, and Sanjoy Sadhukhan

Abstract Airborne Particulate Matter (APM) pollution poses major challenges to public health and the environment thus demands for a structural mitigation alternatives. Phytoremediation on the other hand stands out as a green alternative that promises to tap into the natural cleansing ability of plants and remove pollutants from our environment. It makes a contribution to the general well-being of city and rural groups. In short, plants can capture smoggy air particles in the following ways: by adsorbing them onto their surfaces (similar to how charcoal filters work); absorbing certain gases and converting phytotoxic compounds into friendly forms; and metabolizing atmospheric molecules. Moreover, the method of phytoremediation carries some ancillary advantages such as carbon sequestration, habitat reparation and community appearance in urban centers. Several plant species have shown efficient ways of removing PM from the air, which is affected by its leaf morphology, surface area and biochemical composition. Moreover, the utilization of green infrastructures such as green walls and roofs can increase deposition sites for PMs in addition to enhancing phytoremediation at a large scale in urban areas. However, the phytoremediation works effectively or otherwise by several conditions like type of plants chosen and its grown status, local sources for transferal contaminants in to air and also amount as well as mixing constituents supplied through atmosphere. Among them still several limitations, such as space shortage issues, maintenance requirement and the need for continuously monitoring must be met to fully realize phytoremediation potential in deploying a sustainable application of air quality control. Phytoremediation might be cheap and environmentally benign technology for reducing airborne

D. Saha
Department of Botany, Visva-Bharati, Santiniketan, Birbhum, West Bengal, India
e-mail: debajyotisaha30121999@gmail.com

K. Ball · S. Sadhukhan (✉)
Department of Botany, Raiganj University, Uttar Dinajpur, Raiganj, West Bengal, India
e-mail: sanjoysadhukhan@gmail.com

K. Ball
e-mail: ballkakan@gmail.com

A. Sarkar et al. (eds.), *From Sky to Soil*, https://doi.org/10.1007/978-3-032-31991-3_11

PM pollution in a sustainable manner. Since the benefits of this technology is envisaged to be a solution towards addressing global problem air quality, further research and implementation efforts are required.

Keywords Phytoremediation · Airborne particulate matter · Air pollution · Green solution · Air quality

11.1 Introduction

Airborne Particulate Matter (APM) pollution is a widespread and mounting concern, threatening both of global environmental quality as well as public health. APM is defined as a complex mixture of liquid droplets and solid particles floating in the air that come from multiple sources, such as construction demolition works transportation industrial agriculture Wildfires Volcanic eruptions Particles vary widely in size and source, but among types typically measured are fine particles (PM 2.5), ultrafine particles (PM 0.1), and coarse particles (PM 10) (Şahin et al., 2022). These particles are easily breathed and can penetrate through the blood circulatory system into organs causing respiratory problems, cardiovascular diseases and early mortality as well as systemic inflammation and oxidative stress be major adverse outcomes of combustion-derived air pollution (Mandal et al., 2024a). Beyond its detrimental effects on human health, APM pollution also degrades the environment by causing marine pollution, decreased visibility, ecological degradation, and climate change (Basith et al., 2022). The essence of this chapter is to emphasize the potential employment as a green solution for pollution mitigation using the phytoremediation technique. They are clarifying the mechanisms and principles of phytoremediation, assessing the efficacy of plant species for APM remediation, and talk about case studies, experimental findings, and possible avenues for expanding phytoremediation activities by looking at the definition and classification of APM. Advancements in phytoremediation mechanisms, plant species selection and combining with other APM mitigation strategies will help us to move one step closer both towards a cleaner health life for all humans also pro-sustainable action.

11.2 Understanding Airborne Particulate Matter

APM, also known as fine particulate matter, is a mixture of liquid and solid particles that is suspended in the air. From the composition, through size and shape to its origin, these particles can change a lot, which brings different properties and behaviors (Roy et al., 2024). They can be of natural or anthropogenic origin and vary significantly in size, composition, and properties. APM consists of a wide variety mixture of particles that can contain dust, soot (a fine black powder), pollen, smoke and liquid droplets or bioaerosols. APM particles have a very broad size range of nanometers to micrometers (Kumar et al., 2021).

a. **Coarse APM (with Aerodynamic Size Ranging Between 2.5 and 10 μm)**

   i. ***Sources and Composition***: Coarse APM primarily originates from mechanical processes such as construction, agricultural activities, crushing, grinding, and natural sources like sea spray and pollen. Its composition includes soil particles, pollen, mold spores, road dust, and plant fragments (Jordanova et al., 2012; Kodros et al., 2018).
   ii. ***Impacts***: They can settle on surfaces and cause soiling but are also inhalable, leading to respiratory issues like coughing, asthma, and bronchitis. Their larger size typically means they deposit in the upper respiratory tract, leading to irritation and inflammation. Environmentally, it contributes to the soiling of surfaces, reduces visibility, and affects plant health by obstructing photosynthesis (Rissler et al., 2014).

b. **Fine APM (with Aerodynamic Size Ranging Between 0.1 and 2.5 μm)**

   i. ***Sources and Composition***: Fine APM originates from combustion processes such as vehicle exhaust, industrial emissions, and wildfires. Composed of organic compounds, metals, and sulfates, fine particles are primarily generated through chemical reactions in the atmosphere (Kumar et al., 2021; Şahin et al., 2022).
   ii. ***Impacts***: Fine particles penetrate deep into the lungs and can enter the bloodstream, causing cardiovascular diseases, respiratory disorders, and premature death. They also contribute to reduced visibility and affect climate by altering cloud formation (Thangavel et al., 2022).

c. **Ultrafine APM (with Aerodynamic Size Less Than 0.1 μm)**

   i. ***Sources and Composition***: Ultra-fine particles are primarily produced by combustion processes, such as diesel engines, power plants, and certain industrial activities. They consist of nanoparticles, including metals, carbon, and other toxic compounds (Moreno-Ríos et al., 2022; Nemmar et al., 2013).
   ii. ***Impacts***: Due to their minute size, ultra-fine particles can easily enter the bloodstream through the lungs, causing severe health effects, including cardiovascular diseases, neurotoxicity, and exacerbation of chronic conditions. Environmentally, they contribute to global warming and can interfere with cloud formation and precipitation patterns (Burtscher & Schüepp, 2012; Kwon et al., 2020).

11.3 Phytoremediation: Concept and Mechanisms

Phytoremediation is a clean, cost-effective and eco-friendly technology that utilizes the innate ability of some plants to absorb or accumulate various pollutants (metals, etc.), including APMs. Phytoremediation involves using certain plant species called phytoremediators or hyperaccumulators that can uptake, store and break down environmental pollutants. The green plants serve as natural cleaning agents like the "green machines" and have been able to degrade pollutants through various biochemical and physiological processes, hence their ability to clean up contaminated soil, air and water.

11.3.1 Phytoremediation of APM: General Mechanisms

Phytoremediation offers several mechanisms for removing APM from the environment. While plants are primarily known for mitigating soil and water pollution, they can also significantly reduce APM levels in the atmosphere and improve air quality.

11.3.1.1 Physical Capture

Dry deposition allows plants to take gaseous APM particles up directly from the air. This includes the deposition of particles on plant parts such as leaves, stems and branches. The hydrophobic constituents adsorbed by the waxy cuticle of plant surfaces included organic pollutants and lipophilic compounds derived from APM (Mandal et al., 2024b). Although not a direct input route, the particulate matter is at first trapped by expulsion onto the enter layer which contributes as an initial interception characteristic for APM capture. Subsequently, these particles are washed off with rainfall or intercepted by future lignification cuticles sourced from the apical growth, removing them from the atmosphere. Finally, due to cuticular uptake, the APM particles are captured and kept on plant surfaces which will prevent their redistribution in the atmosphere, therefore facilitating an improved removal efficiency. Furthermore, one way that APMs are released from the atmosphere is by being taken in through stomata, small pores of plant leaves (Das et al., 2024). During the exchange of gases, fine and ultrafine particles can be moved through air to progress into plant tissues via stomatal apertures. APM particles can get into the leaf through a transpiration stream and become immobilized or redeposited to other plant parts, keeping them from returning to the atmosphere. It also helps in decreasing APM levels in the air and probably by modifying the stomatal aperture, it may enhance plant water relations (Arya et al., 2021). Root uptake of contaminants is predominantly by passive processes influenced greatly by the physicochemical properties in soil solution and concentration gradients. In the soil, however, APM particles could dissolve or be suspended in soil water to form aqueous solutions containing

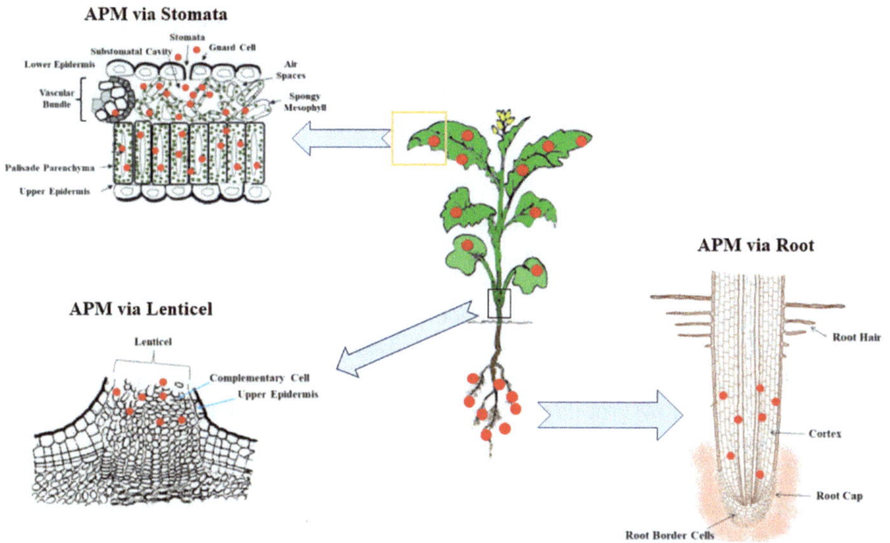

Fig. 11.1 Various mechanisms of phytoremediation through physical capture

dissolved ions and organic compounds. The dissolved substances can then move to the apoplastic and symplastic pathway and subsequently permeate into plant roots, moving from regions of higher concentration to lower concentrations in soil within root tissues (Grantz et al., 2020). Plants not only can passively take up a range of contaminants from the soil but also really actively absorb some other APMs through specific uptake mechanisms—ion channels, transporters and pumps located on root cell membranes. Active transport processes are an example of energy consumption, specific ions or molecules that must be taken up by particular transport proteins. For example, via dedicated transport systems plants can actively import and gather essential nutrients including nitrogen (N), phosphorus (P) or potassium (K). Even APM-related contaminants, the same plant can actively uptake this type of contaminant through targets for specific mechanisms: molecular organic (e.g., heavy elements and metalloids) pollutants (Sembada & Lenggoro, 2024) (Fig. 11.1).

11.3.1.2 Translocation and Sequestration

Translocation—the movement of contaminants, like APM particles, within plant tissues from one place to the other by phenomena not involving pure aqueous mass flow generally accompanied by high metabolic activity. Regarding APM removal, translocation normally occurs after particles are trapped on plant surfaces (including leaves) and have become lodged in the extensive vascular system for plants. After delivered on and into plant surfaces, contaminants can be absorbed through stomatal uptake or cuticular deposition in all parts of the plants such as (root, leaf, stem, or fruit); moreover, some contaminants can translocate from the Uptake site (root surface generally but also Leaf Cuticles, for example) to plant inside via vascular systems. APM particles captured by leaves may translocate within the plant, moving from the leaves to other aerial parts, such as stems and branches, through the xylem. This movement can be passive, powered by water uptake at the roots and governed by transpiration in a network of well-ordered xylem cells that convey sap (Pasricha et al., 2021). The APM particles translocate within the xylem, which can distribute to above-ground biomass of plants and provide a facility for sequestration or storage at different plant organs. In some cases, APM particles may also translocate through the phloem, especially if they have been metabolized or chelated by complexation in the plant tissues. Phloem transport provides a mechanism for the systemic distribution of pollutants and hence redistribution to metabolically active tissues, storage organs or below-ground structures like roots. Involvement of phloem translocation in detoxification and persistence of contaminants within the plant. (Rodrigues et al., 2023). Contaminant sequestration means that contaminants are immobilized and stored in plant tissues where they are isolated from the surrounding environment reducing their bioavailability. These sequestration mechanisms are the greatest benefit as they stop contaminants from getting back out to be able to pollute in restoration size and impacts. Translocation of contaminants within the plant can spread compounds across different parts, allowing for sequestration and containment throughout its biomass. Contaminants translocated within plant tissues can be immobilized in cellular compartments (vacuole, cell wall and intracellular space). Due to these complex interactions plant cells can encapsulate and trap foreign particles within, limiting their bioavailability and thereby reducing toxicity (Mocek-Płóciniak et al., 2023). Plants can keep absorbed contaminants inside several cellular compartments, confining them and keeping them from exposure to the environment where they might harm other organisms. Synonyms for this process are vacuolar accumulation or sequestration—one of the largest plant organelles, in which internally stored toxicants are stably locked up and spatially separated from cytosolic space/metabolism. The degradation, conversion or immobilization of some sequestered contaminants within plant tissues will continue to contribute to the environment's total remediation and detoxification. This allows them to lower their risk of cell toxic contaminants and protect essential cellular functions. They remain inside the plant tissues and cannot be discharged into surroundings hence by immobilization using cell wall binding. The cell wall is a rigid matrix surrounding plant cells and can contribute to cell mechanical support as physical barriers against contaminants mobility making them stay stored over long periods inside the plants (Khan et al., 2023; Priya et al., 2023) (Fig. 11.2).

11.3.2 Microclimatic Modification

Microclimatic modification involves altering local environmental factors, including humidity, wind direction and speed, temperature, and light intensity, within and

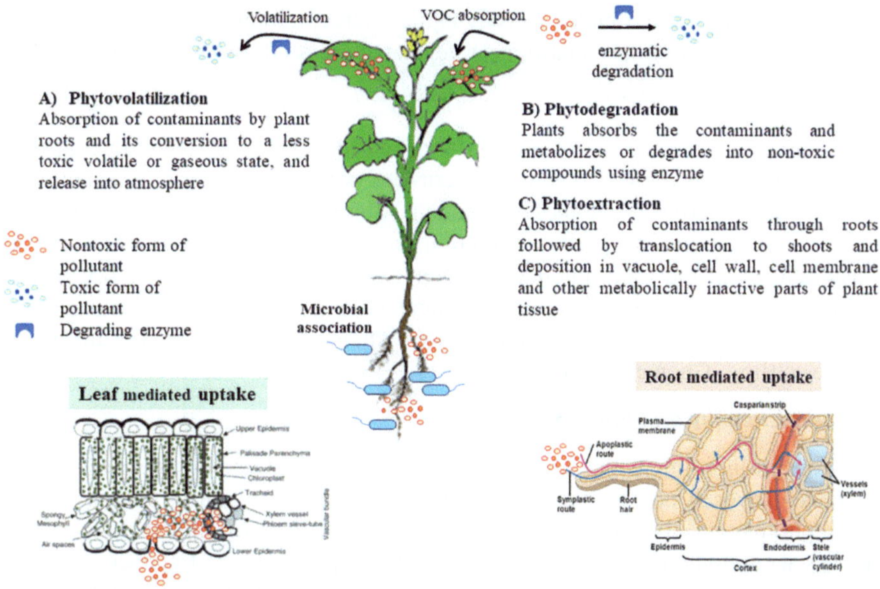

Fig. 11.2 Various pathways for contaminant absorption and degradation in the process of phytoremediation

around vegetation-covered areas. In the context of phytoremediation and the removal of APM, plants can influence microclimatic conditions in their vicinity, leading to changes that enhance APM capture, deposition, and retention.

a. ***Humidity Enhancement***:

One of the primary mechanisms by which plants modify microclimatic conditions is through transpiration—the process by which water is absorbed by plant roots, transported through the plant, and released into the environment through stomata on leaf surfaces. Elevated humidity levels near vegetation can promote the condensation and coagulation of APM particles suspended in the air. Higher humidity increases the likelihood of APM particles coming into contact with water vapor molecules, forming larger aggregates that are more prone to gravitational settling. Increased humidity also reduces the mobility of APM particles, limiting their suspension time in the atmosphere and facilitating their capture by plant surfaces and soil (Caferri & Bassi, 2022). Plant microclimate regulation contributes to the overall effectiveness of phytoremediation in mitigating APM pollution.

b. ***Wind Speed Reduction***:

Slower wind speeds near vegetation enhance the aerodynamic interception of APM particles, increasing their chances of being captured and retained by plant surfaces. Reduced wind speeds decrease the kinetic energy of APM-laden air currents, making it easier for plants to intercept and capture particles through mechanisms such as dry deposition and stomatal uptake. Aerodynamic interception enhances APM removal efficiency and improves air quality near vegetated areas (Barwise & Kumar, 2020).

c. ***Temperature Regulation***:

Shaded areas experience lower temperatures and reduced solar radiation exposure, which can influence APM deposition patterns and particle behavior. Cooler temperatures may affect particle dynamics by reducing volatilization rates, increasing particle cohesion, and promoting particle retention on plant surfaces. Evapotranspiration, transpiration and the total amount of water lost by evaporation from soil have a cooling effect on the surrounding environment. Cooler temperatures near vegetation may promote APM capture and retention by reducing air turbulence and enhancing particle settling (Chen et al., 2023).

d. ***Light Intensity and Solar Radiation***:

Variations in light intensity can affect APM deposition patterns in the rhizosphere, influencing overall phytoremediation efficiency. Light availability plays a crucial role in photosynthesis. Higher light intensity promotes photosynthetic activity and biomass accumulation, increasing plant growth and productivity. Enhanced biomass production contributes to the capture and sequestration of APM particles within plant tissues, reducing their dispersion and environmental impact (Murchie & Burgess, 2022).

11.3.3 Soil Deposition and Retention

a. ***Dry Deposition***:

Dry deposition occurs when APM particles settle directly onto soil surfaces under the influence of gravitational forces. Fine and ultrafine particles suspended in the air gradually settle out onto the ground due to their weight, depositing onto exposed soil surfaces, vegetation, buildings, and other surfaces. Soil surfaces act as passive collectors for APM particles, intercepting particles as they settle out of the atmosphere (Chaichan et al., 2023). Fine particles may adhere to soil particles, aggregates, or organic matter on the soil surface, while coarse particles may accumulate in depressions, cracks, or rough surfaces. Soil texture, structure, moisture content, and vegetation cover influence the efficiency of surface interception, affecting the distribution and retention of APM particles in the soil. Vegetation plays a significant role in enhancing the dry deposition of APM particles onto soil surfaces. Plant canopies intercept and trap airborne particles, preventing re-suspension and promoting deposition onto the ground beneath the vegetation (Emerson et al., 2020).

b. ***Wet Deposition***:

Rain, snow, and fog are examples of precipitation that scavenges particulate matter from the environment, leading to wet deposition. Rainfall events wash APM particles out of the air, transporting them to the ground surface and depositing them onto soil and other surfaces. Wet deposition can efficiently remove APM from the atmosphere, particularly during intense or prolonged precipitation events. Rainwater carries dissolved and suspended particles along flow pathways, redistributing contaminants within the soil matrix and affecting their spatial distribution and mobility. Particle wash-off and soil infiltration processes influence the outcome and conveyance of APM-associated pollutants in the surrounding environment, impacting soil quality and ecosystem health (Casas et al., 2021).

c. ***Soil Retention and Immobilization***:

When placed on soil surfaces, APM particles may interact chemically and physically with organic materials, minerals, and other soil particles. Adsorption involves attaching APM particles to soil colloids and surfaces through electrostatic forces, van der Waals interactions, and chemical bonding. Adsorbed APM particles become incorporated into the soil matrix, where they are immobilized and retained over time. Soil aggregates are crucial in retaining APM particles within the soil matrix. Aggregation processes, such as flocculation, coagulation, and cementation, bind soil particles together into larger, more stable structures. APM particles trapped within soil aggregates are shielded from erosive forces and physical disturbance, reducing their potential for re-entrainment and re-suspension in the atmosphere (Mabagala & Mng'ong'o, 2022). Microbial-mediated processes, such as biodegradation, biotransformation, and mineralization, enhance the long-term sequestration and remediation of APM-contaminated soils, improving soil quality and fertility (Aguilar-Paredes et al., 2023). In phytoremediation, soil deposition and retention are crucial in capturing, immobilizing, and breaking down APM particles in the soil profile. These processes involve physical, chemical, and biological interactions between APM particles, soil components, vegetation, and microbial communities, influencing the fate, transport, and remediation of APM-associated environmental pollutants. Understanding soil deposition and retention mechanisms is essential for designing effective phytoremediation strategies and improving soil quality and ecosystem health in APM-contaminated areas (González Henao & Ghneim-Herrera, 2021).

11.4 Remediating Outdoor and Indoor APM: Selection and Suitability of Plant Species

11.4.1 For Outdoor APM Pollution

Large volumes of airborne particulates are effectively removed from the atmosphere by plants, particularly in urban areas, where it has been reported that trees can dramatically reduce PM10 levels by adsorbing particles on their leaf surfaces or encapsulating them in waxes. (Paulo et al., 2014). The trees along urban roadsides, like *Azadirachta*, *Mangifera*, and *Ficus* spp., effectively reduce automobile particulate matter emissions. Stomatal openings at night allow CAM plants to absorb gases, but C3 and C4 plants absorb more gaseous contaminants during the day. In some plant species, like eucalyptus and pine, polychlorinated biphenyls (PCBs) enter the plant through the cuticle and are processed by cytochrome P450. Metals such as chromium, vanadium, nickel, lead, and iron typically come from the earth into plants; they do not travel to the aerial parts of the plant. Consequently, most metals accumulating in plant aerial parts are absorbed from ambient air (Wolffe et al., 2021). Organic molecules and high-density metals can permeate the wax layer, penetrate the cell's epidermis, and become lodged within the vacuoles and the cell wall. The ability of some bacteria to convert reactive oxygen species into less harmful forms by using their antioxidant properties. This bacterial ability helps plants bio-remediate PM because the PM has been proven to produce reactive oxygen species (ROS), which are detrimental to plants (Pandey & Bajpai, 2019; Mandal et al., 2022).

11.4.2 For Indoor APM Pollution

In indoor environments, plants significantly reduce levels of CO_2, VOCs, PMs, organic carbon, nitrate, sulfate, ammonia, and carbonate (Liu et al., 2022). Green house plants can purify indoor air as a biological filter. *Dracaena deremensis*, *Dracaena marginata*, and *Spathiphyllum* spp. have been shown to effectively remove xylenes, toluene, benzene, and ethylbenzene from indoor environments. The effectiveness of removing benzene from indoor air pollution was examined in eight indoor decorative plants: *Chlorophytum comosum*, *Clitoria ternatea*, *Dracaena sanderiana*, *Euphorbia milli*, *Epipremnum aureum*, *Hedera helix*, *Syngonium podophyllum*, and *Sansevieria trifasciata*. The plant that removes most effectively benzene from the air and water is *C. comosum*. The pre-vegetated frames affixed to walls or other interior structures to create green walls are a relatively new invention (Liu et al., 2022). A green wall system helps in cleaning the air by filtering out contaminants and moderating temperatures within buildings. Plants not just remove carbon-bearing molecules like CO and CO_2 from the air, but also help filter particulate matter out of the air. One of the possible ways to make it is using urban indoor vegetation, which can highly reduce air pollution. One type of green wall is where vegetation can be all or almost part of the walls using a growing medium. It is easy to add to the living infrastructure of an indoor space and measures can be taken that make it possible for green walls, as an example, to be readily available with little construction work but help to reduce particulate matter, carbon dioxide or VOC concentrations. A study has shown potted plants can remove gaseous volatile organic compounds (VOCs) by 10–90% in six hours when sealed (Kumar et al., 2019). It has been shown that eliminating organic pollutants is more successful in root-soil regions exposed to atmospheric conditions. In controlled environmental chambers, the quantities of formaldehyde and BTEX have been dramatically lowered by the plants *Chamaedorea elegans* and *Opuntia microdasys*. In another study, the *Chlorophytum comosum* L. has significantly accumulated indoor PM10, PM2.5, and PM0.2. Furthermore, the researchers observed that the plant wax facilitates the accumulation or adhesion of PM on the leaf surfaces (Vafaie Moghadam et al., 2022). Vertical gardens are becoming more popular in commercial buildings and metropolitan settings because they can provide a space-efficient way to expose additional plant biomass to air pollution. According to reports, indoor green walls considerably increased the total amount of fungal infestation, although the concentrations were still well below the WHO safety recommendations. Thus, the use of botanical air filtration is a potentially effective method of lowering indoor air pollution; however, it is still important to gain more insight into the fundamental mechanism (Mikkonen et al., 2018).

11.4.3 Developments in Biotechnology

In general, biotechnological technologies enable researchers to alter the control of gene expression at specific places, which expedites the finding of new data about the functional genomics of plants. In *Noccaeac aerulescens*, *Arabidopsis* spp., which is a hyperaccumulator of Cd and Zn, *Hirschfeld* spp., which is resistant to Pb toxicity, *Pterisvittata*, and *Brassica* spp., genomes have been sequenced for model phytoremediators (Kurade et al., 2021). It has been found that these phytoremediators effectively remove heavy metals from the surrounding environment. Likewise, overexpression of the t MerC gene resulted in higher metal accumulation in tobacco and Arabidopsis plants, but it also increased the plants' susceptibility to mercury's (Hg) effects. Several additional organic contaminants, such as polychlorinated biphenyls and polycyclic aromatic hydrocarbons, are also included in this approach (Mikkonen et al., 2018). *Atropa belladonna* (gene *rCYP2E1*), poplar (gene *rCYP2E1*), and tobacco (gene *hCYP2E1*) are a few genetically engineered plants used in phytoremediation of organic contaminants. It has been demonstrated that GM plants, such as *P. angustifolia*, *N. tabacum*, and *S. cucubalis*, absorb more heavy metal pollution than their natural counterparts due to the over-expression of glutamylcysteine-converting enzyme. The recombinant *B. juncea* seedlings (*E. coli* gshl gene insertion) showed greater tolerance to non-protein thiols, cadmium, phytochelatins, and glutathione than the wild type (James & Strand, 2009). Genes including *AtNramps*, *AtPcrs*, *CAD1* (*A. thaliana*), *gshI*, *gshII* (*D. innoxia*), *CAX-2*, *NtCBP4* (*N. tabacum*; *A. thaliana*), ferritin (soybean), *merA* (bacteria), and PCS cDNA clone (*B. juncea*) have been shown to promote higher heavy metal tolerance and accumulation. When SRSIp/ArsC and ACT 2p/γ-ECS expressed in combination resulted in greater arsenic (As) tolerance in transgenic *A. thaliana*. Furthermore, in comparison to wild plants, plants acquired 2–3 times more AR and 4–17 times more shoot biomass (Omokhagbor Adams et al., 2020). Through the insertion of *γ-ECS* genes, *ars C* genes, *MerA* genes, and *MerB* genes into Arabidopsis plants, processes of metabolism have been expressed for better phytoremediation, which has led to an increased buildup of pollutants such as mercury, arsenate, and arsenite. Researchers are currently working on genetically modifying common ornamental and houseplants to improve indoor air quality. In this case, the devil's ivy, or pothos ivy, has been genetically altered to eliminate benzene and chloroform by expressing the Mammalian Cytochrome P450 2e1 gene.

11.4.4 Selection Criteria for Suitable Plant Species for Remediation

To remediate Outdoor and Indoor pollutants, have to choose plant species with the following characteristics:

i. ***Tolerance to APM Exposure***:

The susceptibility to infection with the APM varies among plant species. An excess of APM may lead to signs such as leaf chlorosis, decreased growth or wilting in some species whereas other ones are barely affected by the overdose. Plants can employ diverse physiological strategies to respond to or alleviate the stress induced by APM. It means that plants can synthesize some antioxidants (ascorbic acid and glutathione for example) to scavenge the reactive oxygen species produced by APM exposure. In addition, plants may control the stomatal aperture to minimize APM entering leaf tissues or produce protective compounds (flavonoids and phenolics) for plant-environment interactions that mitigate the damages induced by some APMs (Blande et al., 2014).

ii. ***High Biomass Production***:

Fast-growing plants, in addition to containing toxic hydrocarbons that might be useful for breaking down organic pollutants as they degrade, may also be able to out-compete weeds and invaders, reducing the risk of ecosystem degradation while enhancing the efficiency of remediation. Plants that accumulate biomass above and below ground contribute to APM removal and soil stabilization. APM particles are captured by leaf surfaces of above-ground biomass and sequestered within root tissues as well as through adsorption onto rhizosphere soil. The high-biomass species such as bamboo and willow are among the most widely employed hyperaccumulators for APM phytoremediation given their ability to effectively produce biomass and store pollutants (Kumar Rai & Singh, 2020; Zhao et al., 2021).

iii. ***Deep Root Systems***:

Species with deep and extensive root systems can access APM-contaminated soil layers, enhancing pollutant capture and remediation. Plants with a considerable root depth help form channels within the soil, so they play an important role in infiltration to the water table enrichment leachates at all distribution layers of pollutants reaching deep horizons. Moreover, the deep root system stabilizes the soil structure, reducing erosion from shallow erosions and reducing APM re-entrainment and environmental impacts. APM phytoremediation necessitates microbial activity, and the parts of the soil zone that are affected by exudate secretion from the root are known. Plants release organic substances such as amino acids, sugars and organic acids into the rhizosphere which induces microbial metabolism and growth in this region. As a prominent pool of microorganisms in the soil, rhizobacteria and mycorrhizal fungi can contribute to the degradation of pollutants, nutrient cycling, and soil health maintenance with positive effects on both plant (soil root zone) and environmental pollution remelt processes at sites exposed to such contaminants (Priya et al., 2023).

iv. ***Adaptability to Site Conditions***:

Plants should be selected based on their tolerance to site-specific soil and climate conditions. Species adapted to various soil types, pH levels, moisture regimes, and temperature extremes are more likely to thrive in diverse phytoremediation settings. Climate-resilient species, such as drought-tolerant succulents or cold-hardy conifers, can withstand increased fluctuations in temperature and precipitation, allowing them to persist and function in APM-contaminated environments. APM pollution is most often generated in cities and industrial zones due to transport, industry and associated construction dust. To meet these goals, plants destined for phytoremediation in urban environments should be able to tolerate the stress associated with such habitats (e.g. air pollution, soil compaction and impermeable surfaces). In polluted urban landscapes, APM pho-manipulating pit planting of alien foraging bare-root shrub and grass species can be replaced by naturalized or native plants naturally adapted to similar habitats (Ayangbenro & Babalola, 2021; Kisvarga et al., 2023).

v. ***Secondary Metabolite Production***:

Plants in pronunciation emit secondary metabolites such as phytochelatins, glutathione, etc. to detoxify and sequester APM-mediated contaminants. The major phytochelatins and their precursor glutathione (Glu-Cys-Gly) bind to heavy metals such as lead, cadmium, or mercury, leading to the formation of stable complexes. Enzymes such as peroxidases and cytochrome P450s can catabolize organic contaminants like polycyclic aromatic hydrocarbons (PAH) into less harmful compounds. The ability of plants to preferentially take up and accumulate heavy metals, as well as organic pollutants from soil or air is highly variable (Bakshe & Jugade, 2023). Plants that hyperaccumulate metals, including some ferns and mosses, take up such high concentrations of certain heavy metals in their tissues that allow for both phytoextraction and phytostabilization in metal-contaminated soils. Collaborative efforts between ecologists, botanists, environmental engineers, and stakeholders are essential for selecting appropriate plant species and designing successful phytoremediation projects contributing to environmental sustainability and human well-being (Chamba-Eras et al., 2022).

By considering these selection criteria, researchers and practitioners can identify well-suited plant species for APM phytoremediation and tailor remediation strategies to specific environmental challenges and site conditions. Collaborative efforts between ecologists, botanists, environmental engineers, and stakeholders are essential for selecting appropriate plant species and designing successful phytoremediation projects contributing to environmental sustainability and human well-being (Shah et al., 2021).

a. **Examples of Plant Species Used in APM Phytoremediation (Table 11.1)**

Table 11.1 Examples of plant species used in APM phytoremediation

Plant species (scientific name)	Mechanism	References
Indoor plants		
Vetiver Grass (*Vetiveria zizanioides)*	• Utilized for phytoremediation in APM-contaminated areas because of its strong root system, capacity to produce large amounts of biomass, and resistance to toxins • Stabilize soil and prevent erosion	Dorafshan et al. (2023), Otunola et al. (2023), Suelee et al. (2017)
Snake Plant *(Sansevieria trifasciata)*	• Absorbs PM and VOCs like benzene and formaldehyde; efficiently converts carbon dioxide into oxygen, improving indoor air quality	Shalini et al.
Crown of Thorns *(Euphorbia milii)*	• Traps particulate matter on its waxy, thorny surfaces, reducing air pollution by preventing PM from re- entering the air	
Lucky Bamboo *(Dracaena sanderiana)*	• Absorbs and converts toxic substances like formaldehyde into harmless compounds; effective in reducing PM and VOCs through stomatal uptake	
Golden Pothos *(Epipremnum aureum)*	• Captures and stabilizes PM on leaf surfaces; absorbs VOCs such as benzene, formaldehyde, and xylene, reducing overall indoor air pollution	
Spider Plant *(Chlorophytum comosum)*	• Absorbs particulate matter and VOCs through stomata and leaf surfaces; stores them in plant tissues, reducing indoor air pollutants	
Arrowhead Plant *(Syngonium podophyllum)*	• Captures particulate matter and absorbs toxins like formaldehyde through its leaves, contributing to cleaner indoor air	
Outdoor plants		
Common Reed *(Phragmites australis)*	• Helps to stabilize soil, enhance water filtration, and capture airborne particulates • Provide habitat for diverse wildlife and contribute to ecosystem resilience in polluted urban and industrial areas	Milke et al. (2020)
Poplar (*Populus* sp)	• They have deep root systems that penetrate the soil, capturing airborne particulates and stabilizing contaminated soils • They are well-known for reducing greenhouse gas emissions, removing CO_2 from the environment, and improving air quality in polluted urban areas	Simmer and Schnoor (2022)
Willow (*Salix* sp)	• Used for APM remediation in riparian zones, wetlands, and contaminated landscapes • Absorb pollutants from soil and water, reduce erosion, and enhance habitat diversity	Fortin Faubert et al. (2021), Licinio et al.(2022)

(continued)

Table 11.1 (continued)

Plant species (scientific name)	Mechanism	References
Sunflower (*Helianthus* sp)	• Used for APM remediation in contaminated soils and air, particularly in urban and industrial areas • Collect heavy metals like cadmium and lead in their tissues, reducing soil contamination and enhancing soil fertility	Abdullahi et al., (2021), Sabreena et al. (2022)
Indian Mustard (*Brassica juncea)*	• Its strong taproots sift through the soil, drawing out organic substances and heavy metals • It may store large amounts of pollutants in its tissues, it can be used for phytoextraction and phytostabilization of soil contaminated with agricultural pesticides	Dotaniya et al. (2020), Rathore et al. (2019)
Water Hyacinth (*Eichhornia crassipes)*	• Used for APM remediation in contaminated water bodies and wetlands, absorbing pollutants from the water column and stabilising sediments	Churko et al. (2023), Saha et al. (2017)
Water Lettuce (*Pistia stratiotes)*	• Used for APM phytoremediation in ponds, lakes, and waterways, where it captures airborne particulates and absorbs nutrients from the water • Improve water clarity, and reduce algal blooms, contributing to ecosystem health and biodiversity	Gupta et al. (2010)

These examples highlight the diverse range of plant species utilized in APM phytoremediation, each with unique characteristics and remediation capabilities. Researchers and practitioners can effectively mitigate APM pollution, improve air and water quality, and enhance environmental sustainability in polluted urban and industrial landscapes by selecting appropriate plant species and designing tailored remediation strategies.

b. ***Comparative Analysis of Different Plant Species' Efficiency***

A comparative analysis of different plant species' efficiency in APM phytoremediation involves evaluating their performance based on various parameters such as pollutant capture, biomass production, pollutant sequestration, ecological impacts, and economic feasibility. Here's an elaborate discussion of each aspect of the comparative analysis:

i. ***Pollutant Capture Efficiency***:

Greater potential of plants with larger leaf surface areas to remove airborne particulates, including APM through dry deposition. The purpose of comparative studies is to evaluate the architecture and leaf morphology of different plant species to determine their efficiency in trapping particulate material. Stomata are small openings, present on the surfaces of leaves to help in gas exchange and loss of water vapor. Plants that have more stomatal densities can uptake APM particles via the opened orifice of stomata, which allows plants to capture a large amount of air pollutants effectively. These traits are also employed as functional traits of species because comparative studies have shown them to be proxies for APM capture by measuring actual stomatal density and aperture size. The canopy of tall or dense vegetation intercepts more APM particles from the atmosphere than shorter and sparse canopies. The comparative analysis considers factors such as plant height, canopy density, and vertical stratification to assess species' effectiveness in reducing APM concentrations in the air (Park et al., 2022).

ii. ***Biomass Production and Growth Rate***:

Phytoremediation potential is assessed by comparative studies on the biomass production and growth rates of different plant species. High biomass producer species capture APMs storing the pollutants in plant tissues and root systems thus increasing soil stabilization. Rapidly growing species, like those used by photosynthesis, remove carbon dioxide (CO_2) from the air and reduce greenhouse gas emissions as well as improving ambient air quality. In a comparative analysis of plant species, we have to examine their respective carbon sequestration potentials so that the environmental benefits and efforts in minimizing climate change will be realized (Edgar et al., 2021).

iii. ***Pollutant Sequestration and Transformation***:

In this manner, the plants with a significant rate of sequestration and emission mitigated would be expected to cause decreased re-entrainment in the atmosphere because APM-related pollutants could more likely accumulate on soil particles. Root architecture and density are assessed in comparative studies to determine the potential of diverse plant species for pollutant sequestration; soil stabilization, etc. The rhizosphere-associated plant-microbe interactions play a critical role in the fate of pollutants during transformation and immobilization processes. Through comparative analysis, the aim is to understand their involvement in APM phytoremediation and soil health improvement by analyzing the diversity and effect of rhizosphere micro-organisms associated with different plant species (Kell, 2011).

iv. ***Ecological Impacts***:

Habitat and food resources can be generated for wildlife through phytoremediation projects, with benefits to the conservation of biodiversity. Comparative analyses take the ecological impacts of plants on the surrounding ecosystems into account such as habitat diversity, species richness or community composition. Other plant species might be what we think of as invasive and edge out local flora, which causes an inverted profit for native biodiversity. Comparative studies are conducted to categorize plant species based on their invasiveness, so they can be selected for APM phytoremediation in vulnerable ecosystems (De Vitis et al., 2022).

v. ***Economic Feasibility***:

Phytoremediation projects should be economically viable and more affordable than more conventional ways of cleanup. The comparative analysis compares foundation costs, support necessities, and remediation terms of diverse plant species to recognize the most effective and affordable alternatives. This is to assess which options are most cost-effective and sustainable while setting off establishment, maintenance as well as remediation time constraints. Additional economic benefits derived from these plants as a package of ecosystem services, e.g., biomass production, soil stabilization or aesthetic value (if applicable), can increase the total profit when compared to merely considering plant potential for basic pollutant remediation alone. Based on comparative studies, environmental benefits and related resource trading-off in contrast to phytoremediation projects between different plant species are evaluated for optimal deployment (Aghili & Golzary, 2023).

By conducting a comprehensive comparative analysis, researchers and practitioners can identify plant species that are efficient, sustainable, and cost-effective for APM phytoremediation in diverse environmental settings. Integrating multiple criteria and parameters helps inform decision-making, policy development, and landscape management practices, prioritizing environmental protection and human well-being.

11.4.5 Implementation and Optimization of Phytoremediation Techniques

11.4.5.1 Site Selection

Site selection is critical for phytoremediation projects to mitigate APM pollution. This includes the discovery of the best locations for identifying plants that can successfully capture, remediate, and decrease APM levels in the air. Knowledge of the APM pollution sources is a prerequisite to site selection. Airborne pollutants come from sources such as industrial areas, heavy traffic zones and construction sites or agricultural activities to name a few. Due to higher concentrations of APM, sites near these sources are often more readily considered. Determining the concentrations of APM in potential sites for monitoring allows us to evaluate pollution levels and identify priority locations where phytoremediation can contribute to efficient ways of air quality improvement (Kumar et al., 2023). Environmental factors such as topography, soil type and climate arc major determinants of the success or failure of these phytoremediation processes. These include optimal growth temperature, sunlight intensity and duration, well-drained soil with good organic carbon content along with appropriate pH conditions. Ideally, sites are characterized by propitious conditions for plants to grow well and pollute adequately take up natural resources, enabling effective clean-up processes (Mocek-Płóciniak et al., 2023). Places such as residential sites, schools, hospitals and parks that are closer to the sensitive receptors should be treated differently. Phytoremediation processes in these areas are beneficial to human health, gaining a breathing environment, particularly through lowering APM exposure leading to increased air quality for the community. The latter strategy is in line with the public health goal of using phytoremediation to build a more environmentally and socially beneficial array of polluted sites, based on where sensitive receptors are located (Bikis, 2023). Site selection should not be ignored as it includes consideration of existing land use patterns and involves a process where stakeholder engagement is key. Working with local communities, government agencies and landowners also helps to ensure that phytoremediation projects are consistent with the needs of the community. Ensuring that you involve stakeholders early on in the project development process will give your projects a better chance of support and successful implementation. Determining land ownership, access rights and regulatory requirements will aid in the identification of suitable sites as well as streamline the project planning process (Adhi & Muslim, 2023)

The selection of the site also depends on its accessibility and available infrastructure. Not only should the sites be accessible, and convenient in transport and logistics to facilitate project management as well as monitoring and maintenance. Abundant water sources meet the requirements of irrigation, equipment storage, and support facilities like nurseries and greenhouses that add to project feasibility coupled with economical operations. Assessing infrastructure requirements and limitations will point out which locations are feasible, but also aid in planning the logistics of the project (Dhanaraju et al., 2022). Land use history, contamination associated with the site and specific extent of many environmental attributes are important to be considered in choosing sites. If the soil is a clean clay or sand, no further effort beyond planting should be necessary once negotiated but sites with prior industrial use will need remediation of any existing contamination before starting phytoremediation. Assessment of site-specific environmental factors like topography, drainage patterns and microclimates also brings about potential demands that assist in determining both challenges as well as solutions toward the successful realization of a project (Padhye et al., 2023).

11.4.5.2 Design Considerations

Proper design is required for the efficient implementation of phytoremediation projects to address APM pollution. These include planning and design of the project layout, vegetation arrangements, species choice and maintenance strategies to maximize APM capture; remediation effectiveness during peak loads and achieve other environmental benefits. Where and how to strategically place vegetation are critical choices that can either maximize APM capture or offset remediation efficiency. Therefore, strategically planting high APMs such as industrial areas, roads, and construction sites can help intercept and filter these pollutants before their transport into the area. When utilized along emission sources of APMs, vegetative barriers create physical walls that catch flying pollutant souls and incarcerate them. Ground cover acts as a green filter, preventing the spread of pollutants and improving the air quality in nearby areas. APM capture and remediation effectiveness is maximized when vegetation is optimally spaced out in a field (Kumar et al., 2023; Priya et al., 2023).

The specific plant species with high APM capture and tolerance potentials are necessary for the success of phytoremediation projects. Plant materials with large leaf surface areas, abundant foliage or deep root systems are preferred for APM capture and pollutant uptake. The best choice for plants is native species that have adapted to local environmental conditions, need less maintenance and are more tolerant of stressors. This is useful in increasing the ability of your model to remove pollutants efficiently—with hyperaccumulator and bioaccumulator species capable of storing large amounts of toxins inside their tissues. spaced out in a field. The sequestration of pollutants by these species in their biomass aids the removal and reduces APM concentrations (Mocek-Płóciniak et al., 2023).

High-density planting patterns encourage plant competition for nutrients, water and light resources ensuring rapid growth of plants which helps the faster uptake rate of pollutants as mentioned above. Optimizing plant density with respect to site conditions, species characteristics and project goals is integral in effective capture of APMs or remediation. High-density planting could increase the uptake and biomass, which caused a higher APM removal rate promoting air pollution mitigation. greater variety of high-quality habitats (Katumo et al., 2022). Careful watering, fertilizing, pruning and pest control help ensure flora's health and longevity. Optimal growth, uptake of contaminants and performance in a remediation process are supported via good maintenance practices. It is important to monitor APM concentrations, plant growth and soil quality in order to evaluate the success of projects as well as determine areas for improvement. While this weblog patently encourages continuous monitoring for evaluating project performance and tracking changes in component concentrations, remember that it also flags the huge outcome benefit within the management actions as one optimizes remediation strategies (Wang et al., 2024). Competently designing based on microclimatic and other unique site conditions, including temperature range or specific climactic precipitations like fog drip allows better plant placement in turn bringing about long-term project success. Further aid in selecting plant species well-suited to local climatic conditions will result in the best growth, so APM capture and solutions are successful. Assessing soil properties such as texture, structure, pH, and nutrient content helps determine plant suitability and growth requirements. Amending soils with organic matter, compost, and other soil amendments improves soil fertility, water retention, and microbial activity, creating favorable circumstances for the growth of plants and pollutant uptake (Vitasse et al., 2021).

- **Notable Phytoremediation Projects for APM**

Urban initiatives like the Urban Greening Initiative, Rooftop Garden Installation, Urban Park Renovation Initiative, and Urban Forest Expansion Initiative focus on using vegetation such as Vetiver Grass, Poplar Trees, Succulents, Common Reed, Water Hyacinth, Native Trees to capture PM10, PM2.5, PAHs and VOCs, improving air quality, urban sustainability, to reduce urban heat island effect, improve building energy efficiency, and promote green infrastructure development (Alsalama et al., 2021; Kowarik et al., 2019). Industrial and brownfield remediation projects employ hyperaccumulator species like Indian Mustard and Sunflowers to extract heavy metals, stabilizing contaminated soils (Megharaj & Naidu, 2017; Yan et al., 2020). Community-focused efforts, including roadside greening programs and schoolyard projects, plant grasses, shrubs, and trees to reduce pollution exposure, promote biodiversity, promote environmental stewardship, and inspire future generations to engage in sustainable practices (Bikomeye et al., 2021; Muthu et al., 2021). These projects collectively illustrate how phytoremediation integrates plant-based solutions to address environmental challenges, enhance urban ecosystems, and foster community engagement.

11.5 Future Perspectives

Although numerous studies on APM phytoremediation have been reported, many study gaps still exist which provide insight for future research to improve remediation efficiency as well as tackle new emerging challenges in the field of phytoremediation. Characterizing microbial communities and their enzymatic pathways that colonize phytoremediation plants could thus be useful to define new strategies for bettering the plant-microbe collaboration in pollutant degradation in soil. Furthermore, few studies provide the most full and continuous consideration of long-term effectiveness from all currently operating phytoremediation applications. In addition, research should also center on the discovery of new plant species more efficiently in APM remediation and adaptation process of phytoremediation to a wide range of environmental conditions as well as different types of pollutants, hence making phytoremediation an applicable method within specific situations and thus increasing its effectiveness.

The future of APM phytoremediation looks promising as modern trends and technologies are developed to widen their potential scale. One important trend in using real-time sensor networks and remote sensing technology for monitoring APM levels is the timely management of sensors distributed on-site phytoremediation strategies. Drones and satellite imaging technology advances provide exciting opportunities for wide-scale APM hotspot mapping and the planting of vegetation in high pollution burden zones. Additionally, implementation of bioengineering strategies including genetic modification and gene editing can further improve plant resistance to APM pollutants, resulting in heightened removal potential.

The policy should encourage more implementation by offering incentives for green infrastructure, funding research and demonstrations, and environmental regulation guidelines/standards on phytoremediation and land use planning. Such collaborative efforts among governmental agencies, educational institutions, industrial stakeholders and community organizations are indispensable to facilitate the establishment of a comprehensive APM management strategy that effectively addresses phytoremediation with other pollution control measures. Finally, public outreach and education can inform the public on phytoremediation benefits to promote citizen involvement in green actions ultimately allowing for informed citizens to advocate sustainable practices and policy change. These issues are interconnected and by taking a multi-sectoral approach utilizing all stakeholder knowledge, policymakers can lead the transition to greener healthier environments as well as eliminate some of these APM pollution challenges.

11.6 Conclusion

Phytoremediation presents one of the most efficient strategies for APM elimination by allowing natural plants to remove, adsorb or degrade them from ecosystems. Plants are natural systems that can be used to capture and remediate APM pollutants through phytoextraction, rhizofiltration, phytostabilization as well as phytodegradation strategies for the improvement of air quality, ecosystem health and human welfare. Secondly, phytoremediation has several advantages like lower cost in comparison to conventional methods and it works largely on the principle of managing pollution as well. Hence, future APM management strategies shall be based on a combined system approach integrating phytoremediation with other pollution control measures where possible involving policymakers, researchers and practitioners as well as community stakeholders. Therefore, phytoremediation has a lot of potential to minimize APM contamination and construct more sustainable environments hence healthier for the present and future generations by promoting innovations, and knowledge exchange requisite for advancing stakeholder collaborations.

References

Abdullahi, M. B., Animashaun, I. M., Garba, Y., Adamu, A., & Yisa, P. B. (2021). Phytoremediation of lead and chromium using sunflower (Helianthus Annuus L.) in contaminated soils of IBB University, Lapai, Nigeria. *Journal of Applied Sciences and Environmental Management, 25*(6), 1003–1007. https://doi.org/10.4314/jasem.v25i6.18

Adams, O., Godleads, P. T. F., Okoro, S. E., & Ehinomen, I. (2020). Bioremediation, biostimulation and bioaugmention: A review. *International Journal of Environmental Bioremediation & Biodegradation, 3*(1), 28–39. https://doi.org/10.12691/ijebb-3-1-5

Adhi, A. B., & Muslim, F. (2023). Development of stakeholder engagement strategies to improve sustainable construction implementation based on lean construction principles in Indonesia. *Sustainability, 15*(7), 6053. https://doi.org/10.3390/su15076053

Aghili, S., & Golzary, A. (2023). Greening the earth, healing the soil: A comprehensive life cycle assessment of phytoremediation for heavy metal contamination. *Environmental Technology & Innovation, 32*, Article 103241. https://doi.org/10.1016/j.eti.2023.103241

Aguilar-Paredes, A., Valdés, G., Araneda, N., Valdebenito, E., Hansen, F., & Nuti, M. (2023). Microbial community in the composting process and its positive impact on the soil biota in sustainable agriculture. *Agronomy, 13*(2), 542. https://doi.org/10.3390/agronomy13020542

Alsalama, T., Koç, M., & Isaifan, R. J. (2021). Mitigation of urban air pollution with green vegetation for sustainable cities: A review. *International Journal of Global Warming, 25*(3/4), 498. https://doi.org/10.1504/IJGW.2021.119014

Arya, G. C., Sarkar, S., Manasherova, E., Aharoni, A., & Cohen, H. (2021). The plant cuticle: An ancient guardian barrier set against long-standing rivals. *Frontiers in Plant Science, 12*, Article 663165. https://doi.org/10.3389/fpls.2021.663165

Ayangbenro, A. S., & Babalola, O. O. (2021). Reclamation of arid and semi-arid soils: The role of plant growth-promoting archaea and bacteria. *Current Plant Biology, 25*, Article 100173. https://doi.org/10.1016/j.cpb.2020.100173

Bakshe, P., & Jugade, R. (2023). Phytostabilization and rhizofiltration of toxic heavy metals by heavy metal accumulator plants for sustainable management of contaminated industrial sites: A comprehensive review. *Journal of Hazardous Materials Advances, 10*, Article 100293. https://doi.org/10.1016/j.hazadv.2023.100293

Barwise, Y., & Kumar, P. (2020). Designing vegetation barriers for urban air pollution abatement: A practical review for appropriate plant species selection. *Npj Climate and Atmospheric Science, 3*(1), 12. https://doi.org/10.1038/s41612-020-0115-3

Basith, S., Manavalan, B., Shin, T. H., Park, C. B., Lee, W.-S., Kim, J., & Lee, G. (2022). The impact of fine particulate matter 2.5 on the cardiovascular system: A review of the invisible killer. *Nanomaterials, 12*(15), 2656. https://doi.org/10.3390/nano12152656

Bikis, A. (2023). Urban air pollution and greenness in relation to public health edited by Q. Wang. *Journal of Environmental and Public Health, 2023*, 1–18. https://doi.org/10.1155/2023/8516622

Bikomeye, J., Balza, J., & Beyer, K. (2021). The impact of schoolyard greening on children's physical activity and socioemotional health: A systematic review of experimental studies. *International Journal of Environmental Research and Public Health, 18*(2), 535. https://doi.org/10.3390/ijerph18020535

Blande, J. D., Holopainen, J. K., & Niinemets, Ü. (2014). Plant volatiles in polluted atmospheres: Stress responses and signal degradation. *Plant, Cell & Environment, 37*(8), 1892–1904. https://doi.org/10.1111/pce.12352

Burtscher, H., & Schüepp, K. (2012). The occurrence of ultrafine particles in the specific environment of children. *Paediatric Respiratory Reviews, 13*(2), 89–94. https://doi.org/10.1016/j.prrv.2011.07.004

Caferri, R., & Bassi, R. (2022). Plants and water in a changing world: A physiological and ecological perspective. *Rendiconti Lincei. Scienze Fisiche e Naturali, 33*(3), 479–487. https://doi.org/10.1007/s12210-022-01084-7

Casas, G., Martinez-Varela, A., Vila-Costa, M., Jiménez, B., & Dachs, J. (2021). Rain amplification of persistent organic pollutants. *Environmental Science & Technology* acs.est.1c03295. https://doi.org/10.1021/acs.est.1c03295

Chaichan, M. T., Kazem, H. A., Al-Waeli, A. H. A., Sopian, K., Fayad, M. A., Alawee, W. H., Dhahad, H. A., Isahak, W. N. R. W., & Al-Amiery, A. A. (2023). Sand and dust storms' impact on the efficiency of the photovoltaic modules installed in Baghdad: A review study with an empirical investigation. *Energies, 16*(9), 3938. https://doi.org/10.3390/en16093938

Chamba-Eras, I., Griffith, D. M., Kalinhoff, C., Ramírez, J., & Gázquez, M. J. (2022). Native hyperaccumulator plants with differential phytoremediation potential in an artisanal gold mine of the Ecuadorian amazon. *Plants, 11*(9), 1186. https://doi.org/10.3390/plants11091186

Chen, H., Liu, R., & Zhang, Y. (2023). The impact of vegetation canopy on the outdoor thermal environment in cold winter and spring. *Sustainability, 15*(17), 12818. https://doi.org/10.3390/su151712818

Churko, E. E., Nhamo, L., & Chitakira, M. (2023). Phytoremediation capacity of water hyacinth (Eichhornia Crassipes) as a nature-based solution for contaminants and physicochemical characterization of lake water. *Water, 15*(14), 2540. https://doi.org/10.3390/w15142540

Das, S., Barman, C., Roy, A., Mandal, M., Popek, R., Adit, A., & Sarkar, A. (2024). Air pollution may alter reproductive dynamics/efficacy in plants: Correlative evidences from an urban sprawl in Eastern Indo-Gangetic plain. *Aerobiologia, 1–19*. https://doi.org/10.1007/s10453-024-09811-3

Dhanaraju, M., Chenniappan, P., Ramalingam, K., Pazhanivelan, S., & Kaliaperumal, R. (2022). Smart Farming: Internet of Things (IoT)-based sustainable agriculture. *Agriculture, 12*(10), 1745. https://doi.org/10.3390/agriculture12101745

Dorafshan, M. M., Abedi-Koupai, J., Eslamian, S., & Amiri, M. J. (2023). Vetiver grass (Chrysopogon Zizanoides L.): A hyper-accumulator crop for bioremediation of unconventional water. *Sustainability, 15*(4), 3529. https://doi.org/10.3390/su15043529

Dotaniya, M. L., Rajendiran, S., Meena, V. D., Vassanda Coumar, M., Saha, J. K., Bhogal, N. S., & Patra, A. K. (2020). Comparative evaluation of phytoremediation potential of indian mustard (Brassica Juncea) varieties under sewage irrigated sites. *Journal of the Indian Society of Soil Science, 68*(4), 450–457. https://doi.org/10.5958/0974-0228.2020.00036.5

Edgar, V.-N., Fabián, F.-L., Mario, P.-C., & Ileana, V.-R. (2021). Coupling plant biomass derived from phytoremediation of potential toxic-metal-polluted soils to bioenergy production and high-value by-products—A review. *Applied Sciences, 11*(7), 2982. https://doi.org/10.3390/app11072982

Emerson, E. W., Hodshire, A. L., DeBolt, H. M., Bilsback, K. R., Pierce, J. R., McMeeking, G. R., & Farmer, D. K. (2020). Revisiting particle dry deposition and its role in radiative effect estimates. Proceedings of the National Academy of Sciences, 117(42), 26076–26082. https://doi.org/10.1073/pnas.2014761117

Faubert, F., Maxime, D. D., Hijri, M., & Labrecque, M. (2021). Willows used for phytoremediation increased organic contaminant concentrations in soil surface. *Applied Sciences, 11*(7), 2979. https://doi.org/10.3390/app11072979

Grantz, D. A., Karr, M., & Burkhardt, J. (2020). Heterogeneity of stomatal pore area is suppressed by ambient aerosol in the homobaric species, vicia faba. *Frontiers in Plant Science, 11*, 897. https://doi.org/10.3389/fpls.2020.00897

Gupta, R., Tripathi, P., Kumar, R., Sharma, A., & Mishra, A. (2010). Pistia stratiotes (Jalkumbhi). *Pharmacognosy Reviews, 4*(8), 153. https://doi.org/10.4103/0973-7847.70909

González Henao, S., & Ghneim-Herrera, T. (2021). Heavy metals in soils and the remediation potential of bacteria associated with the plant microbiome. *Frontiers in Environmental Science, 9*, 604216. https://doi.org/10.3389/fenvs.2021.604216

James, C. A., & Strand, S. E. (2009). Phytoremediation of small organic contaminants using transgenic plants. *Current Opinion in Biotechnology, 20*(2), 237–241. https://doi.org/10.1016/j.copbio.2009.02.014

Jordanova, D., Jordanova, N., Lanos, P., Petrov, P., & Tsacheva, T. (2012). Magnetism of outdoor and indoor settled dust and its utilization as a tool for revealing the effect of elevated particulate air pollution on cardiovascular mortality. *Geochemistry, Geophysics, Geosystems, 13*(8), 2012GC004160. https://doi.org/10.1029/2012GC004160

Kumar, M., Muthusamy, A., Kumar, V., & Bhalla-Sarin, N. (2019). *In Vitro plant breeding towards novel agronomic traits: Biotic and abiotic stress tolerance*. Singapore: Springer Singapore.

Kumar Rai, P., & Singh, J. S. (2020). Invasive alien plant species: Their impact on environment, ecosystem services and human health. *Ecological Indicators, 111*, 106020. https://doi.org/10.1016/j.ecolind.2019.106020

Katumo, D. M., Liang, H., Ochola, A. C., Lv, M., Wang, Q.-F., & Yang, C.-F. (2022). Pollinator diversity benefits natural and agricultural ecosystems, environmental health, and human welfare. *Plant Diversity, 44*(5), 429–435. https://doi.org/10.1016/j.pld.2022.01.005

Kell, D. B. (2011). Breeding crop plants with deep roots: Their role in sustainable carbon, nutrient and water sequestration. *Annals of Botany, 108*(3), 407–418. https://doi.org/10.1093/aob/mcr175

Khan, S., Masoodi, T. H., Pala, N. A., Murtaza, S., Mugloo, J. A., Sofi, P. A., Zaman, M. U., Kumar, R., & Kumar, A. (2023). Phytoremediation prospects for restoration of contamination in the natural ecosystems. *Water, 15*(8), 1498. https://doi.org/10.3390/w15081498

Kisvarga, S., Horotán, K., Wani, M. A., & Orlóci, L. (2023). Plant responses to global climate change and urbanization: Implications for sustainable urban landscapes. *Horticulturae, 9*(9), 1051. https://doi.org/10.3390/horticulturae9091051

Kodros, J. K., Volckens, J., Jathar, S. H., & Pierce, J. R. (2018). Ambient particulate matter size distributions drive regional and global variability in particle deposition in the respiratory tract. *GeoHealth, 2*(10), 298–312. https://doi.org/10.1029/2018GH000145

Kowarik, I., Hiller, A., Planchuelo, G., Seitz, B., Von Der Lippe, M., & Buchholz, S. (2019). Emerging urban forests: Opportunities for promoting the wild side of the urban green infrastructure. *Sustainability, 11*(22), 6318. https://doi.org/10.3390/su11226318

Kumar, P., Kalaiarasan, G., Porter, A. E., Pinna, A., Kłosowski, M. M., Demokritou, P., Chung, K. F., Christopher Pain, D. K., Arvind, R. A., Adcock, I. M., & Dilliway, C. (2021). An overview of methods of fine and ultrafine particle collection for physicochemical characterisation and toxicity assessments. *Science of the Total Environment, 756*, Article 143553. https://doi.org/10.1016/j.scitotenv.2020.143553

Kumar, R., Verma, V., Thakur, M., Singh, G., & Bhargava, B. (2023). A systematic review on mitigation of common indoor air pollutants using plant-based methods: A phytoremediation approach. *Air Quality, Atmosphere & Health, 16*(8), 1501–1527. https://doi.org/10.1007/s11869-023-01326-z

Kurade, M. B., Ha, Y.-H., Xiong, J.-Q., Govindwar, S. P., Jang, M., & Jeon, B.-H. (2021). Phytoremediation as a green biotechnology tool for emerging environmental pollution: A step forward towards sustainable rehabilitation of the environment. *Chemical Engineering Journal, 415*, Article 129040. https://doi.org/10.1016/j.cej.2021.129040

Kwon, H.-S., Ryu, M. H., & Carlsten, C. (2020). Ultrafine particles: Unique physicochemical properties relevant to health and disease. *Experimental & Molecular Medicine, 52*(3), 318–328. https://doi.org/10.1038/s12276-020-0405-1

Licinio, A., Laur, J., Pitre, F. E., & Labrecque, M. (2022). Willow and herbaceous species' phytoremediation potential in Zn-contaminated farm field soil in eastern Québec, Canada: A greenhouse feasibility study. *Plants, 12*(1), 167. https://doi.org/10.3390/plants12010167

Liu, H., Luo, L., Jiang, G., Li, G., Zhu, C., Meng, W., Zhang, J., Jiao, Q., Pengqiang, D., & Li, X. (2022). Sulfur enhances cadmium bioaccumulation in Cichorium Intybus by altering soil properties, heavy metal availability and microbial community in contaminated alkaline soil. *Science of the Total Environment, 837*, Article 155879.

Megharaj, M., & Naidu, R. (2017). Soil and brownfield bioremediation. *Microbial Biotechnology, 10*(5), 1244–1249. https://doi.org/10.1111/1751-7915.12840

Mikkonen, A., Li, T., Vesala, M., Saarenheimo, J., Ahonen, V., Kärenlampi, S., Blande, J. D., Tiirola, M., & Tervahauta, A. (2018). Biofiltration of airborne VOCs with green wall systems-microbial and chemical dynamics. *Indoor Air, 28*(5), 697–707. https://doi.org/10.1111/ina.12473

Milke, J., Gałczyńska, M., & Wróbel, J. (2020). The importance of biological and ecological properties of phragmites Australis (Cav.) Trin. Ex Steud., in phytoremediation of aquatic ecosystems—The review. *Water, 12*(6), 1770. https://doi.org/10.3390/w12061770

Mocek-Płóciniak, A., Mencel, J., Zakrzewski, W., & Roszkowski, S. (2023). Phytoremediation as an effective remedy for removing trace elements from ecosystems. *Plants, 12*(8), 1653. https://doi.org/10.3390/plants12081653

Moreno-Ríos, A. L., Tejeda-Benítez, L. P., & Bustillo-Lecompte, C. F. (2022). Sources, characteristics, toxicity, and control of ultrafine particles: An overview. *Geoscience Frontiers, 13*(1), Article 101147. https://doi.org/10.1016/j.gsf.2021.101147

Murchie, E. H., & Burgess, A. J. (2022). Casting light on the architecture of crop yield. *Crop and Environment, 1*(1), 74–85. https://doi.org/10.1016/j.crope.2022.03.009

Muthu, M., Gopal, J., Kim, D.-H., & Sivanesan, I. (2021). Reviewing the impact of vehicular pollution on road-side plants—Future perspectives. *Sustainability, 13*(9), 5114. https://doi.org/10.3390/su13095114

Mabagala, F. S., & Mng'ong'o, M. E. (2022). On the tropical soils; the influence of Organic Matter (OM) on phosphate bioavailability. *Saudi Journal of Biological Sciences, 29*(5), 3635–3641. https://doi.org/10.1016/j.sjbs.2022.02.056.

Mandal, M., Sarkar, M., Khan, A., Biswas, M., Masi, A., Rakwal, R., Agrawal, G. K., Srivastava, A., & Sarkar, A. (2022). Reactive oxygen species (ROS) and reactive nitrogen species (RNS) in plants–maintenance of structural individuality and functional blend. *Advances in Redox Research, 5*, Article 100039. https://doi.org/10.1016/j.arres.2022.100039

Mandal, M., Roy, A., & Sarkar, A. (2024a). Understanding the possible cellular responses in plants under micro (nano)-plastic (MNPs): Balancing the structural harmony with functions. *Science of the Total Environment, 957*, Article 177732. https://doi.org/10.1016/j.scitotenv.2024.177732

Mandal, M., Roy, A., Singh, P., & Sarkar, A. (2024b). Quantification and characterization of airborne microplastics and their possible hazards: A case study from an urban sprawl in Eastern India. *Frontiers in Environmental Chemistry, 1499873*. https://doi.org/10.3389/fenvc.2024.1499873

Moghadam, V., Ameneh, A. I., Saadatmand, S., Ebadi, M., & Ardebili, Z. O. (2022). *New insights into the transcriptional, epigenetic, and physiological* responses to zinc oxide nanoparticles in datura stramonium; potential species for phytoremediation. *Journal of Plant Growth Regulation, 41*(1), 271–281. https://doi.org/10.1007/s00344-021-10305-6

Nemmar, A., Holme, J. A., Rosas, I., Schwarze, P. E., & Moreno, E. (2013). Recent advances in particulate matter and nanoparticle toxicology: A review of the *in vivo* and *in vitro* studies. *BioMed Research International, 2013*, 1–22. https://doi.org/10.1155/2013/279371

Otunola, B. O., Aghoghovwia, M. P., Thwala, M., Arias, A., Jordaan, R., Hernandez, J. C., & Ololade, O. O. (2023). Improving capacity for phytoremediation of vetiver grass and indian mustard in heavy metal (Al and Mn) contaminated water through the application of clay minerals. *Environmental Science and Pollution Research, 30*(18), 53577–53588. https://doi.org/10.1007/s11356-023-26083-5

Padhye, L. P., Srivastava, P., Jasemizad, T., Bolan, S., Hou, D., Shaheen, S. M., Rinklebe, J., O'Connor, D., Lamb, D., Wang, H., Siddique, K. H. M., & Bolan, N. (2023). Contaminant containment for sustainable remediation of persistent contaminants in soil and groundwater. *Journal of Hazardous Materials, 455*, Article 131575. https://doi.org/10.1016/j.jhazmat.2023.131575

Pandey, V. C., & Bajpai, O. (2019). Phytoremediation. In *Phytomanagement of polluted sites* (pp 1–49). Elsevier.

Park, S., Lee, J. K., Kwak, M. J., Lim, Y. J., Handong Kim, S., Jeong, G., Son, J.-a, Chang-Young, O., Je, S. M., Chang, H., Kim, K., & Woo, S. Y. (2022). Relationship between leaf traits and pm-capturing capacity of major urban-greening species. *Horticulturae, 8*(11), 1046. https://doi.org/10.3390/horticulturae8111046

Pasricha, S., Mathur, V., Garg, A., Lenka, S., Verma, K., & Agarwal, S. (2021). Molecular mechanisms underlying heavy metal uptake, translocation and tolerance in hyperaccumulators—an analysis. *Environmental Challenges, 4*, Article 100197. https://doi.org/10.1016/j.envc.2021.100197

Priya, A. K., Muruganandam, M., Ali, S. S., & Kornaros, M. (2023). Clean-up of heavy metals from contaminated soil by phytoremediation: A multidisciplinary and eco-friendly approach. *Toxics, 11*(5), 422. https://doi.org/10.3390/toxics11050422

Paulo, J. C., Pratas, J., Varun, M., Dsouza, R., & Manoj, S. (2014). Phytoremediation of soils contaminated with metals and metalloids at mining areas: Potential of native flora. In M. C. Hernandez Soriano (Ed.), *Environmental risk assessment of soil contamination.* InTech.

Rathore, S. S., Shekhawat, K., Dass, A., Kandpal, B. K., & Singh, V. K. (2019). Phytoremediation mechanism in Indian mustard (Brassica Juncea) and its enhancement through agronomic interventions. *Proceedings of the National Academy of Sciences, India Section B: Biological Sciences, 89*(2), 419–27. https://doi.org/10.1007/s40011-017-0885-5

Rissler, J., Nordin, E. Z., Eriksson, A. C., Nilsson, P. T., Frosch, M., Sporre, M. K., Wierzbicka, A., Svenningsson, B., Löndahl, J., Messing, M. E., Sjogren, S., Hemmingsen, J. G., Loft, S., Pagels, J. H., & Swietlicki, E. (2014). Effective density and mixing state of aerosol particles in a near-traffic urban environment. *Environmental Science & Technology, 48*(11), 6300–6308. https://doi.org/10.1021/es5000353

Rodrigues, C. I. D., Brito, L. M., & Nunes, L. J. R. (2023). Soil carbon sequestration in the context of climate change mitigation: A review. *Soil Systems, 7*(3), 64. https://doi.org/10.3390/soilsystems7030064

Roy, R., & D'Angiulli, A. (2024). Air pollution and neurological diseases, current state highlights. *Frontiers in Neuroscience, 18*, 1351721. https://doi.org/10.3389/fnins.2024.1351721

Roy, A., Das, S., Singh, P., Mandal, M., Kumar, M., Rajlaxmi, A., & Sarkar, A. (2024). Summer1125 time monitoring and source apportionment study of both coarse, fine, and ultra-fine particulate pollution in Eastern Himalayan Darjeeling: A hint to health risk during peak tourist season. Mapan, 1–15. https://doi.org/10.1007/s12647-024-00776-9

Sabreena, S. H., Bhat, S. A., Kumar, V., Ganai, B. A., & Ameen, F. (2022). Phytoremediation of heavy metals: An indispensable contrivance in green remediation technology. *Plants, 11*(9), 1255. https://doi.org/10.3390/plants11091255

Saha, P., Shinde, O., & Sarkar, S. (2017). Phytoremediation of industrial mines wastewater using water hyacinth. *International Journal of Phytoremediation, 19*(1), 87–96. https://doi.org/10.1080/15226514.2016.1216078

Şahin, Ü. A., Harrison, R. M., Alam, M. S., Beddows, D. C. S., Bousiotis, D., Shi, Z., Crilley, L. R., Bloss, W., Brean, J., Khanna, I., & Verma, R. (2022). Measurement report: Interpretation of wide-range particulate matter size distributions in Delhi. *Atmospheric Chemistry and Physics, 22*(8), 5415–5433. https://doi.org/10.5194/acp-22-5415-2022

Sembada, A. A., & Wuled Lenggoro, I. (2024). Transport of nanoparticles into plants and their detection methods. *Nanomaterials, 14*(2), 131. https://doi.org/10.3390/nano14020131

Shah, A., Nazari, M., Antar, M., Msimbira, L. A., Naamala, J., Lyu, D., Rabileh, M., Zajonc, J., & Smith, D. L. (2021). PGPR in agriculture: A sustainable approach to increasing climate change resilience. *Frontiers in Sustainable Food Systems, 5*, Article 667546. https://doi.org/10.3389/fsufs.2021.667546

Simmer, R. A., & Schnoor, J. L. (2022). Phytoremediation, bioaugmentation, and the plant microbiome. *Environmental Science & Technology, 56*(23), 16602–16610. https://doi.org/10.1021/acs.est.2c05970

Suelee, A. L., Hasan, S. N. M. S., Kusin, F. M., Yusuff, F. M., & Ibrahim, Z. Z. (2017). Phytoremediation potential of vetiver grass (vetiveria zizanioides) for treatment of metal-contaminated water. *Water, Air, & Soil Pollution, 228*(4), 158. https://doi.org/10.1007/s11270-017-3349-x

Thangavel, P., Park, D., & Lee, Y.-C. (2022). Recent insights into particulate matter (PM2.5)-mediated toxicity in humans: An overview. *International Journal of Environmental Research and Public Health, 19*(12), 7511. https://doi.org/10.3390/ijerph19127511

Vitasse, Y., Baumgarten, F., Zohner, C. M., Kaewthongrach, R., Fu, Y. H., Walde, M. G., & Moser, B. (2021). Impact of microclimatic conditions and resource availability on spring and autumn phenology of temperate tree seedlings. *New Phytologist, 232*(2), 537–550. https://doi.org/10.1111/nph.17606

Vitis, D., Marcello, K. H., Barak, R. S., Egerton-Warburton, L., Ernst, A. R., Evans, M., Fant, J. B., Foxx, A. J., Hadley, K., Jabcon, J., O'Shaughnessey, J., Ramakrishna, S., Sollenberger, D., Taddeo, S., Casanova, R., Woolridge, C., Lan, X., Zeldin, J., & Kramer, A. T. (2022). Why are some plant species missing from restorations? A diagnostic tool for temperate grassland ecosystems. *Frontiers in Conservation Science, 3*, 1028295. https://doi.org/10.3389/fcosc.2022.1028295

Wang, T., Jiaxin, X., Chen, J., Liu, P., Hou, X., Yang, L., & Zhang, L. (2024). Progress in microbial fertilizer regulation of crop growth and soil remediation research. *Plants, 13*(3), 346. https://doi.org/10.3390/plants13030346

Wolffe, M. C., Wild, O., Long, S. P., & Ashworth, K. (2021). Temporal variability in the impacts of particulate matter on crop yields on the north China plain. *Science of the Total Environment, 776*, Article 145135. https://doi.org/10.1016/j.scitotenv.2021.145135

Yan, A., Wang, Y., Tan, S. N., Yusof, M. L. M., Ghosh, S., & Chen, Z. (2020). Phytoremediation: A promising approach for revegetation of heavy metal-polluted land. *Frontiers in Plant Science, 11*, 359. https://doi.org/10.3389/fpls.2020.00359

Zhao, W., Liu, T., Liu, Y., Wang, H., Wang, R., Ma, Q., Dong, H., & Bi, X. (2021). The significance of biomass allocation to population growth of the invasive species ambrosia artemisiifolia and ambrosia trifida with different densities. *BMC Ecology and Evolution, 21*(1), 175. https://doi.org/10.1186/s12862-021-01908-4

Chapter 12
Advancing Urban Resilience and Sustainable Development Through Green Infrastructure

Rahul Kumar and Suchismita Roy

Abstract Sustainable development is an overall socioeconomic as well as environmental change that pertains to the prerequisites of the present circumstance, without conceding the ability of future generations to meet their needs. Unsustainable rapid urban growth has led to major increase in pollutants of the air. Particulate matters have become available for many human diseases and environmental issues. Urban greening is an approach or method to build sustainable designs creating eco-friendly cities, using only sustainable methods and means for its construction or growth. It is a great way to mitigate carbon emissions, reduce carbon footprints, futuristic planning of recycling and reusing materials and making the world a better place for generations to come. In the scope of this chapter, we will discuss ways and tools about how to usher urban greening for the betterment of our existing urban culture. In this chapter, we tried to deal with the essential issues associated with urbanization with special focus on the issue of particulate matters and how urban greening can be used as a mitigation measure against it.

Keywords Anthroposphere · Biosphere · Ecological gentrification · Green infrastructure · Particulate matter · Sustainable urbanization · Gentrification · Urban pollution island

12.1 Introduction

"When the last tree is cut down, the last fish eaten, and the last stream poisoned, you will realize that you cannot eat money."–Native American saying (source: Wasserman, 1983, ("The Oxford Dictionary of Proverbs," 2004).

R. Kumar
Department of Zoology, Sheodeni Sao College (Magadh University), Kaler, India
e-mail: rahuldayanand33@gmail.com

S. Roy (✉)
Department for Cell and Molecular Medicine, University of California, San Diego, CA, USA
e-mail: roy.suchismita2011@gmail.com

A. Sarkar et al. (eds.), *From Sky to Soil*, https://doi.org/10.1007/978-3-032-31991-3_12

Rate of urbanization is directly allied to the extensive usage of natural resources. Urbanization comes with a cost (Richardson, 1987). Forests are vanishing, water bodies are getting polluted, air is becoming unbreathable, food security is threatened due to shrinking farmlands and increasing population size, shortage of drinking water is rampant, rate of waste production is way higher than the rate of waste recycling and management, rise in the number of infectious, lifestyle and environment related diseases and morbidities is manifold, decline in biodiversity is significant, and so on (Kuddus et al., 2020; Tong et al., 2018). Economic growth is taking a heavy toll on environmental health. According to the World Bank, about 56% of the global population reside in cities which is roughly equivalent to 4.4 billion individuals. This value is supposed to double by 2050 making 7 out of every 10 persons living in cities. With rapid economic, demographic, and spatial growth, the number of urban hubs is increasing at an unwavering level. Though urbanization is a crucial part of economic growth, it is adversely affecting our environment. Reduction in green spaces and increase in air pollution are results of unsustainable urban planning. Suspended particulate matter (PM) is foremost air pollutants in urban areas (Chávez-García & González-Méndez, 2021) having deleterious effects. In 2015, member countries of the United Nations adopted the 2030 Agenda for Sustainable Development at the core of which are 17 Sustainable Development Goals (SDGs) (worldtop20.org, 2022). Out of these 17 SDGs, the 11th goal is concerned with "sustainable cities and communities." SDG-11 has ten key targets of which inclusive and sustainable urbanization, reducing the adverse impact of urbanization on environment, preserving urban natural heritage and universal access to urban green spaces are the ones which form an integral part of the idea of urban greening in sustainable urbanization. Present chapter deals with various issues associated with urbanization with special reference to PM and urban greening as a measure to alleviate these concerns to promote an infrastructure of sustainable urbanization.

12.2 Uprising of Urban Ecosystem

Urban ecosystem is largely a complex socio-ecological system (Mollashahi & Szymura, 2022) with regular feedback between them. Humans are the main drivers of the change in an urban ecosystem (Anees et al., 2023). Knowledge, ideology, technology, values, health, nutrition, language, personality, population, biophysical characteristics, resource exploitation pattern, etc., are directly associated with the actions of urban inhabitants and are a part of the social system. Ecological system includes microorganisms, air, water, weeds, crops, climate, pests, trees, soils, livestock, human-built structures, etc. There is a flow of energy, material, and information from the ecological to the social system and there develops a system of coadaptation with the influence of both the systems (Levin et al., 2013).

Humans interact with all components belonging to the biosphere, viz. atmosphere, lithosphere, and hydrosphere. The effect of human influence in an urban ecosystem is so huge that a separate component is recognized as the anthroposphere.

Biosphere at a large scale comprises of atmosphere, lithosphere (including pedosphere) and hydrosphere. Humans are a part of the same biosphere. With the over-influencing effect of mankind in the urban ecosystem, with human presence, man-made structures, a fourth component has gained predominance as the anthroposphere (Fig. 12.1). Anthroposphere impacts biosphere as it includes all kinds of human activities involved with severe material flow and impacted decision making, such as in land usage pattern, migration, dispersal and traveling pattern and urban man-made structures (Donahue, 2015). Human impacts are also in habitat destruction and fragmentation of lands by processes as deforestation, urbanization, and more industrial infrastructure development. These activities impact is in the form of pollution of air, water, and land. Emission of greenhouse gasses, unusual fluctuations in temperature are directly linked to human induced climate change (Mandal et al., 2023a). Human activities such as overfishing, hunting, poaching, overharvesting of trees for timber and non-timber forest items and overexploitation of natural resources are the side effects of urbanization. Increased use of synthetic fertilizers and fossil fuels have also altered biogeochemical cycles of nitrogen, phosphorus, and carbon. Due to increase in demand for food, transport facilities, new infrastructures, and heavy usage of electricity, change in land use is widespread. Biodiversity rich natural forest areas are being diverted to activities like farming and mining. Such ongoing human activities and interventions in urban areas have negative impacts on the biosphere (Manisalidis et al., 2020). Interaction between biosphere and anthroposphere can be compared with a negative–positive ecological interaction like parasitism or predation where human beings are largely benefitted from the biosphere, but biosphere is negatively impacted by human beings (Darimont et al., 2023).

Fig. 12.1 Components of the urban ecosystem

12.3 Suspended Particulate Matter and Mechanism of Their Formation

Suspended particulate matters (PM) or particulate pollution (Gokul et al., 2023) are airborne particles of different sizes and varying chemical compositions. These particles may be solid as well as liquid in nature. Table 12.1 enlists types, size, of PMs and its primary affecting tract on humans. It is to be noted that the smaller particles such as particulate matter are more harmful than larger sized particles (Souza et al., 2021).

Particle size has a significant influence on how a particular PM affects the body. Most coarse PMs are easily removed by mucociliary clearance from the respiratory system after inhalation. Conversely, fine PMs can reach alveoli, where the lung parenchyma retains it before it can enter the bloodstream. Ultrafine PMs can pass past pulmonary barriers and respiratory mucosae, where it can be taken up by lung cells such as alveolar macrophages (Keenan et al., 2023). $PM_{2.5}$ is known to have more deleterious effects on human health than PM_{10}. According to the Global Burden of Disease (GBD) study, $PM_{2.5}$ is estimated to be the cause of 5 million deaths per year (Thangavel et al., 2022).

Airborne particulate matter (PM) is a complex mixture of solids and aerosols constituted of small droplets of liquid, dry solid fragments, and solid cores with liquid coatings. Particles differ extensively in size, shape, and chemical arrangement, and may contain inorganic ions, metallic compounds, elemental carbon, organic compounds, and compounds from the earth's crust (Mandal et al., 2023b). Air pollutants in the air are existing as a suspension of a complex mixture of several gasses and PMs. There are two kinds of PMs based on the source of emission. If the PM is directly emitted to the atmosphere, it is primary PM. Otherwise PM can also form in the atmosphere by conversion of different gasses to particles, it is called secondary PM. Both primary and secondary PM may undergo various chemical and physical

Table 12.1 PMs can be classified into following broad categories based on its aerodynamic diameter (a relative quantification of particle mass and shape) (Gokul et al., 2023; Newell et al., 2021)

Types	Subtypes	Size	Parts affected	Reference
Coarse PMs	PM10, PM2.5	2.5 to 10 μm	Upper respiratory tract	Chen et al. (2019) and Keet et al. (2018)
Fine PMs	PM2.5	less than 2.5 μm	Lung parenchyma	Li et al. (2018)
Ultrafine PMs		less than 0.1 μm	Alveolar macrophages	Buonanno et al. (2015) and Keenan et al. (2023)
Quasi ultrafine PMs	PM0.3	< 0.3 μm	Infiltrate deep into organs and tissues by entering the blood circulation via alveoli	Habre et al. (2021)

transformations in the atmosphere depending upon the surrounding conditions (R. Zhang et al., 2015). Urban primary PM is resourced from anthropogenic sources (R. Zhang et al., 2015).

12.4 Urbanization as a Source to Rising Airborne Particulate Matters

Urbanization comes with a pinch of salt. The entire process of urbanization involves mass movement of population from a village to a city, which certainly involves challenges with economic, social, and mental agility. This urban area with localized high PM concentrations in the atmosphere is known as urban pollution island (Fig. 5.2) (UPI) (Singh et al., 2022) (Fig. 5.2). In satellite images, UPI appears as a denser region marked by reduced Aerosol Optical Depth (AOD). AOD is a smart measure which appears as a dimensionless parameter used to quantify the reduction of solar radiation due to scattering or absorbance by the atmospheric aerosols between the observation site and the top of the atmosphere (Daxini & Wu, 2024). Apart from its impacts on human health, high concentration of PM in UPI may affect precipitation patterns, energy (heat) budget, soil health, crop production, outdoor visibility, and performance of photovoltaic devices (Barat & Parth Sarthi, 2023; Daxini & Wu, 2024). Radiative forcing (RF) of PM contributes toward climate change (Pulselli & Marchi, 2015). RF as defined by the Intergovernmental Panel on Climate Change quantifies the impact of a particular climatic element on the quantity of the retained downward-directed radiant energy reaching Earth surface (Anthropogenic and Natural Radiative Forcing, 2013; Intergovernmental Panel on Climate Change (IPCC), 2023; Pinault, 2018). Positive RF value suggests the possible role of a pollutant in global warming whereas negative RF value implies its role in global cooling (Kumar & Sharma, 2020). Fine PM has a cooling effect on the atmosphere, whereas the coarse PM has a warming effect (S. L. Chen et al., 2021). With context to climatic change, some constituents of the ambient PM mixture enhance climate warming (e.g., black carbon released as black sooty material discharged from gas and diesel engines, power plants which coal-fired), while others have a cooling influence (e.g., nitrate and sulfate). Apart from these impacts, particulate air pollution impacts on psychological well-being and has been found to affect visual aesthetic preferences (Zhao et al., 2020). High PM concentrations at UPI creates haze by blocking sunlight (Y. Li et al., 2019). PM is involved in formation of smog during winters which also reduces outside visibility and affects daily work (Geddes & Murphy, 2012). Prolonged lack of sunshine and gloomy weather has been found to have correlation with symptoms of depression. High PM levels also disrupt insect sensory functions thereby impacting insect pollinators directly linked to agriculture and food security (Jankielsohn, 2023).

With these aspects, there have been challenges with employment, accommodation, and space problems, along with numerous impending challenges, many of which are interconnected and complex and ultimately affecting health and life expectancy

(K. H. Kim et al., 2015). In the scope of this chapter we are highlighting some key challenges which have developed in terms of affecting the ecosystem mainly PMs. There are various factors contributing to the degradation of air quality in areas which are urban (Goossens et al., 2021). These factors have been discussed as below.

12.5 Infrastructure Strain on Environment

Rapid urbanization methods have put immense pressure on existing infrastructure-based constructions. This sudden shoot up in the number of high-rise buildings has caused a serious threat to the environment (Das et al., 2023). There is an upsurge in of industrial belts rising the rate of combustion of coal, oil, gasoline; transformation products of NOx, SO_2, and organics including biogenic organics, e.g., terpenes; high temperature processes; smelters, and steel mills. For example, the most prevalent metal on earth, when studied by material flow analysis (MFA), showed that since 1950–2007, end-user claim for only the metal aluminum (Al) has increased to about 30-fold. Broadly, Al is used in vehicles, industrial equipment, construction accessories, and metal products but the contemporary global anthropogenic Al through its transformation from bauxite ore to a final product is not environment friendly PM existing in the urban areas has a substantial amount of Al from natural and human reinforced activities (Al-Thani et al., 2018; Roshan et al., 2019).

12.5.1 Deforestation

Due to expanding urban population, due to migration, expansion of urban limits has become inevitable. This expansion is mostly at the cost of surrounding green areas. Ecologically rich forest areas are now acting as a sink of harmful pollutants. Forests are the living lungs of mankind, toward purifying air and maintaining its breathability. Therefore, deforestation leads to stagnation of air pollutants in the atmosphere leading to rise in PMs (Roy et al., 2023). The process of deforestation to clear up land space and for reusing or reclaiming agricultural lands necessitates the urge of burning down of green belt, which releases huge amount of smoke and harmful gases into the air. This not only contributes to air PM locally, but the smoke can travel long distances, affecting air quality in regions far from the site of deforestation. The biggest health threat from smoke is from fine particles, $PM_{2.5.}$ Biomass burning produces a variety of toxic gaseous pollutants, such as CO, CH_4, VOCs, PAH, and more importantly several heavy metals (Pb, Cu, Fe, Zn, and Hg, etc.). These glasses and metals account for about one-third of the particulate matter mass that is emitted into the urban atmosphere (Akbari et al., 2021).

12.5.2 *Power Plants and Power Generation*

Power in the form of electricity is a necessity in contemporary times but with the release of power there is significant take up by fossil fuels. Burning fossil fuels is a major source of fine $PM_{2.5}$. Burning fossil fuels also releases other pollutants, including sulfur dioxide (SO_2), nitrogen oxides (NO_X), and carbon dioxide (CO_2). These pollutants react to form more PMs, as well as ground-level ozone. Energy-related fossil-fuel combustion in high-mid income countries and biomass burning in low-income countries accounts for most of the universal air pollution, has the potential to generate 85% of respirable particulate pollution generating almost all sulfur dioxide and nitrogen oxide content of PMs. Other harmful chemicals which get emitted are black carbon, polycyclic aromatic hydrocarbons (PAH), nitrogen and sulfur dioxides, mercury, and volatile chemicals that affect ozone at ground level (Perera, 2018).

12.5.3 *Excessive Concrete Usage*

From building newer concrete buildings to more humans working in these construction zones, the urban jungle does pose a serious threat. From PM10 to PM2.5, there is serious upsurge of these overall particulate matters. Suspended PMs such as cement, dust, and gypsum in the air in construction zone pose serious threats with every day of their work (Siebert et al., 2001; Tong et al., 2018).

12.6 Environmental Degradation Due to Traffic Congestion

Urbanization is vehemently contributing to environmental degradation through increased pollution rate, deforestation programs, habitat loss, and the expansion of impervious surfaces(Lindbom et al., 2006).The influx of people and vehicles in urban areas leads to traffic congestion and air pollution, which have detrimental effects on public health, productivity, and quality of life (Lindbom et al., 2006). Laboratory based research, using both animals or mammalian cells, have confirmed the presence of varied fractions and types of PMs from these areas. In a normal routine life, the impact of traffic-related particles to the total particle burden in urban areas, the particles generated by the wear and tear of tires and road surface are huge. et al. showed the granite roadway made about 1.7 times as high PM_{10} than quartzite roadway and ultrafine particles seem to be produced from the tires and not the roadways. These PMs have induced proinflammatory cytokines and interleukins, and that this cytokine is part of oxidative stress in humans. On the other hand, subway particles had a higher cell toxicity effect than granite and quartzite, which may affect the capacity to secrete cytokines. This is probably due to the high iron content in the subway particles (Lindbom et al., 2006). Urban areas require heavy transportation facilities, relying mostly on diesel and petrol. Although compressed natural gas (CNG) and electricity-based vehicles have gained ground. Numerous studies have evaluated the associations between PM or traffic-related air pollutants and cognitive outcomes in younger people from throughout the world.

12.7 Occupational Particulate Dust Exposure

Industrialization is one of the major reasons behind air pollution. Industries require big input of power, and power generation requires large scale combustion of fossil fuels. Apart from this, many industries themselves release toxic chemicals and PMs to the atmosphere. Accidental release of toxic gasses from some industries has also been reported claiming lives of many people living around it. The Bhopal gas tragedy of 1984 is a notable example. Exposure to ambient particulate air pollution is a recognized risk factor for cardiovascular disease; however, occupational particulate exposure has become limitless due to urbanization and demand for jobs, and its side effects are immense. Relatively less attention has been given to the effects of occupational exposures, which depends on various parameters as rate of exposure, duration of exposure, composition of the PM and concentration and dosage of PM (Fang et al., 2010). Based on a large dataset, Vinnikov et al. (Vinnikov et al., 2023) have reported highest $PM_{2.5}$ concentrations to be found in the concrete-mixing unit (median 1180 $\mu g/m^3$), followed by metalworking (510 $\mu g/m^3$), armature workshop (375 $\mu g/m^3$) and molding site (245 $\mu g/m^3$), which is largely diff from normal office (29.5 $\mu g/m^3$) (Vinnikov et al., 2023).

12.8 Urbanization: Polluting Every Other Sources

Particles can be hauled to long distances by wind and then they can act as sedimentation on ground or water sources. Depending on their composition, these settling may include making lakes and streams acidic, changing the nutrient balance in waters bodies such as in backwaters and coastal areas and large basins of rivers, draining the nutrients in soil, harming sensitive forests and farm crops, affecting the diversity of ecosystems, contributing to acid rain formation (Mandal et al., 2025). A study on snail by Hartono et al., 2017 (Hartono et al., 2017) showed that high $PM_{2.5}$ concentration (7.75 mg/L) and chemical components related to $PM_{2.5}$, including Al and acidity (pH 5.0) induced a significant decrease in snails' movement by reducing its activity. In contrast, a low concentration of $PM_{2.5}$ (3.88 mg/L) such as lead (Pb) and zinc (Zn) did not affect snails' behavior. The results suggest that high PM2.5 deposition in water bodies, and associated features as acidification and some metals toxicity, can have an adverse effect on aquatic species. In another study, from downtown Beijing the washing effect of rain on PMs showed that the concentration of

$PM_{2.5}$ was negatively correlated to the magnitude of rainfall which accumulated, showing that $PM_{2.5}$ reduction was from over 400 $\mu g/m^3$ to 10–30 $\mu g/m^3$ in a period with 1.6 mm precipitation (Ouyang et al., 2015). Its ecological effect stems from the acidification of water bodies through acid rains with pH value which is at the lower range (often < 4.5) (Haines, 1981). Studies have correlated sulfate and nitrate aerosols, in PM2.5 (Gao et al., 2016) causing a pH decrease in affected water bodies (Haines, 1981). Research on effects of $PM_{2.5}$ on aquatic ecosystems and organisms is pertinent for aquatic ecology (Hartono et al., 2017; Huang et al., 2014)[1].

12.9 Effect of Urbanization in Historical Assets

PMs have the tenacity to stain and damage and corrode stone and other building materials, which even includes cultural heritages such as statues and monuments. These effects are related to acid rain. The most common and prominent occurrence noted on stone structures and monuments is the accumulation of pollutants on their surface causing the formation of dark color patinas known as "black crusts." Black crusts are made by the cementing material gypsum, inside which atmospheric PM is embedded (Török et al., 2011). The carbonaceous fraction, i.e., OC (organic carbon) and EC (elemental carbon), is one of PM's constituents (Comite et al., 2021) of these crusts as well.

12.9.1 Effect of Urbanization in Indoors Setups

Use of fossil fuels in household appliances like cooking stoves, ovens, etc., hydrocarbon-based wall paints, air-conditioners, refrigerators, aerosols, tobacco smoking, asbestos, various makeup products, etc., also contribute toward air pollution. Particles of indoor origin are derived from biological sources, such as allergens, pollens grains, spores and molds, dust mites and cockroaches. Other sources are from smoking tobacco, cooking, and burning wood, candles, or incense. PMs are also generated from complex reactions of gaseous pollutants emitted from sources as household cleaning products, humidifiers, and fresheners. The effect of indoor air pollutant and health can surpass the outdoor pollutant as people spend most time in an indoor environment, which made it one of the weightiest topics of air pollution research (Vardoulakis et al., 2020).

12.9.2 Effects of PM on Human Health

The percentage of people dying from non-communicable diseases globally significantly increased from 54% in 1990 to 74% in 1991. 20% of all these incidents is attributed to environmental risk factors out of which 50% is attributed singly to air pollution (Pozzer et al., 2023). According to World Health Organization estimates, exposure to fine particles PMs in contaminated air results in the deaths of almost 7 million people annually.

PM exposure can negatively affect human health and may cause many life-threatening disorders (Table 12.2). PMs can affect our body through contact (or adsorption) with the external body surface (eyes and skin) or through inhalation. Many eyes related issues like ocular surface disease, glaucoma, cataract, intraocular inflammation, retinal vessel disease, and refractive error result from exposure to PMs (Ehsanifar & Ghadim, 2022). PM exposure to skin leads to atopic eczema, extrinsic aging, skin inflammation (Dijkhoff et al., 2020). The aerodynamic diameter of PMs determines its anatomical reach, which in turn determines how the inhaled PM would impact a certain organ. For instance, PM_{10} affects upper respiratory tract, $PM_{2.5}$ may enter lower respiratory tract and $PM_{0.1}$ may even infiltrate deep into organs and tissues by entering the blood circulation via alveoli (Grant-Jacob & Mills, 2022; Loaiza-Ceballos et al., 2022).

PMs are responsible for various respiratory disorders like pneumonia, chronic obstructive pulmonary disease, asthma (Kyung & Jeong, 2017). It may also irritate the nasal epithelium and may induce allergic rhinitis and rhinosinusitis (Mariani et al., 2021; Mo, 2019). Prolonged exposure to PMs also affects immunity and makes the person susceptible to bacterial (e.g., *Listeria, S. pneumoniae, K. pneumoniae, S. aureus, P. aeruginosa*) and viral infections (e.g., Influenza virus) (L. Yang et al., 2020). Entry and accumulation of PMs to blood circulation may have life threatening consequences. It may result in cardiovascular disorders like acute ischemia and thrombosis, cardiac arrhythmias, acute heart failure, atherosclerosis, hypertension, and chronic heart failure (Hamanaka & Mutlu, 2018). Scientists have found correlation between exposure to PM and diabetes mellitus pathogenesis (Ccami-Bernal et al., 2023; He et al., 2017; R. Li et al., 2021; Mozafarian et al., 2022). Chronic effects of PM exposure are reported to cause DNA damage (Nemmar et al., 2013). Various cancers like lung cancer, pancreatic cancer, etc., have been reported due to exposure to PMs (Kyung & Jeong, 2017). Fine PMs can even cross blood–brain barrier and affect normal brain functions leading to cognitive decline, brain stroke, neuroinflammation, and neurodegenerative disorders. PM exposure leads to multiple reproductive disorders and childbirth defects as well (Bell et al., 2010; Johnson et al., 2021; Park et al., 2021; Sun et al., 2015; Wieczorek et al., 2024). Therefore, we find that exposure to PM has multiple severe impacts on human health (Fig. 12.2).

Due to extensive industrial, transportation, and household activities, levels of PM in densely populated urban regions are quite higher compared to rural areas. It leads to the formation of an urban pollution island.

Table 12.2 Diseases caused due to suspended PMs in the surrounding environment

No	Organ affected/disorder type	Diseases	References
1	Ocular disorders	Ocular surface disease, Glaucoma, Cataract, Intraocular inflammation, Retinal vessel disease, Refractive error	Ehsanifar and Ghadim (2022)
2	Respiratory disorders	Pneumonia, Lung cancer, Chronic obstructive pulmonary disease, Asthma	Kyung and Jeong (2017)
3	Host defense	Increased susceptibility to bacterial (e.g., *Listeria, S. pneumoniae, K. pneumoniae, S. aureus, P. aeruginosa*) and viral infections (e.g., Influenza virus)	Yang et al. (2020)
4	Skin	Atopic eczema, Extrinsic aging, Skin inflammation	Dijkhoff et al. (2020)
5	Cardiovascular diseases	Acute ischemic/thrombotic events, Cardiac arrhythmias, Acute heart failure, Atherosclerosis, Hypertension, Chronic heart failure	Hamanaka and Mutlu (2018)
6	Nose	Allergic rhinitis and Rhinosinusitis	Mariani et al. (2021)
7	Pancreas	Pancreatic cancer, Diabetes mellitus type 1, Diabetes mellitus type 2	Ccami-Bernal et al. (2023), He et al. (2017), Li et al. (2021) and Mozafarian et al. (2022)
8	Brain	Cognitive decline, brain stroke, neuroinflammation, neurodegenerative disorders like dementia, Alzheimer's disease	Li et al. (2022) and You et al. (2022)
9	Reproductive disorders	Hormonal disruption, male infertility, decreased sperm count, female infertility, decrease in ovarian reserve, early pregnancy loss, endometrial tissue damage	Park et al. (2021) and Wieczorek et al. (2024)
10	Childbirth defects	Low birth weight, Premature birth, Miscarriage/stillbirth, Infant mortality	Bell et al. (2010), Johnson et al. (2021) and Sun et al. (2015)

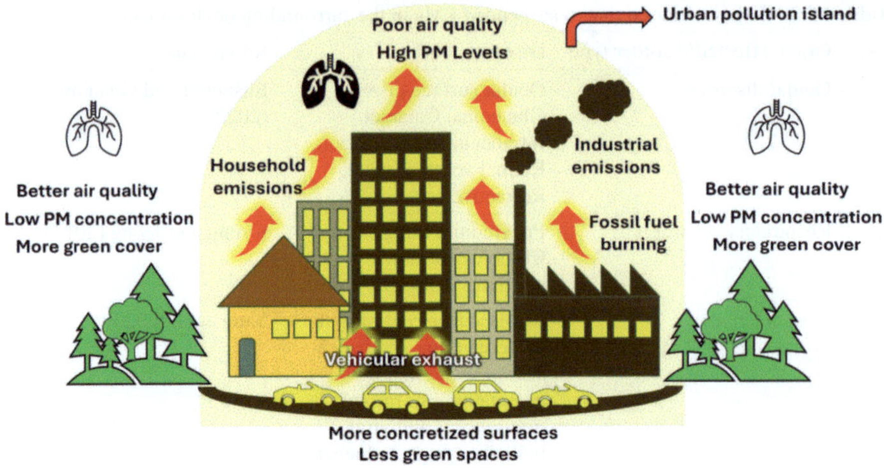

Fig. 12.2 Concept of Urban pollution island

12.9.3 Genomic Imprint of the Health Benefits of Urban Greening

As already discussed, urban greening is known to be associated with reduction in risk of many diseases but there is limited knowledge regarding the biological basis for the positive health effects of having a high amount of greenery in the surrounding environment (Xu et al., 2021). Apart from reducing PM levels from urban atmosphere and reducing the risk of diseases caused due to PM exposure, urban greening also has an independent impact on the human health. In this regard, very recently, scientists have started exploring the effects of urban greening on cellular epigenetic signatures (Caizergues et al., 2022; Dockx et al., 2022a, 2022b; Kim et al., 2015; Nwanaji-Enwerem et al., 2024). Epigenetics is the study of heritable chemical changes which can modulate gene expression without any changes to the nucleotide sequence itself (Hughes, 2014). These changes produce epigenetic signatures and play important roles in controlling cellular functions (Margueron et al., 2005). Epigenetic signatures primarily include histone modifications, DNA methylation patterns and non-coding RNAs (Petronis, 2010). Post-translational modifications of histone proteins regulate gene expression (Khan, 2015). The process of acetylation is linked to the activation of transcription, while deacetylation is related with the transcriptional inactivation. The effects of histone methylation vary depending on the specific site of methylation (Kouzarides, 2007). DNA methylation primarily suppresses gene expression. Methyltransferases are enzymes responsible for DNA methylation, while demethylases are enzymes that perform DNA demethylation (Kumar et al., 2023). Non-coding RNAs regulate gene expression at both the transcriptional and translational levels. Long non-coding RNA (lncRNA) and micro-RNA (miRNA) are prominent types of regulatory non-coding RNAs (Kumar et al., 2023). The regulatory functions of these RNAs are also associated with DNA and histone modifications. Cross-talks exist between different types of epigenetic signatures (Y. Kondo, 2009; Y. Li, 2021). So far, the effects of urban greening on DNA methylation patterns have only been explored but as different epigenetic signatures crosstalk, there is possibility of the influence of urban greening on other types of epigenetic signatures as well (Fig. 5.4).

12.9.4 Concept of Ecological Gentrification

Environmental, ecological, or green gentrification is a process in which cleaning up pollution or providing green amenities increases local property values and attracts wealthier residents to a previously polluted or disenfranchised neighborhood. Ecological gentrification is the process of enhancing green infrastructure in urban localities. It originated from a large environmental movement to increase green infrastructure and sustainability efforts in urban cities. Green infrastructure plannings include planting trees to purify air quality. Ecological gentrification projects typically target communities that have experienced environmental degradation because of decades of exposure to pollutants (chemical, water, and air)[3].

Ecological gentrification often leads to environmental improvement but has an impact on society. This increases property values of an area. The improved environmental quality and aesthetics make these neighborhoods more desirable. As a result, property values and rents increase, causing an imbalance of low- and high-income residents. Higher costs can lead to the displacement of long-time, lower-income residents who can no longer afford to live in these areas causing economic polarization. This can create a socio-economic drift, often resulting in a more affluent population moving in. This in turn causes changes in community character and culture.

To address ecological gentrification and nullify the negative impacts, we can bring in ecological changes in the cities itself through strategies like inclusionary zoning for people with low to moderate incomes. The society can be highly regulated through rent control and tenant protections acts. There is a need for community and land Trustees for long-term affordability and community control over land use. There is also the need for developing job opportunities for all groups of income people. By balancing environmental goals with social equity considerations, cities can work toward achieving sustainable development that benefits all residents.

Understanding the ecology of gentrification requires examining how these various factors intersect and shape the lived experiences of residents, both those who benefit from the process and those who are marginalized or displaced. It also involves recognizing the uneven distribution of power and resources that underlie gentrification and exploring strategies for creating more equitable and inclusive urban environments. Hutchings et al., 2023 (Hutchings et al., 2023) also showed evidence that gentrified areas did not have better air quality. They also suspected that the no change in air quality is likely associated with the growth of new high risers, sports arenas and stores and demolitions of the older existing buildings.

12.9.5 Urban Greening

Urban greening is about integrating nature and urbanization. Urban greening is the process of incorporating green spaces and elements into the urban environment (ref). Trees, shrubs, grass covered lands, and other vegetation are the fundamental components of urban green spaces (De Haas et al., 2021). It promotes sustainable urban development. It stresses upon protecting and including natural green spaces as well as creating new green spaces while urban planning and landscaping. There are public green spaces as well as private green spaces. Public green spaces are accessible to everyone and includes roadside trees, sport fields, urban parks and nature reserves, riparian areas (rivers, lakes, and small streams), farmlands, greenways and trails, nature conservation areas, garden, or natural forests inside of public educational and research institutions, gardens in the office premises, green walls, cemeteries, sacred groves, zoos, etc. Private green spaces include gardens and vegetation growing in the premises of corporate offices, private firms, private educational institutions, home gardens, rooftop gardens, balcony gardens, etc. (Grace & Jeuland, 2018; Roy et al., 2012; Wolch et al., 2014). Urban green spaces are necessary for both the urban environment as well as human health. Urban greening is a broad concept. It includes development of green infrastructure, ecosystem-based adaptation, and nature-based solutions (Dorst et al., 2019).

Green infrastructure (GI) is defined as a "strategically planned network of natural and semi-natural areas with other environmental features designed and managed to deliver a wide range of ecosystem services" by the European commission (European Commission, 2013). According to Convention of Biological Diversity (David Cooper & Noonan-Mooney, 2013), Ecosystem Based Adaptation (EBA) is "the sustainable use of biodiversity and ecosystem services into an overall adaptation strategy to help people to adapt to the adverse effects of climate change that can be cost-effective and generate social, economic and cultural co-benefits and contribute to the conservation of biodiversity." Nature-based solutions (NBS) has been defined by Albert et. Al. (Albert et al., 2019) as "actions that alleviate a well-defined societal challenge, utilize ecosystem processes of spatial, blue and green infrastructure networks, and are embedded within viable governance or business models for implementation." Here, blue and green infrastructure refers to water bodies and vegetation related spaces. All these three concepts are interconnected and have a common goal of integrating social, environmental, and economic development for sustainable urban development. In the present chapter, the term GI has been used frequently to refer to an urban design with abundant green spaces. Well planned green spaces and urban infrastructure together constitute GI.

12.9.6 Selection of Plant Species for Urban Greening

As already discussed, urban greening includes both preservation of existing green spaces in urban areas as well as establishment of new green spaces, thus for any urban greening project, selection of suitable plant species is pertinent. Many such studies have been conducted which deal with identifying the suitable selection criteria for choosing the right plant species for any urban greening project (Asgarzadeh et al., 2014; Beck et al., 1994; Darabi et al., 2023; Goodness, 2018; Jiménez et al., 2014; Maya-Manzano et al., 2017; Miller, 1996; Núñez-Florez et al., 2019; Pauleit, 2003; Radhakrishnan et al., 2019; Rahnema et al., 2019; Sæbø et al., 2003; Wei et al., 2017). For creating an urban ecosystem certain criterion such as easy growing and low maintenance plants, suitable for growing in that soil and faster growing, cost effective perennial plants are chosen. Apart from these some important selection criteria are also considered as discussed below.

12.9.7 Endemic Plant Species Help in Preserving Biodiversity

Endemic or local species of a particular environment are preferred over exotic species for urban greening. It helps both ways in conservation of our native plants and associated fauna (Berthon et al., 2021). Combination of plant species is selected which support other life forms and help in preserving biodiversity (Sercombe et al., 2011). However, care should be taken as urbanization modifies local amenities and may make an unsuitable area for native plant species (Sjöman et al., 2016).

12.9.8 Shape, Size, and Pollinationn Factors of Plant Species

Trees in the urban zone provides all round services, such as mitigating heat island of the environment, refining air quality, arresting rainwater, and acting as topsoil for carbon storage. Shade providing plant species are preferred for the roadside, which also provide cooling effects during harsher weathers. But care should also be taken for pollination factors, as it is a health issue, for example, London plane trees (*Platanus* × *acerifolia*, Platanaceae) have good canopy coverage and provide shade when used as a street tree but are also stands out as an odd for producing high dose pollen and irritating trichome fibers as allergens (Sercombe et al., 2011). Nectar bearing plants are preferred for support of pollinator growth in the city. Pollinators play an important role in fruit and seed production.

12.9.9 Resilience/Stress Tolerant Nature of Plant Species

Climatic changes topped with urbanization have changed the overall factors of the environment, thus, plant's resilience to abiotic stresses has also to be considered for urban green restructuring. Such species of plants are selected are selected which readily adapt to the local climatic conditions and withstand changes in temperature and precipitation. Drought tolerant/resistant plant species are selected for sustained plant growth under dry and arid conditions where availability of water is an issue. Disease resistant plant species are preferred for long-lasting green spaces. Plants resistant against attack of insect-pests are also most preferred. Many plant species are good at absorbing water at fast rate and help in managing floods and restoration.

12.9.10 Aesthetics and Aesthetics and Religious Considerations

Plant species are selected which enhances the beauty, visual appeal, and aesthetic value of the city. Sacred groves revered by local communities can aid both in conservation of cultural practices as well as urban greening. Off late, studies have pointed the attention on trees species with pigments, such as plants harboring anthocyanins have gained prominence as ornamental plants (Lo Piccolo & Landi, 2021). Species such as *Acer platanoides*, *A. pseudoplatanus*, *A. palmatum*, *Fagus sylvatica*, *Corylus maxima*, *Prunus cerasifera*, *Cercis canadensis* and *Cotinus coggygria* are widely used as permanent, red leafy species for urban areas. These plants with higher anthocyanin content are also abiotic stress tolerant which are beneficial for rapid urbanization pattern (Dadea et al., 2017).

12.9.11 Phytoremediation and Invasiveness

Phytoremediation is a "soft" remediation tool, that makes use of the ability of some plant species to accumulate certain elements. Such plant species are preferred for urban greening for phytoremediation, phytostabilization, as well as bioindicators (Dadea et al., 2017). Fast-growing tree species, such as Betula pendula and Robinia pseudoacacia, can be ideal low-cost candidates for phytoremediation applications, due to their physiological properties of high rate of transpiration rates, quicker growth rate after transplantation, and high biomass production. Other helpful species which aid in phytoremediation of heavy metals are Ailanthus altissima, Acer pseudoplatanus. Trees as Willows and poplars are valuable resources for phytoremediation. On the other hand, species like the eucalyptus, black locust, birch, and paulownia are good candidates for phytoextraction owing to their high metal bioconcentration ability and high yield (Pajević et al., 2016).

Among the native species of any area, invasive species have the power for phytoremediation due to their high adaptive and hyperaccumulation capability. Invasive plant species as Japanese hops, honeysuckles, English Ivy are a fear factor for other plant species because of their mass growth. There are listed management plans such as National Invasive Species Management Plan, which provide major goals and instructions for future plant introductions.

Based on all the good characteristics of a plant species, it is difficult or almost impossible for a plant species to fulfill all the criteria. Therefore, a combination of different plant species is selected which share many (if not all) for the greening project. Different governmental organizations and research institutions keep upgrading the list of plant species from time to time which are suitable for urban greening projects in their respective countries. It would also be beneficial to develop tools and techniques for selecting species that are well-adapted and suitable to urban conditions. From the point of view of phytosanitary, it is pertinent to use the species after analyzing an inventory of trees and shrubs species in various urban environments. Several research groups have proposed strategies for Urban Green Space (USG). The United States Department of Agriculture developed software that enables the selection of tree species according to the desired environmental effects or functional benefits. Garcia-Garcia et al., 2016 proposed a type of weighted index for each species based on how much severity the plants can endure as anthropogenic, biotic, and abiotic being the factors. They calculated from a measure which they defined as observed Species per Green Area per Year (SAYs) and concluded on the scale that he higher the number of damaged SAYs, the less prone a species should be for outdoor planting (Garcia-Garcia et al., 2016).

12.9.12 Effect of Urban Greening on Environment

Urban greening is an effective method for remediation of polluted urban air. Many plants, especially trees have great PM retention capabilities, by virtue of which these help in improvement of urban air quality (Fig. 12.3) (Baraldi et al., 2019; Han et al., 2020; Paull et al., 2020; L. Zhang et al., 2019). Every aerial part of a plant can captures PM to different extant, but leaves stand out to be the most efficient. Leaves work as a biofilter and natural air purifiers (Bikis, 2023). Leaf trichomes increase the surface area of the leaves and help in capturing PM. Trichomes also prevent the re-release of the captured PM. Other microstructures present on leaf surface such as folds, ridges, waxes, oils, furrows, folds, and stomata also play important role in PM retention. It has been found that high stomatal density leads to more PM retention in comparison to low stomatal density of the leaf (Yan et al., 2018). It has been found that trees with lanceolate leaves are more effective in PM capture than elliptical and obovate leaves (Leonard et al., 2016). Even trees with small leaves but denser canopies are efficient PM capturers and retainers (S. L. Chen et al., 2021; Hwang et al., 2011).

Increasing urban green spaces can help in drastically reducing PM concentrations from the urban atmosphere. Plants have abilities to capture and retain PM from

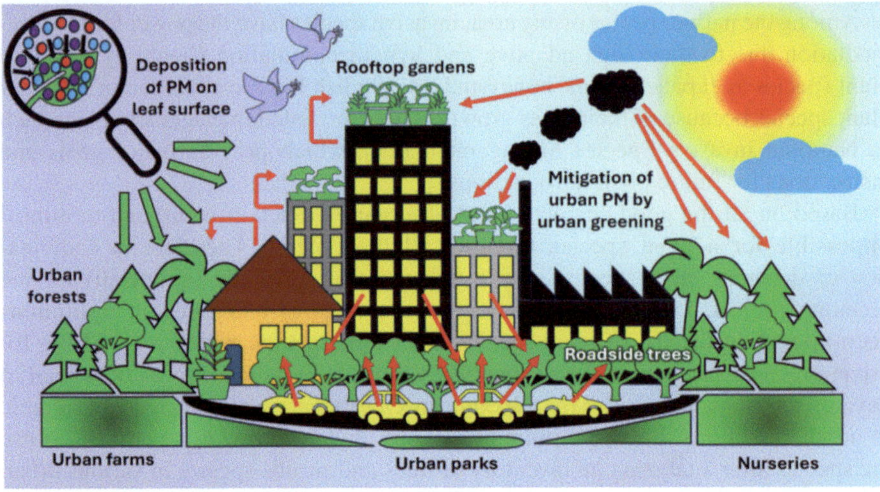

Fig. 12.3 Urban greening in mitigation of urban PM

surrounding air. Conservation of existing green spaces and inclusion of new green spaces in the city is crucial for sustainable urban development.

Studies have shown that roadside trees are effective bio-monitors of air quality, and they efficiently capture PMs released from vehicular exhausts (Rai, 2016; Sevik et al., 2020). Similarly green walls have been found to reduce PM concentration on busy roads of the city (Ysebaert et al., 2021). In other words, urban greening significantly reduces PM concentrations in the urban atmosphere (Chaudhuri & Kumar, 2022; Junior et al., 2022). Some of the changes one can expect in the environment due to urban greening are discussed broadly here (Edeigba et al., 2024; Martens et al., 2022; C. Yang et al., 2017).

12.9.13 Regulation of Urban Environmental Pollution

Urban areas produce huge amount of heat. Urban greening has a cooling effect on the cities. urban greening improves air quality and makes it breathable for humans as well as other organisms. Clean and fresh air promotes growth of more green spaces and promotes sustenance of other life forms creating biodiversity. Apart from capturing PM, urban greening helps in sequestering carbon from atmospheric carbon dioxide and helps in reducing urban heating and global warming. experimental and modeling studies show the effect of green space on air pollutant concentrations in urban settings is highly variable and context specific (Venter et al., 2024).

Urban green spaces act as a buffer against noise pollution. Reduction in noise pollution helps in survival and sustenance of various animal species in the city. Residential noise is considered one of the most omnipotent and intrusive pollutants in the Modern World. A study carried out in the cities of Brazil based on the biophysical characteristics of green areas proved that vegetation proved to be efficient in lessening noise. The density and number of trees was proportional to the blocked sound decibel units with the reduction of noise propagation (de Oliveira et al., 2022).

Urban green areas amend building energy capacity and coupled emissions from larger power plants by shading buildings, circulating, and cooling air temperatures and varying wind speeds around buildings. From energy modeling and field sampling data of urban connected green areas relative to residential buildings, annual energy saving due to trees were reported to have minimized the cost (Nowak et al., 2017). Modeling tree effects on residential building energy use in urban areas in the United States itself, revealed annual energy savings of about 7.2%, worth at $7.8 billion (Nowak et al., 2017), which is a huge efficient cost saving.

12.9.14 Rise of Urban Biotechnology

Urbanization is heavily impacting evolutionary traits, as species are facing challenges with mutation, genetic drifts, adaptive and non-adaptive evolutionary processes, phenotypic plasticity, and epigenetic modifications, ultimately the genome getting affected. For plant species favoring urban greening, their traits such as flower development, water retention capacity, stress tolerance, and such traits are of prime importance. Introgression of such genes for these desired traits into these valuable plant species is a foundation for the future development of urban biotechnology and urban forestry. Genetic engineering has the potential to create new varieties of urban flowering plants. Calli induction with favored traits, micropropagation of plants, and plant regeneration technologies need to be developed for urban plant species (Gladkov et al., 2024). For example, the method of cell selection was used for the species Agrostis stolonifera, which could combine better heavy metal retention capacity as well as resistant to soil contamination. The tool for micropropagation or clonal propagation was used for white oaks, which lead to the fruitful production of stress tolerant oak species (Sax et al., 2019). There is a huge list of usually colors of decorative plants such as blue roses, chrysanthemums, lily, and petunia (Qi et al., 2013) which have already been obtained via engineering the genome.

12.9.15 Effect of Urban Greening on Human Health

Urban greening has a positive impact on human health (Kondo et al., 2018). Based on the deleterious effects of PM exposure on human health as discussed earlier in the present chapter, it can easily be inferred that by mitigating urban PM problem, urban population can be saved from the harmful effects and lethal consequences of PM. Along with this, urban greening has many direct and indirect effects on human health (Q. Li et al., 2023). Urban greening improves and restores mental health

(Y. Zhang et al., 2020), reduces mortality rate (Rojas-Rueda et al., 2019), reduce events of cardiac arrest, works on improving obesity problem (Luo et al., 2020), reduces tendency of type 2 diabetes (Ccami-Bernal et al., 2023), and better reach in pregnancy outcomes (Hu et al., 2021). Greenness at school and home has a positive impact on the cognitive development of children. Exposing kids to childhood green space has been related with reduced hyperactive behavior, better attention and mental soundness, better visual memory-related outcomes, and improved social skills as well (Dockx et al., 2022a, 2022b). In case of adults, inverse relationship has been found between green spaces and insulin resistance in adolescents. Urban greening, therefore, helps in reducing the prevalence of diabetes mellitus type. A research group from China in their study concluded the link between abundant residential green space was with lower risk of T2 Diabetes and decreased fasting blood glucose levels (R. Li et al., 2021). Through effective PM capturing and retention, urban green spaces play an important role in enhancing aerial visibility by reducing haze and smog in the city. By reducing noise pollution, urban green spaces help in lowering mental stress and disturbances and promote calmness and mental peace (de Oliveira et al., 2022). Urban green spaces reduce the risk of preterm birth. In nutshell, urban greening "Harm mitigation, capacity instoration, immunity improvement and capacity restoration" (Nwanaji-Enwerem et al., 2024).

12.9.16 Urban Greening and Sustainable Development

Urban greening incorporates ecological, environmental, and cultural benefits into the process of urban design, planning and management, ultimately fostering sustainable development (Paudel & States, 2023). Sustainable development is the concept of achieving development that fulfills the requirements of the current generation without jeopardizing the capacity of future generations to fulfill their own needs. As already discussed, urban greening involves preserving existing green spaces and creating new ones. Urban greening aims at protecting the urban environment for the present as well as future generations. Urban greening has the capacity to restore nature and increase biodiversity while collaborating with various stakeholders with the purpose of achieving sustainable development. In a recent study, it has been found that cities are warming 29% faster than urban areas. Urban areas contribute significantly toward global warming. By employing mitigation strategies like urban greening, cities can significantly cut their carbon emissions along with reducing PM load. Local initiatives like urban greening have global impacts.

12.9.17 Biological Basis of the Health Benefits of Urban Greening

As already discussed, urban greening is known to be associated with reduction in risk of several vital diseases, but there is limited knowledge regarding the biological basis for the positive health effects of having a high amount of greenery in the surrounding environment (Xu et al., 2021). Apart from reducing PM levels from urban atmosphere and reducing the risk of diseases caused due to PM exposure, urban greening also has an independent impact on the human health. In this regard, very recently, scientists have started exploring the effects of urban greening on cellular epigenetic signatures (Caizergues et al., 2022; Kim et al., 2023; Nwanaji-Enwerem et al., 2024; Xu et al., 2021). Epigenetics is the study of heritable chemical changes which can modulate gene expression without any changes to the nucleotide sequence itself (Hughes, 2014). These changes produce epigenetic signatures and play important roles in controlling cellular functions (Margueron et al., 2005). Epigenetic signatures primarily include histone modifications, DNA methylation patterns and non-coding RNAs (Petronis, 2010). Post-translational modifications of histone proteins regulate gene expression (Khan, 2015). The process of acetylation is linked to the activation of transcription, while deacetylation is related with the transcriptional inactivation. The effects of histone methylation vary depending on the specific site of methylation (Kouzarides, 2007). DNA methylation primarily suppresses gene expression. Methyltransferases are enzymes responsible for DNA methylation, while demethylases are enzymes that perform DNA demethylation (Kumar et al., 2023). Non-coding RNAs regulate gene expression at both the transcriptional and translational levels. Long non-coding RNA (lncRNA) and micro-RNA (miRNA) are prominent types of regulatory non-coding RNAs (Kumar et al., 2023). The regulatory functions of these RNAs are also associated with DNA and histone modifications. Cross-talks exist between different types of epigenetic signatures (Y. Kondo, 2009; Y. Li, 2021) (So far, the effects of urban greening on DNA methylation patterns have only been explored but as different epigenetic signatures crosstalk, there is possibility of the influence of urban greening on other types of epigenetic signatures as well) (Fig. 12.4).

Urban greening affects DNA methylation patterns but its influence on histone modifications and non-coding RNAs is still unknown (indicated with "?"). Histone acetylation is associated with the activation of transcription, whereas deacetylation is connected to the transcriptional inactivity. The consequences of histone methylation differ based on the position of methylation. The main function of DNA methylation is to inhibit gene expression. Methyltransferases are enzymatic catalysts that facilitate the addition of a methyl group to DNA, resulting in DNA methylation. Conversely, demethylases are enzymatic catalysts that facilitate the removal of a methyl group from DNA, resulting in DNA demethylation. Non-coding RNAs control gene expression at both the transcriptional and translational stages. Long non-coding RNA (lncRNA) and micro-RNA (miRNA) are significant categories of regulatory non-coding RNAs. The regulatory roles of these RNAs are also linked to DNA and

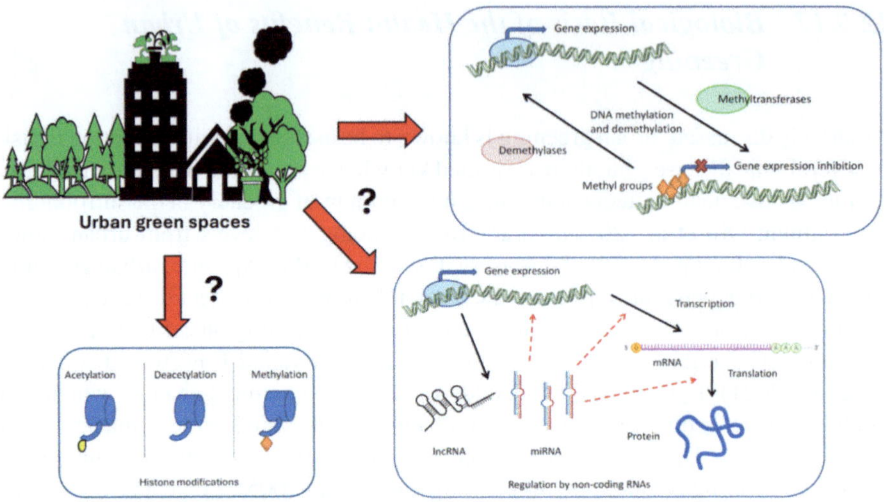

Fig. 12.4 Epigenetic effects of urban greening

histone modification. As various epigenetic signatures crosstalk, there is a potential for urban greening to also impact other types of epigenetic signatures.

Using DNA methylation profiling (K. Kim et al., 2023) have found positive relationship between greenness and slower epigenetic aging in an American biracial urban cohort. Epigenetic aging refers to acquisition of DNA methylation pattern associated with aging. Another study found that maternal exposure to green spaces around the home during pregnancy was linked to the methylation of the imprinting control region of IGF2/H19 gene (IGF2/H19 ICR) in the offspring. Likewise, maternal exposure to green spaces has been found to be positively correlated with the methylation state of HTR2A gene in placental tissue. This finding has important implications for serotonin levels and the neurodevelopment of children (Dockx et al., 2022a, 2022b). Through blood DNA methylation profiling, Xu et al. (Xu et al., 2021) has identified various genetic loci with specific DNA methylation patterns in populations residing nearby urban green spaces. Many of these loci are mapped to genes associated with various diseases. There are only few such studies which specifically investigate the epigenetic impacts of urban greening. This field of study is still in its early stages and there are more fascinating discoveries yet to come.

12.9.18 Conclusion

Polluted urban atmosphere is home to a complex mixture of gasses and PMs that negatively impact urban environment and human health. Urban greening plays a significant role in reducing the negative effects of PM on the environment and human health. Although addressing these challenges requires holistic approaches, that involve collaboration between governments, urban planners, civil society organizations, the private sector, and local communities at large. Sustainable urban development strategies focusing on equity, resilience, and environmental sustainability are essential for creating inclusive and livable cities for all residents. Along with the plans for urban greening., there should be strict regulatory measures by the government. Under law by the Environmental Protection Agency (EPA) (LaKind et al., 2023) by the United States of America, in 2024 EPA is planning to lower the annual standard of PM to 9 μg per cubic meter. One another federal law by the US government exists as the Clean Air Act (Yuan et al., 2022), which sets limits of toxicity for certain pollutants and promotes it not to be spread. Personal precaution against PMs is another thing which can be taken care off. It has been seen that in regular face mask wearers, the overall exposure is reduced by less than 20. As a strategy of self-protection from long-term exposure to particulate matter, wearing facemasks outdoors is inferior to the installation of indoor air purifiers (Shen et al., 2021).

Urban greening is by far the best method, including the development of green infrastructure, ecosystem-based adaptation, and nature-based solutions. Urban greening is about integrating nature and urbanization, incorporating green spaces and elements into the urban environment. It promotes sustainable urban development by protecting and including natural green spaces and creating new green spaces during urban planning and landscaping and beautification. Urban greening contributes to sustainable development, aiming to protect the urban environment for present and future generations. In conclusion, urban greening plays a crucial role in improving the quality of life in urban settings and fostering sustainable development. By incorporating ecological, environmental, and cultural benefits into urban design, planning, and management, cities can protect their environment and contribute to a more sustainable future. Though some recent studies report and claim that reduction in anthropogenic emissions instead of urban infrastructure should usher in better air quality (Venter et al., 2024), but it's the cumulative culture which will bring the green change.

References

Akbari, M. Z., Thepnuan, D., Wiriya, W., Janta, R., Pansompong, P., Hemwan, P., Charoenpanyanet, A., & Chantara, S. (2021). Emission factors of metals bound with PM2.5 and ashes from biomass burning simulated in an open-system combustion chamber for estimation of open burning emissions. *Atmospheric Pollution Research, 12*(3). https://doi.org/10.1016/j.apr.2021.01.012

Albert, C., Schröter, B., Haase, D., Brillinger, M., Henze, J., Herrmann, S., Gottwald, S., Guerrero, P., Nicolas, C., & Matzdorf, B. (2019). Addressing societal challenges through nature-based solutions: How can landscape planning and governance research contribute? *Landscape and Urban Planning, 182*. https://doi.org/10.1016/j.landurbplan.2018.10.003

Al-Thani, H., Koc, M., & J. Isaifan, R. (2018). Investigations on deposited dust fallout in urban Doha: Characterization, source apportionment and mitigation. *Environment and Ecology Research, 6*(5). https://doi.org/10.13189/eer.2018.060510

Anees, M. M., Mann, D., & Mahato, S. (2023). Urban ecosystems research in India: Advances and opportunities. *Current Landscape Ecology Reports, 8*(1). https://doi.org/10.1007/s40823-022-00083-6

Anthropogenic and natural radiative forcing. (2013). In *Climate Change 2013 the Physical Science Basis: Working Group I Contribution to the Fifth Assessment Report of the Intergovernmental Panel on Climate Change* (Vol. 9781107057999). https://doi.org/10.1017/CBO9781107415324.018

Asgarzadeh, M., Vahdati, K., Lotfi, M., Arab, M., Babaei, A., Naderi, F., Pir Soufi, M., & Rouhani, G. (2014). Plant selection method for urban landscapes of semi-arid cities (a case study of Tehran). *Urban Forestry and Urban Greening, 13*(3). https://doi.org/10.1016/j.ufug.2014.04.006

Baraldi, R., Chieco, C., Neri, L., Facini, O., Rapparini, F., Morrone, L., Rotondi, A., & Carriero, G. (2019). An integrated study on air mitigation potential of urban vegetation: From a multi-trait approach to modeling. *Urban Forestry and Urban Greening, 41*. https://doi.org/10.1016/j.ufug.2019.03.020

Barat, A., & Parth Sarthi, P. (2023). Characteristics of remotely sensed urban pollution island (UPI) & its linkage with surface urban heat island (SUHI) over Eastern India. *Aerosol Science and Engineering, 7*(2). https://doi.org/10.1007/s41810-023-00176-7

Beck, H. W., Gilman, E. F., & Fowler, P. A. (1994). Expert system for tree selection in urban forestry. *Applied Engineering in Agriculture, 10*(5).

Bell, M. L., Belanger, K., Ebisu, K., Gent, J. F., Lee, H. J., Koutrakis, P., & Leaderer, B. P. (2010). Prenatal exposure to fine particulate matter and birth weight: Variations by particulate constituents and sources. *Epidemiology, 21*(6). https://doi.org/10.1097/EDE.0b013e3181f2f405

Berthon, K., Thomas, F., & Bekessy, S. (2021). The role of 'nativeness' in urban greening to support animal biodiversity. *Landscape and Urban Planning, 205*. https://doi.org/10.1016/j.landurbplan.2020.103959

Bikis, A. (2023). Urban air pollution and greenness in relation to public health. *Journal of Environmental and Public Health, 202*. https://doi.org/10.1155/2023/8516622

Buonanno, G., Giovinco, G., Morawska, L., & Stabile, L. (2015). Lung cancer risk of airborne particles for Italian population. *Environmental Research, 142*. https://doi.org/10.1016/j.envres.2015.07.019

Caizergues, A. E., Le Luyer, J., Grégoire, A., Szulkin, M., Senar, J. C., Charmantier, A., & Perrier, C. (2022). Epigenetics and the city: Non-parallel DNA methylation modifications across pairs of urban-forest Great tit populations. *Evolutionary Applications, 15*(1). https://doi.org/10.1111/eva.13334

Ccami-Bernal, F., Soriano-Moreno, D. R., Fernandez-Guzman, D., Tuco, K. G., Castro-Díaz, S. D., Esparza-Varas, A. L., Medina-Ramirez, S. A., Caira-Chuquineyra, B., Cortez-Soto, A. G., Yovera-Aldana, M., & Rojas-Rueda, D. (2023). Green space exposure and type 2 diabetes mellitus incidence: A systematic review. *Health and Place, 82*. https://doi.org/10.1016/j.healthplace.2023.103045

Chaudhuri, S., & Kumar, A. (2022). Urban greenery for air pollution control: A meta-analysis of current practice, progress, and challenges. In *Environmental monitoring and assessment* (Vol. 194, No. 3). https://doi.org/10.1007/s10661-022-09808-w

Chávez-García, E., & González-Méndez, B. (2021). Particulate matter and foliar retention: Current knowledge and implications for urban greening. In *Air quality, atmosphere and health* (Vol. 14, No. 9). https://doi.org/10.1007/s11869-021-01032-8

Chen, C. H., Wu, C. Da, Chiang, H. C., Chu, D., Lee, K. Y., Lin, W. Y., Yeh, J. I., Tsai, K. W., & Guo, Y. L. L. (2019). The effects of fine and coarse particulate matter on lung function among the elderly. *Scientific Reports, 9*(1). https://doi.org/10.1038/s41598-019-51307-5

Chen, S. L., Chang, S. W., Chen, Y. J., & Chen, H. L. (2021). Possible warming effect of fine particulate matter in the atmosphere. *Communications Earth and Environment, 2*(1). https://doi.org/10.1038/s43247-021-00278-5

Comite, V., Miani, A., Ricca, M., La Russa, M., Pulimeno, M., & Fermo, P. (2021). The impact of atmospheric pollution on outdoor cultural heritage: an analytic methodology for the characterization of the carbonaceous fraction in black crusts present on stone surfaces. *Environmental Research, 201*. https://doi.org/10.1016/j.envres.2021.111565

Dadea, C., Russo, A., Tagliavini, M., Mimmo, T., & Zerbe, S. (2017). Tree species as tools for biomonitoring and phytoremediation in urban environments: A review with special regard to heavy metals. *Arboriculture and Urban Forestry, 43*(4). https://doi.org/10.48044/jauf.2017.014

Darabi, H., Moarrab, Y., Balist, J., & Naroei, B. (2023). Resilient plant species selection for urban green infrastructure development in arid regions: a case of Qom, Iran. *Urban Ecosystems, 26*(6). https://doi.org/10.1007/s11252-023-01410-3

Darimont, C. T., Cooke, R., Bourbonnais, M. L., Bryan, H. M., Carlson, S. M., Estes, J. A., Galetti, M., Levi, T., MacLean, J. L., McKechnie, I., Paquet, P. C., & Worm, B. (2023). Humanity's diverse predatory niche and its ecological consequences. *Communications Biology, 6*(1). https://doi.org/10.1038/s42003-023-04940-w

Das, S., Roy, A., Masiwal, R., Mandal, M., Popek, R., & Sarkar, A. (2023). Comprehensive analysis of PM1 composition in the Eastern Indo-Gangetic Basin: A three-year urban study. *Sustainability, 15*(20), 14894. https://doi.org/10.3390/su152014894

David Cooper, H., & Noonan-Mooney, K. (2013). Convention on biological diversity. In *Encyclopedia of Biodiversity*, 2nd Ed. https://doi.org/10.1016/B978-0-12-384719-5.00418-4

Daxini, R., & Wu, Y. (2024). Review of methods to account for the solar spectral influence on photovoltaic device performance. *Energy, 286*. https://doi.org/10.1016/j.energy.2023.129461

Dijkhoff, I. M., Drasler, B., Karakocak, B. B., Petri-Fink, A., Valacchi, G., Eeman, M., & Rothen-Rutishauser, B. (2020). Impact of airborne particulate matter on skin: A systematic review from epidemiology to in vitro studies. *Particle and Fibre Toxicology, 17*(1). https://doi.org/10.1186/s12989-020-00366-y

Dockx, Y., Bijnens, E. M., Luyten, L., Peusens, M., Provost, E., Rasking, L., Sleurs, H., Hogervorst, J., Plusquin, M., Casas, L., & Nawrot, T. S. (2022). Early life exposure to residential green space impacts cognitive functioning in children aged 4 to 6 years. *Environment International, 161*. https://doi.org/10.1016/j.envint.2022.107094

Dockx, Y., Bijnens, E., Saenen, N., Aerts, R., Aerts, J. M., Casas, L., Delcloo, A., Dendoncker, N., Linard, C., Plusquin, M., Stas, M., Van Nieuwenhuyse, an, Van Orshoven, J., Somers, B., & Nawrot, T. (2022). Residential green space in association with the methylation status in a CpG site within the promoter region of the placental serotonin receptor HTR2A. *Epigenetics, 17*(13). https://doi.org/10.1080/15592294.2022.2088464

Donahue, C. J. (2015). The anthroposphere, material flow analysis, and chemical education. *Journal of Chemical Education, 92*(4). https://doi.org/10.1021/ed500510j

Dorst, H., van der Jagt, A., Raven, R., & Runhaar, H. (2019). Urban greening through nature-based solutions—Key characteristics of an emerging concept. *Sustainable Cities and Society, 49*. https://doi.org/10.1016/j.scs.2019.101620

Edeigba, B. A., Ashinze, U. K., Umoh, A. A., Biu, P. W., & Daraojimba, A. I. (2024). Urban green spaces and their impact on environmental health: A Global Review. *World Journal of Advanced Research and Reviews, 21*(2). https://doi.org/10.30574/wjarr.2024.21.2.0518

Ehsanifar, M., & Ghadim, B. K. (2022). Exposure to urban air pollution particulate matter and ocular disorders. *Journal of Biomedical Research & Environmental Sciences, 3*(6). https://doi.org/10.37871/jbres1506

European Commission. (2013). Green Infrastructure (GI)—Enhancing Europe's Natural Capital. In *Communication from the Commission to the European Parliament, the Council, the European Economic and Social Committee and the Committee of the Regions, Brussels, Belgium.*

Fang, S. C., Cassidy, A., & David, C. C. (2010). A systematic review of occupational exposure to particulate matter and cardiovascular disease. *International Journal of Environmental Research and Public Health, 7*(4). https://doi.org/10.3390/ijerph7041773

Gao, J. F., Fan, X. Y., Pan, K. L., Li, H. Y., & Sun, L. X. (2016). Diversity, abundance and activity of ammonia-oxidizing microorganisms in fine particulate matter. *Scientific Reports, 6*. https://doi.org/10.1038/srep38785

Garcia-Garcia, M. J., Sánchez-Medina, A., Alfonso-Corzo, E., & Gonzalez Garcia, C. (2016). An index to identify suitable species in urban green areas. *Urban Forestry and Urban Greening, 16*. https://doi.org/10.1016/j.ufug.2016.01.006

Geddes, J. A., & Murphy, J. G. (2012). The science of smog: A chemical understanding of ground level ozone and fine particulate matter. In *Metropolitan Sustainability: Understanding and Improving the Urban Environment*. https://doi.org/10.1533/9780857096463.3.205

Gladkov, E.A., & Gladkova, O.V., (2024). *Prospects of urban biotechnology for ornamental plants.*

Gokul, T., Kumar, K. R., Prema, P., Arun, A., Balaji, P., & Faggio, C. (2023). Particulate pollution and its toxicity to fish: An overview. In *Comparative Biochemistry and Physiology Part—C: Toxicology and Pharmacology* (Vol. 270). https://doi.org/10.1016/j.cbpc.2023.109646

Goodness, J. (2018). Urban landscaping choices and people's selection of plant traits in Cape Town, South Africa. *Environmental Science and Policy, 85*. https://doi.org/10.1016/j.envsci.2018.02.010

Goossens, J., Jonckheere, A. C., Dupont, L. J., & Bullens, D. M. A. (2021). Air pollution and the airways: Lessons from a century of human urbanization. *Atmosphere, 12*(7). https://doi.org/10.3390/atmos12070898

Grace, D., & Jeuland, M. (2018). Preferences for attributes of sacred groves and temples along an urbanization gradient in the national capital Region of India. *Ecological Economics, 152*. https://doi.org/10.1016/j.ecolecon.2018.06.008

Grant-Jacob, J. A., & Mills, B. (2022). Deep learning in airborne particulate matter sensing: a review. *Journal of Physics Communications, 6*(12). https://doi.org/10.1088/2399-6528/aca45e

De Haas, W., Hassink, J., & Stuiver, M. (2021). The Role of Urban Green Space in Promoting Inclusion: Experiences From the Netherlands. *Frontiers in Environmental Science, 9*. https://doi.org/10.3389/fenvs.2021.618198

Habre, R., Girguis, M., Urman, R., Fruin, S., Lurmann, F., Shafer, M., Gorski, P., Franklin, M., McConnell, R., Avol, E., & Gilliland, F. (2021). Contribution of tailpipe and non-tailpipe traffic sources to quasi-ultrafine, fine and coarse particulate matter in southern California. *Journal of the Air and Waste Management Association, 71*(2). https://doi.org/10.1080/10962247.2020.1826366

Haines, T. A. (1981). Acidic precipitation and its consequences for aquatic ecosystems: A review. *Transactions of the American Fisheries Society, 110*(6). https://doi.org/10.1577/1548-8659(1981)110<669:apaicf>2.0.co;2

Hamanaka, R. B., & Mutlu, G. M. (2018). Particulate matter air pollution: Effects on the cardiovascular system. *Frontiers in Endocrinology, 9*. https://doi.org/10.3389/fendo.2018.00680

Han, D., Shen, H., Duan, W., & Chen, L. (2020). A review on particulate matter removal capacity by urban forests at different scales. In *Urban Forestry and Urban Greening* (Vol. 48). https://doi.org/10.1016/j.ufug.2019.126565

Hartono, D., Lioe, B., Zhang, Y., Li, B., & Yu, J. (2017). Impacts of particulate matter (PM2.5) on the behavior of freshwater snail Parafossarulus striatulus. *Scientific Reports, 7*(1). https://doi.org/10.1038/s41598-017-00449-5

He, D., Wu, S., Zhao, H., Qiu, H., Fu, Y., Li, X., & He, Y. (2017). Association between particulate matter 2.5 and diabetes mellitus: A meta-analysis of cohort studies. *Journal of Diabetes Investigation, 8*(5). https://doi.org/10.1111/jdi.12631

Hu, C. Y., Yang, X. J., Gui, S. Y., Ding, K., Huang, K., Fang, Y., Jiang, Z. X., & Zhang, X. J. (2021). Residential greenness and birth outcomes: A systematic review and meta-analysis of observational studies. *Environmental Research, 193*. https://doi.org/10.1016/j.envres.2020.110599

Huang, X. H. H., Bian, Q., Ng, W. M., Louie, P. K. K., & Yu, J. Z. (2014). Characterization of PM2.5 major components and source investigation in suburban Hong Kong: A one year monitoring study. *Aerosol and Air Quality Research, 14*(1). https://doi.org/10.4209/aaqr.2013.01.0020

Hughes, V. (2014). Epigenetics: The sins of the father. *Nature, 507*(7490). https://doi.org/10.1038/507022a

Hutchings, H., Zhang, Q., Grady, S., Mabe, L., & Okereke, I. C. (2023). Gentrification and air quality in a large urban county in the United States. *International Journal of Environmental Research and Public Health, 20*(6). https://doi.org/10.3390/ijerph20064762

Hwang, H. J., Yook, S. J., & Ahn, K. H. (2011). Experimental investigation of submicron and ultrafine soot particle removal by tree leaves. *Atmospheric Environment, 45*(38). https://doi.org/10.1016/j.atmosenv.2011.09.019

Intergovernmental Panel on Climate Change (IPCC). (2023). Climate change 2021—The physical science basis. In *Climate Change 2021—The Physical Science Basis*. https://doi.org/10.1017/9781009157896

Jankielsohn, A. (2023). Sustaining insect biodiversity in agricultural systems to ensure future food security. *Frontiers in Conservation Science, 4*. https://doi.org/10.3389/fcosc.2023.1195512

Jiménez, E. I. T., Castrejón, A. F., & Sánchez, M. G. (2014). Criteria for selection of plant species for its deployment in thermally efficient architectural green roofs in the Metropolitan Area of Mexico City: Methodological guidelines. *Energy Procedia, 57*. https://doi.org/10.1016/j.egypro.2014.10.043

Johnson, N. M., Hoffmann, A. R., Behlen, J. C., Lau, C., Pendleton, D., Harvey, N., Shore, R., Li, Y., Chen, J., Tian, Y., & Zhang, R. (2021). Air pollution and children's health—A review of adverse effects associated with prenatal exposure from fine to ultrafine particulate matter. *Environmental Health and Preventive Medicine, 26*(1). https://doi.org/10.1186/s12199-021-00995-5

Junior, D. P. M., Bueno, C., & da Silva, C. M. (2022). The effect of urban green spaces on reduction of particulate matter concentration. *Bulletin of Environmental Contamination and Toxicology, 108*(6). https://doi.org/10.1007/s00128-022-03460-3

Keenan, C. E., Kelleher, R., & Gray, S. G. (2023). Environmental pollution, epigenetics, and cancer. In *Epigenetic Cancer Therapy* (2nd Edn.). https://doi.org/10.1016/B978-0-323-91367-6.00023-4

Keet, C. A., Keller, J. P., & Peng, R. D. (2018). Long-term coarse particulate matter exposure is associated with asthma among children in medicaid. *American Journal of Respiratory and Critical Care Medicine, 197*(6). https://doi.org/10.1164/rccm.201706-1267OC

Khan, S. A. (2015). Global histone post-translational modifications and cancer: Biomarkers for diagnosis, prognosis and treatment? *World Journal of Biological Chemistry, 6*(4). https://doi.org/10.4331/wjbc.v6.i4.333

Kim, K. H., Kabir, E., & Kabir, S. (2015). A review on the human health impact of airborne particulate matter. *Environment International, 74*. https://doi.org/10.1016/j.envint.2014.10.005

Kim, K., Joyce, B. T., Nannini, D. R., Zheng, Y., Gordon-Larsen, P., Shikany, J. M., Lloyd-Jones, D. M., Hu, M., Nieuwenhuijsen, M. J., Vaughan, D. E., Zhang, K., & Hou, L. (2023). Inequalities in urban greenness and epigenetic aging: Different associations by race and neighborhood socioeconomic status. *Science Advances, 9*(26). https://doi.org/10.1126/sciadv.adf8140

Kondo, M. C., Fluehr, J. M., McKeon, T., & Branas, C. C. (2018). Urban green space and its impact on human health. *International Journal of Environmental Research and Public Health, 15*(3). https://doi.org/10.3390/ijerph15030445

Kondo, Y. (2009). Epigenetic cross-talk between DNA methylation and histone modifications in human cancers. *Yonsei Medical Journal, 50*(4). https://doi.org/10.3349/ymj.2009.50.4.455

Kouzarides, T. (2007). Chromatin modifications and their function. In *Cell* (Vol. 128, No. 4). https://doi.org/10.1016/j.cell.2007.02.005

Kuddus, M. A., Tynan, E., & McBryde, E. (2020). Urbanization: A problem for the rich and the poor? *Public Health Reviews, 41*(1). https://doi.org/10.1186/s40985-019-0116-0

Kumar, R., & Sharma, A. K. (2020). A review of atmospheric changes amidst COVID19 pandemic in the light of IPCC special report 2018 and COP24 conference. *Uttar Pradesh Journal of Zoology, 41*(14).

Kumar, R., Keshari, P., Mishra, Y., & Sharma, A. K. (2023). Modern approaches to study the effects of phytochemicals on cellular epigenome. In *Phytochemicals as an Epigenetic Modifier in Cancer Prevention*. https://doi.org/10.1088/978-0-7503-5252-9ch12

Kyung, S. Y., & Jeong, S. H. (2017). Adverse health effects of particulate matter. *Journal of the Korean Medical Association, 60*(5). https://doi.org/10.5124/jkma.2017.60.5.391

LaKind, J. S., Burns, C. J., Johnson, G. T., & Lange, S. S. (2023). Epidemiology for risk assessment: The US Environmental Protection Agency quality considerations and the Matrix. In *Hygiene and Environmental Health Advances* (Vol. 6). https://doi.org/10.1016/j.heha.2023.100059

Leonard, R. J., McArthur, C., & Hochuli, D. F. (2016). Particulate matter deposition on roadside plants and the importance of leaf trait combinations. *Urban Forestry and Urban Greening, 20*. https://doi.org/10.1016/j.ufug.2016.09.008

Levin, S., Xepapadeas, T., Crépin, A. S., Norberg, J., De Zeeuw, A., Folke, C., Hughes, T., Arrow, K., Barrett, S., Daily, G., Ehrlich, P., Kautsky, N., Mäler, K. G., Polasky, S., Troell, M., Vincent, J. R., & Walker, B. (2013). Social-ecological systems as complex adaptive systems: Modeling and policy implications. *Environment and Development Economics, 18*(2). https://doi.org/10.1017/S1355770X12000460

Li, T., Hu, R., Chen, Z., Li, Q., Huang, S., Zhu, Z., & Zhou, L. F. (2018). Fine particulate matter (PM2.5): The culprit for chronic lung diseases in China. *Chronic Diseases and Translational Medicine, 4*(3). https://doi.org/10.1016/j.cdtm.2018.07.002

Li, Y., Zheng, C., Ma, Z., & Quan, W. (2019). Acute and cumulative effects of haze fine particles on mortality and the seasonal characteristics in Beijing, China, 2005–2013: A time-stratified case-crossover study. *International Journal of Environmental Research and Public Health, 16*(13). https://doi.org/10.3390/ijerph16132383

Li, R., Chen, G., Jiao, A., Lu, Y., Guo, Y., Li, S., Wang, C., & Xiang, H. (2021). Residential green and blue spaces and type 2 diabetes mellitus: A population-based health study in China. *Toxics, 9*(1). https://doi.org/10.3390/toxics9010011

Li, W., Lin, G., Xiao, Z., Zhang, Y., Li, B., Zhou, Y., Ma, Y., & Chai, E. (2022). A review of respirable fine particulate matter (PM2.5)-induced brain damage. In *Frontiers in Molecular Neuroscience* (Vol. 15). https://doi.org/10.3389/fnmol.2022.967174

Li, Q., Liu, Y., Yang, L., Ge, J., Chang, X., & Zhang, X. (2023). The impact of urban green space on the health of middle-aged and older adults. *Frontiers in Public Health, 11*. https://doi.org/10.3389/fpubh.2023.1244477

Li, Y. (2021). Modern epigenetics methods in biological research. In *Methods* (Vol. 187). https://doi.org/10.1016/j.ymeth.2020.06.022

Lindbom, J., Gustafsson, M., Blomqvist, G., Dahl, A., Gudmundsson, A., Swietlicki, E., & Ljungman, A. G. (2006). Exposure to wear particles generated from studded tires and pavement induces inflammatory cytokine release from human macrophages. *Chemical Research in Toxicology, 19*(4). https://doi.org/10.1021/tx0503101

Lo Piccolo, E., & Landi, M. (2021). Red-leafed species for urban "greening" in the age of global climate change. *Journal of Forestry Research, 32*(1). https://doi.org/10.1007/s11676-020-01154-2

Loaiza-Ceballos, M. C., Marin-Palma, D., Zapata, W., & Hernandez, J. C. (2022). Viral respiratory infections and air pollutants. In *Air quality, atmosphere and health* (Vol. 15, No. 1). https://doi.org/10.1007/s11869-021-01088-6

Luo, Y. N., Huang, W. Z., Liu, X. X., Markevych, I., Bloom, M. S., Zhao, T., Heinrich, J., Yang, B. Y., & Dong, G. H. (2020). Greenspace with overweight and obesity: A systematic review and meta-analysis of epidemiological studies up to 2020. In *Obesity reviews* (Vol. 21, No. 11). https://doi.org/10.1111/obr.13078

Mandal, M., Das, S., Roy, A., Rakwal, R., Jones, O. A., Popek, R., Agrawal, G. K., & Sarkar, A. (2023a). Interactive relations between plants, phyllosphere microbial community, and particulate matter pollution. *Science of the Total Environment, 890*, Article 164352. https://doi.org/10.1016/j.scitotenv.2023.164352

Mandal, M., Popek, R., Przybysz, A., Roy, A., Das, S., & Sarkar, A. (2023b). Breathing fresh air in the city: Implementing avenue trees as a sustainable solution to reduce particulate pollution in urban agglomerations. *Plants, 12*(7), 1545. https://doi.org/10.3390/plants12071545

Mandal, M., Roy, A., & Sarkar, A. (2025). Micro (nano) plastics pollution in an agri-ecosystem: Uncovering the hidden threats to soil, crop health and performance. *Pedosphere.* https://doi.org/10.1016/j.pedsph.2025.11.007

Manisalidis, I., Stavropoulou, E., Stavropoulos, A., & Bezirtzoglou, E. (2020). Environmental and health impacts of air pollution: A review. In *Frontiers in Public Health* (Vol. 8). https://doi.org/10.3389/fpubh.2020.00014

Margueron, R., Trojer, P., & Reinberg, D. (2005). The key to development: Interpreting the histone code? In *Current Opinion in Genetics and Development* (Vol. 15, No. 2). https://doi.org/10.1016/j.gde.2005.01.005

Mariani, J., Iodice, S., Cantone, L., Solazzo, G., Marraccini, P., Conforti, E., Bulsara, P. A., Lombardi, M. S., Howlin, R. P., Bollati, V., & Ferrari, L. (2021). Particulate matter exposure and allergic rhinitis: The role of plasmatic extracellular vesicles and bacterial nasal microbiome. *International Journal of Environmental Research and Public Health, 18*(20). https://doi.org/10.3390/ijerph182010689

Martens, D., Öztürk, Ö., Rindt, L., Twarok, J., Steinhardt, U., & Molitor, H. (2022). Supporting biodiversity: Structures of participatory actions in urban green spaces. *Frontiers in Sustainable Cities, 4.* https://doi.org/10.3389/frsc.2022.952790

Maya-Manzano, J. M., Fernández-Rodríguez, S., Monroy-Colín, A., Silva-Palacios, I., Tormo-Molina, R., & Gonzalo-Garijo, Á. (2017). Allergenic pollen of ornamental plane trees in a Mediterranean environment and urban planning as a prevention tool. *Urban Forestry and Urban Greening, 27.* https://doi.org/10.1016/j.ufug.2017.09.009

Miller, R. W. (1996). *Urban forestry: Planning and managing urban greenspaces.* Urban Forestry: Planning and Managing Urban Greenspaces.

Mo, J. H. (2019). Association of particulate matter with ENT diseases. *Clinical and Experimental Otorhinolaryngology, 12*(3). https://doi.org/10.21053/ceo.2019.00752

Mollashahi, H., & Szymura, M. (2022). Urban ecosystem: An interaction of biological and physical components. In *Biodiversity of Ecosystems.* https://doi.org/10.5772/intechopen.97742

Mozafarian, N., Hashemipour, M., Yazdi, M., Hani Tabaei Zavareh, M., Hovsepian, S., Heidarpour, M., & Taheri, E. (2022). The association between exposure to air pollution and type 1 diabetes mellitus: A systematic review and meta-analysis. *Advanced Biomedical Research, 11*(1). https://doi.org/10.4103/abr.abr_80_21

Nemmar, A., Holme, J. A., Rosas, I., Schwarze, P. E., & Alfaro-Moreno, E. (2013). Recent advances in particulate matter and nanoparticle toxicology: A review of the in vivo and in vitro studies. *BioMed Research International, 2013.* https://doi.org/10.1155/2013/279371

Newell, K., Cusack, R. P., Kartsonaki, C., Chaudhary, N., & Kurmi, O. P. (2021). Household air pollution and associated health effects in low and middle income countries. *Encyclopedia of Respiratory Medicine, 6.* https://doi.org/10.1016/B978-0-12-801238-3.11494-1

Nowak, D. J., Appleton, N., Ellis, A., & Greenfield, E. (2017). Residential building energy conservation and avoided power plant emissions by urban and community trees in the United States. *Urban Forestry and Urban Greening, 21.* https://doi.org/10.1016/j.ufug.2016.12.004

Núñez-Florez, R., Pérez-Gómez, U., & Fernández-Méndez, F. (2019). Functional diversity criteria for selecting urban trees. *Urban Forestry and Urban Greening, 38.* https://doi.org/10.1016/j.ufug.2019.01.005

Nwanaji-Enwerem, U., McGeary, J. E., & Grigsby-Toussaint, D. S. (2024). Greenspace, stress, and health: How is epigenetics involved? *Frontiers in Public Health, 12.* https://doi.org/10.3389/fpubh.2024.1333737

de Oliveira, J. D., Biondi, D., & Dos Reis, A. R. N. (2022). The role of urban green areas in noise pollution attenuation. *DYNA (Colombia), 89*(220). https://doi.org/10.15446/dyna.v89n220.95822

Ouyang, W., Guo, B., Cai, G., Li, Q., Han, S., Liu, B., & Liu, X. (2015). The washing effect of precipitation on particulate matter and the pollution dynamics of rainwater in downtown Beijing. *Science of the Total Environment, 505*. https://doi.org/10.1016/j.scitotenv.2014.09.062

Pajević, S., Borišev, M., Nikolić, N., Arsenov, D. D., Orlović, S., & Župunski, M. (2016). Phytoextraction of heavy metals by fast-growing trees: A review. *Phytoremediation: Management of Environmental Contaminants 3*. https://doi.org/10.1007/978-3-319-40148-5_2

Park, S. R., Lee, J. W., Kim, S. K., Yu, W. J., Lee, S. J., Kim, D., Kim, K. W., Jung, J. W., & Hong, I. S. (2021). The impact of fine particulate matter (PM) on various beneficial functions of human endometrial stem cells through its key regulator SERPINB2. *Experimental and Molecular Medicine, 53*(12). https://doi.org/10.1038/s12276-021-00713-9

Paudel, S., & States, S. L. (2023). Urban green spaces and sustainability: Exploring the ecosystem services and disservices of grassy lawns versus floral meadows. *Urban Forestry and Urban Greening, 84*. https://doi.org/10.1016/j.ufug.2023.127932

Pauleit, S. (2003). Urban street tree plantings: Identifying the key requirements. *Proceedings of the Institution of Civil Engineers: Municipal Engineer, 156*(1). https://doi.org/10.1680/muen.2003.156.1.43

Paull, N. J., Krix, D., Irga, P. J., & Torpy, F. R. (2020). Airborne particulate matter accumulation on common green wall plants. *International Journal of Phytoremediation, 22*(6). https://doi.org/10.1080/15226514.2019.1696744

Perera, F. (2018). Pollution from fossil-fuel combustion is the leading environmental threat to global pediatric health and equity: Solutions exist. *International Journal of Environmental Research and Public Health, 15*(1). https://doi.org/10.3390/ijerph15010016

Petronis, A. (2010). Epigenetics as a unifying principle in the aetiology of complex traits and diseases. *Nature, 465*(7299). https://doi.org/10.1038/nature09230

Pinault, J. L. (2018). Anthropogenic and natural radiative forcing: Positive feedbacks. *Journal of Marine Science and Engineering, 6*(4). https://doi.org/10.3390/jmse6040146

Pozzer, A., Anenberg, S. C., Dey, S., Haines, A., Lelieveld, J., & Chowdhury, S. (2023). Mortality attributable to ambient air pollution: A review of global estimates. *GeoHealth, 7*(1). https://doi.org/10.1029/2022GH000711

Pulselli, F. M., & Marchi, M. (2015). Global Warming Potential and the Net Carbon Balance. *Reference Module in Earth Systems and Environmental Sciences*. https://doi.org/10.1016/b978-0-12-409548-9.09526-9

Qi, Y., Lou, Q., Quan, Y., et al. (2013). Flower-specific expression of the Phalaenopsis flavonoid 3′, 5′-hydoxylase modifies flower color pigmentation in Petunia and Lilium. *Plant Cell Tiss Organ Cult, 115*, 263–273.

Radhakrishnan, M., Kenzhegulova, I., Eloffy, M. G., Ibrahim, W. A., Zevenbergen, C., & Pathirana, A. (2019). Development of context specific sustainability criteria for selection of plant species for green urban infrastructure: The case of Singapore. *Sustainable Production and Consumption, 20*. https://doi.org/10.1016/j.spc.2019.08.004

Rahnema, S., Sedaghathoor, S., Allahyari, M. S., Damalas, C. A., & Bilali, H. El. (2019). Preferences and emotion perceptions of ornamental plant species for green space designing among urban park users in Iran. *Urban Forestry and Urban Greening, 39*. https://doi.org/10.1016/j.ufug.2018.12.007

Rai, P. K. (2016). Impacts of particulate matter pollution on plants: Implications for environmental biomonitoring. *Ecotoxicology and Environmental Safety, 129*. https://doi.org/10.1016/j.ecoenv.2016.03.012

Richardson, H. W. (1987). The costs of urbanization: a four-country comparison. *Economic Development & Cultural Change, 35*(3). https://doi.org/10.1086/451604

Rojas-Rueda, D., Nieuwenhuijsen, M. J., Gascon, M., Perez-Leon, D., & Mudu, P. (2019). Green spaces and mortality: a systematic review and meta-analysis of cohort studies. *The Lancet Planetary Health, 3*(11). https://doi.org/10.1016/S2542-5196(19)30215-3

Roshan, D. R., Koc, M., Isaifan, R., Shahid, M. Z., & Fountoukis, C. (2019). Aerosol optical thickness over large urban environments of the Arabian Peninsula-speciation, variability, and distributions. *Atmosphere, 10*(5). https://doi.org/10.3390/atmos10050228

Roy, A., Mandal, M., Das, S., Kumar, M., Popek, R., Awasthi, A., Giri, B. S., Mondal, K. C., & Sarkar, A. (2023). Non-exhaust particulate pollution in Asian countries: A comprehensive review of sources, composition, and health effects. *Environmental Engineering Research, 29*(3), Article 230384. https://doi.org/10.4491/eer.2023.384

Roy, S., Byrne, J., & Pickering, C. (2012). A systematic quantitative review of urban tree benefits, costs, and assessment methods across cities in different climatic zones. *Urban Forestry and Urban Greening, 11*(4). https://doi.org/10.1016/j.ufug.2012.06.006

Sæbø, A., Benedikz, T., & Randrup, T. B. (2003). Selection of trees for urban forestry in the Nordic countries. *Urban Forestry and Urban Greening, 2*(2). https://doi.org/10.1078/1618-8667-00027

Sax, M.S., Bassuk, N., Bridgen, M.(2019). Tissue Culture Clonal Propagation of Hybrid White Oaks for the Urban Environment. *HortSci, 54*(12), 2214–2223.

Sercombe, J. K., Green, B. J., Rimmer, J., Burton, P. K., Katelaris, C. H., & Tovey, E. R. (2011). London Plane Tree bioaerosol exposure and allergic sensitization in Sydney, Australia. *Annals of Allergy, Asthma and Immunology, 107*(6). https://doi.org/10.1016/j.anai.2011.08.011

Sevik, H., Cetin, M., Ozel, H. B., Akarsu, H., & Zeren Cetin, I. (2020). Analyzing of usability of tree-rings as biomonitors for monitoring heavy metal accumulation in the atmosphere in urban area: a case study of cedar tree (Cedrus sp.). *Environmental Monitoring and Assessment, 192*(1). https://doi.org/10.1007/s10661-019-8010-2

Shen, H., Liu, B., Chen, Y., Zhu, X., Yun, X., Meng, W., Lu, C., Shen, G., Hu, Y., Russell, A. G., Smith, K. R., & Tao, S. (2021). Individual and population level protection from particulate matter exposure by wearing facemasks. *Environment International, 146*. https://doi.org/10.1016/j.envint.2020.106026

Siebert, U., Rothenbacher, D., Daniel, U., & Brenner, H. (2001). Demonstration of the healthy worker survivor effect in a cohort of workers in the construction industry. *Occupational and Environmental Medicine, 58*(12). https://doi.org/10.1136/oem.58.12.774

Singh, J., Payra, S., Mishra, M. K., & Verma, S. (2022). An analysis of particulate pollution using urban aerosol pollution island intensity over Delhi, India. *Environmental Monitoring and Assessment, 194*(12). https://doi.org/10.1007/s10661-022-10573-z

Sjöman, H., Morgenroth, J., Sjöman, J. D., Sæbø, A., & Kowarik, I. (2016). Diversification of the urban forest—Can we afford to exclude exotic tree species? *Urban Forestry and Urban Greening, 18*. https://doi.org/10.1016/j.ufug.2016.06.011

Souza, I. da C., Morozesk, M., Mansano, A. S., Mendes, V. A. S., Azevedo, V. C., Matsumoto, S. T., Elliott, M., Monferrán, M. V., Wunderlin, D. A., & Fernandes, M. N. (2021). Atmospheric particulate matter from an industrial area as a source of metal nanoparticle contamination in aquatic ecosystems. *Science of the Total Environment, 753*. https://doi.org/10.1016/j.scitotenv.2020.141976

Sun, X., Luo, X., Zhao, C., Chung Ng, R. W., Lim, C. E. D., Zhang, B., & Liu, T. (2015). The association between fine particulate matter exposure during pregnancy and preterm birth: A meta-analysis. *BMC Pregnancy and Childbirth, 15*(1). https://doi.org/10.1186/s12884-015-0738-2

Thangavel, P., Park, D., & Lee, Y. C. (2022). Recent Insights into Particulate Matter (PM2.5)-Mediated toxicity in humans: An overview. *International Journal of Environmental Research and Public Health, 19*(12). https://doi.org/10.3390/ijerph19127511

The Oxford dictionary of proverbs. (2004). *Choice Reviews Online, 42*(03). https://doi.org/10.5860/choice.42-1286

Tong, R., Cheng, M., Zhang, L., Liu, M., Yang, X., Li, X., & Yin, W. (2018). The construction dust-induced occupational health risk using Monte-Carlo simulation. *Journal of Cleaner Production, 184*. https://doi.org/10.1016/j.jclepro.2018.02.286

Török, Á., Licha, T., Simon, K., & Siegesmund, S. (2011). Urban and rural limestone weathering; the contribution of dust to black crust formation. *Environmental Earth Sciences, 63*(4). https://doi.org/10.1007/s12665-010-0737-6

Vardoulakis, S., Giagloglou, E., Steinle, S., Davis, A., Sleeuwenhoek, A., Galea, K. S., Dixon, K., & Crawford, J. O. (2020). Indoor exposure to selected air pollutants in the home environment: A systematic review. *International Journal of Environmental Research and Public Health, 17*(23). https://doi.org/10.3390/ijerph17238972

Venter, Z. S., Hassani, A., Stange, E., Schneider, P., & Castell, N. (2024). Reassessing the role of urban green space in air pollution control. *Proceedings of the National Academy of Sciences of the United States of America, 121*(6). https://doi.org/10.1073/pnas.2306200121

Vinnikov, D., Abenova, A., Raushanova, A., & Rapisarda, V. (2023). Occupational exposure to fine particulate matter in the reinforced concrete production and its association with respiratory symptoms and lung function. BMC Public Health, 23(1). https://doi.org/10.1186/s12889-023-16753-x

Wei, H., Huang, Z., & Lin, M. (2017). A decision support system for plant optimization in urban areas with diversified solar radiation. *Sustainability (Switzerland), 9*(2). https://doi.org/10.3390/su9020215

Wieczorek, K., Szczęsna, D., Radwan, M., Radwan, P., Polańska, K., Kilanowicz, A., & Jurewicz, J. (2024). Exposure to air pollution and ovarian reserve parameters. *Scientific Reports, 14*(1). https://doi.org/10.1038/s41598-023-50753-6

Wolch, J. R., Byrne, J., & Newell, J. P. (2014). Urban green space, public health, and environmental justice: The challenge of making cities "just green enough." *Landscape and Urban Planning, 125*. https://doi.org/10.1016/j.landurbplan.2014.01.017

worldtop20.org. (2022). *The Global Challenge for Government Transparency: The Sustainable Development Goals (SDG) 2030 Agenda*. New Jersey Minority Educational Development.

Xu, R., Li, S., Li, S., Wong, E. M., Southey, M. C., Hopper, J. L., Abramson, M. J., & Guo, Y. (2021). Residential surrounding greenness and DNA methylation: An epigenome-wide association study. *Environment International, 154*. https://doi.org/10.1016/j.envint.2021.106556

Yan, G., Liu, J., Zhu, L., Zhai, J., Cong, L., Ma, W., Wang, Y., Wu, Y., & Zhang, Z. (2018). Effectiveness of wetland plants as biofilters for inhalable particles in an urban park. *Journal of Cleaner Production, 194*. https://doi.org/10.1016/j.jclepro.2018.05.168

Yang, C., He, X., Wang, R., Yan, F., Yu, L., Bu, K., Yang, J., Chang, L., & Zhang, S. (2017). The effect of urban green spaces on the urban thermal environment and its seasonal variations. *Forests, 8*(5). https://doi.org/10.3390/f8050153

Yang, L., Li, C., & Tang, X. (2020). The impact of PM2.5 on the host defense of respiratory system. *Frontiers in Cell and Developmental Biology, 8*. https://doi.org/10.3389/fcell.2020.00091

You, R., Ho, Y. S., & Chang, R. C. C. (2022). The pathogenic effects of particulate matter on neurodegeneration: a review. *Journal of Biomedical Science, 29*(1). https://doi.org/10.1186/s12929-022-00799-x

Ysebaert, T., Koch, K., Samson, R., & Denys, S. (2021). Green walls for mitigating urban particulate matter pollution—A review. In *Urban Forestry and Urban Greening* (Vol. 59). https://doi.org/10.1016/j.ufug.2021.127014

Yuan, M., Barron, A. R., Selin, N. E., Picciano, P. D., Metz, L. E., Reilly, J. M., & Jacoby, H. D. (2022). Meeting U.S. greenhouse gas emissions goals with the international air pollution provision of the clean air act. *Environmental Research Letters, 17*(5). https://doi.org/10.1088/1748-9326/ac6227

Zhang, R., Wang, G., Guo, S., Zamora, M. L., Ying, Q., Lin, Y., Wang, W., Hu, M., & Wang, Y. (2015). Formation of urban fine particulate matter. *Chemical Reviews, 115*(10). https://doi.org/10.1021/acs.chemrev.5b00067

Zhang, L., Zhang, Z., Chen, L., & McNulty, S. (2019). An investigation on the leaf accumulation-removal efficiency of atmospheric particulate matter for five urban plant species under different rainfall regimes. *Atmospheric Environment, 208*. https://doi.org/10.1016/j.atmosenv.2019.04.010

Zhang, Y., Mavoa, S., Zhao, J., Raphael, D., & Smith, M. (2020). The association between green space and adolescents mental well-being: A systematic review. *International Journal of Environmental Research and Public Health, 17*(18). https://doi.org/10.3390/ijerph17186640

Zhao, J., Huang, Y., Tang, T., & Li, Y. (2020). Effects of particulate air pollution on visual aesthetic preference. *Proceedings of the Institution of Civil Engineers: Urban Design and Planning, 173*(3). https://doi.org/10.1680/jurdp.19.00040

MIX
Papier aus verantwortungsvollen Quellen
Paper from responsible sources
FSC® C105338

If you have any concerns about our products,
you can contact us on
ProductSafety@springernature.com

In case Publisher is established outside the EU,
the EU authorized representative is:
Springer Nature Customer Service Center GmbH
Europaplatz 3, 69115 Heidelberg, Germany

Printed by Libri Plureos GmbH
in Hamburg, Germany